Asia in Transition

Volume 21

This book series, indexed in Scopus, is an initiative in conjunction with Springer under the auspices of the Universiti Brunei Darussalam – Institute of Asian Studies (http://ias.ubd.edu.bn/). It addresses the interplay of local, national, regional and global influences in Southeast, South and East Asia and the processes of translation and exchange across boundaries and borders. The series explores a variety of disciplinary and interdisciplinary perspectives.

Submission and Peer Review:

Proposal submissions are to be sent to the Series Editor-in-Chief, Prof Dr Hoon Chang Yau: changyau.hoon@ubd.edu.bn and Springer Publishing Editor Alex Westcott Campbell: alexandra.campbell@springernature.com. Please contact the publishing editor to obtain the correct proposal form. All proposals will undergo peer review by the editorial board members. If accepted, the final manuscript will be peer reviewed internally by the editorial board and may also be externally reviewed (single blind) by Springer ahead of acceptance and publication.

Jose Monfred C. Sy · Ananeza P. Aban
Editors

Reimagining Development in Southeast Asia

Alternative Practices from the Grassroots and Social Movements

ubd INSTITUTE OF ASIAN STUDIES

Springer

Editors
Jose Monfred C. Sy
Department of Filipino and Philippine Literature
University of the Philippines
Quezon City, Philippines

Program on Alternative Development
University of the Philippines Center for Integrative and Development Studies
Quezon City, Philippines

Ananeza P. Aban
Department of Conflict and Development Studies
Ghent University
Ghent, Belgium

Program on Alternative Development
University of the Philippines Center for Integrative and Development Studies
Quezon City, Philippines

ISSN 2364-8252 ISSN 2364-8260 (electronic)
Asia in Transition
ISBN 978-981-97-9139-2 ISBN 978-981-97-9140-8 (eBook)
https://doi.org/10.1007/978-981-97-9140-8

This Springer imprint is published by the registered company Springer Nature Singapore Pte Ltd.
The registered company address is: 152 Beach Road, #21-01/04 Gateway East, Singapore 189721, Singapore

Acknowledgments

This book would not have been possible without the generous support and collaboration of numerous individuals and organizations. We extend our heartfelt gratitude to a range of entities whose contributions have been indispensable to this project.

First and foremost, we acknowledge the invaluable support of the following organizations: Movement for Alternatives and Solidarity in Southeast Asia (MASSA), the University of the Philippines College of Social Work and Community Development, 11.11.11 of Belgium, Comité Catholique Contre La Faim Et Pour Le Développement (CCFD)—Terre Solidaire, and the Taiwan Foundation for Democracy. We are also deeply grateful to the Rosa Luxemburg Stiftung, Global Tapestry of Alternatives, Focus on the Global South, ASEAN SOGIE Caucus, Women's Legal and Human Rights Bureau—Philippines, Vietnam Peace and Development Foundation, Participatory Development Training Centre—Laos, Save Our Schools Network, and Asian Regional Exchange for New Alternatives (ARENA) for their unwavering support and commitment to our research program.

We also wish to express our sincere appreciation to the individuals who have contributed their expertise and dedication to this work. Our gratitude extends to Teresa S. Encarnacion Tadem, Ph.D., former executive director of the University of the Philippines Center for Integrative and Development Studies. We are thankful for the support of Naruemon Thabchumpon, Ph.D., Karl Arvin Hapal, Jenito Jewata Santana, Ryan Silverio, and Fernando 'Ato Lekinawa' Costa, as well as Chalida Tajaroensuk, Khamphoui Saythalat, and Valentin da Costa Pinto.

Special thanks are due to Dr. May Shi Sho, Shantoy Hades, Sokunthy Ros, and Antonio Miguel da Costa. We thank Ma. Stephanie Joy Andaya as well for her assistance in proofreading the first draft of the book. Our appreciation extends to Ashish Kothari, Gaudencio Souza, Geraldine Yam, Dong Huy Cuong, Jelen Paclarin, Nelia Menezes, and R. D. Benyamin Daud and the Diocese of Larantuka. Lastly, we honor the memory of Maureen Pagaduan and thank her for her support.

Each of you has played a crucial role in the development and realization of this work. We are deeply appreciative of your expertise, dedication, and support. Thank you for contributing to this collective endeavor and helping us bring this project to fruition.

Contents

Part VI Towards a Peoples' Solidarity Economy

Part VII Conclusion

Editors and Contributors

About the Editors

Jose Monfred C. Sy is an assistant professor of Philippine studies at the Department of Filipino and Philippine Literature, University of the Philippines (UP) Diliman. He received his Master of Arts (Philippine Studies) and Bachelor of Arts (Comparative Literature), summa cum laude, from the same university. He is currently a research fellow of the Program on Alternative Development of the UP Center for Integrative and Development Studies, where he is a co-editor of the multi-volume monograph series *Alternative Practices across Southeast Asia* (2020, 2023).

His most recent monograph, *Mga Mapa-kuwento ng Bakwit School: Paglalakbay, Pag-aaral, at Pakikibaka ng Kabataang Lumad* (*Map-Stories from the Bakwit School: The Sojourn, Study, and Struggle of Lumad Youth for the Ancestral Land*, Sentro ng Wikang Filipino, 2024), explores the intersections of education and resistance among indigenous youth. He also co-edited the volumes *Suring Supling: Kalipunan ng Rebyu ng mga Akdang Pambata sa Pilipinas* (*Collection of Reviews of Children's Books in the Philippines*, Sentro ng Wikang Filipino, 2022) and *Mula Palengke Patungong Paaralan: Critical, Nationalist, and Democratic Pedagogy in the Philippines* (*From the Marketplace to the School*, IBON Foundation, 2024). His research on social movements, anticolonialism and decoloniality, environmental humanities, and children's literature also appears in *Kritika Kultura, Humanities Diliman, Social Science Diliman, Daluyan, Suvannabhumi, AlterNative*, and *Philippine Studies.*

He is a member of the Congress of Teachers/Educators for Nationalism and Democracy (CONTEND), an anti-imperialist group of academics that advocates for a nationalist, scientific, and mass-oriented education, as well as the Alliance of Concerned Teachers (ACT), the largest non-traditional teachers' organization in the country that advances the sector's economic and political welfare. Aside from his academic and activist work, he is the author of several children's books. e-mail: jose_monfred.sy@up.edu.ph

Ananeza P. Aban is currently a Ph.D. researcher of the Department of Conflict and Development Studies and a recipient of the Special Research Fund doctoral scholarship of Ghent University in Belgium. She also serves as a senior research analyst of UP Center for Integrative and Development Studies Program on Alternative Development. Among her major tasks are coordinating the field research across Southeast Asia, co-editing the *Alternative Practices* monograph series (2020, 2023), and ensuring the international linkages of the Program. She finished her Master of Community Development degree from the University of the Philippines in Diliman.

Prior to her work in the academe as a full-time researcher, she headed the Secretariat Team of the ASEAN Civil Society Conference/ASEAN Peoples' Forum in 2017, the broadest regional platform of civil society groups that engages issues on ASEAN. Her interest on ASEAN and Southeast Asian issues began in 2002 during the Southeast Asia People's Festival in Phnom Penh, Cambodia, where she was part of.

While working in the academe, she still joins the social movement in the Philippines and at the Southeast Asia level. She has lived and worked with Metro Manila urban poor communities as well as indigenous peoples in Mindanao, where she practiced participatory development planning and action research.

She has conducted field research in many parts of Southeast Asia, including the borderland communities. Her fields of interest are globalization and development studies, critical agrarian studies, resistance and solidarity, radical political geography, gender, and development. Many of her papers were published by UP CIDS while some of her recent essays appeared in the websites: the Global Tapestry of Alternatives and Governance in Conflict Network. e-mail: Ananeza.Aban@ugent.be

Contributors

Ananeza P. Aban Ghent University, Ghent, Belgium

Josie Abuque Ayta Mag-indi Community, Porac, Pampanga, Philippines

Jeroen Adam Ghent University, Ghent, Belgium

Andrew Aeria Persatuan Pemangkin Kesedaran Sosial, Kuching, Malaysia

Raphael Baladad Focus on the Global South, Manila, Philippines

Carl Eugene Balbas University of the Philippines, Diliman, Quezon City, Philippines

Nathaniel Punongbayan Candelaria Department of Political Science, College of Social Sciences and Philosophy, University of the Philippines Diliman, Quezon City, Philippines

Marivic Capuno Ayta Mag-indi Community, Porac, Pampanga, Philippines

Rita Capuno Ayta Mag-indi Community, Porac, Pampanga, Philippines

Rafael V. Dimalanta University of the Philippines, Diliman, Quezon City, Philippines

Bianca Ysabelle E. Franco Independent Scholar, Quezon City, Philippines

Evelyn D. Galangera Ma-Igting na Samahan ng Panlipunang Negosyante ng Towerville, San Jose del Monte, Bulacan, Philippines

Jennifer A. Macabudbud Ma-Igting na Samahan ng Panlipunang Negosyante ng Towerville, San Jose del Monte, Bulacan, Philippines

Micah Magaro University of the Philippines, Diliman, Quezon City, Philippines

Angeli Fleur G. Nuque University of the Philippines Center for Integrative and Development Studies (UP CIDS), Quezon City, Philippines

Rowena E. Osal Ma-Igting na Samahan ng Panlipunang Negosyante ng Towerville, San Jose del Monte, Bulacan, Philippines

Venarica Papa University of the Philippines, Diliman, Quezon City, Philippines

Josephine Parilla PATAMABA Workers in the Informal Sector Enterprise, Angono, Rizal, Philippines

Rosalinda Pineda Ofreneo University of the Philippines, Diliman, Quezon City, Philippines

Regie Anne D. Placido Center for Asian Mission for the Poor (CAMP)—Asia, San Jose del Monte, Bulacan, Philippines

Supatsak Pobsuk Focus on the Global South, Bangkok, Thailand

Benjamin R. Quiñones Jr. Asian Solidarity Economy Council, Quezon City, Philippines

Ma. Victoria R. Raquiza National College of Public Administration and Governance, University of the Philippines, Diliman, Quezon City, Philippines

Justin Jette Roldan University of the Philippines, Diliman, Quezon City, Philippines

Roja Estrelita Salvador Social Watch Philippines, Diliman, Quezon City, Philippines

Jose Monfred C. Sy University of the Philippines (UP) Diliman, Quezon City, Philippines

Honey B. Tabiola University of the Philippines Center for Integrative and Development Studies, Diliman, Quezon City, Philippines

Eduardo C. Tadem University of the Philippines (UP) Diliman, Quezon City, Philippines

Donald Takhell Indigenous Perspectives, Manipur, India

Violeta Valenton Ayta Mag-indi Community, Porac, Pampanga, Philippines

Benjamin B. Velasco School of Labor and Industrial Relations, University of the Philippines, Diliman, Quezon City, Philippines

Part I
Introduction

Chapter 1
Finding Alternatives Among the Grassroots and Social Movements

Jose Monfred C. Sy and Eduardo C. Tadem

Abstract This introductory chapter examines the complexities surrounding the concept of "development" and its entanglement with power dynamics, colonial legacies, and capitalist imperatives. From Truman's 1949 address to the emergence of the era of development amidst colonialism, the discourse on development has been fraught with challenges. The chapter highlights how mainstream paradigms, exemplified by the Association of Southeast Asian Nations (ASEAN)'s initiatives, often prioritize corporate interests over community well-being. Critiques of ASEAN's model underscore the need for alternative approaches rooted in principles of equality, solidarity, and sustainability. Various alternative visions, from grassroots initiatives to civil society movements, challenge neoliberal capitalism and advocate for a pluriversal approach that embraces diverse ways of being and knowing. By foregrounding marginalized voices, these alternatives offer pathways toward a more inclusive future. The chapter sets the stage for exploring decolonial alternatives in Southeast Asia beyond the geographical, political, and socio-economic scope imposed by ASEAN.

Keywords Alternatives · Alternative development · Development discourse · Neoliberal capitalism · Pluriverse

Like other broad terms such as "freedom" and "democracy," "development" often evokes an optimistic connotation despite the difficulty of pinning it down. Therein lies the problem. On the one hand, it can easily be associated with the pursuit of a better life for all (Peet and Hartwick 2015). Unlike "growth," which may also describe a detriment, development is commonly equated by scholars to structural changes and improvements that encompass economic progress (Boothroyd and Davis 1993; Green and Haines 2002). On the other, the term's broadness lends itself to different

J. M. C. Sy (✉) · E. C. Tadem
University of the Philippines (UP) Diliman, Quezon City, Philippines
e-mail: jose_monfred.sy@up.edu.ph

E. C. Tadem
e-mail: ectadem@up.edu.ph

J. M. C. Sy and A. P. Aban (eds.), *Reimagining Development in Southeast Asia*, Asia in Transition 21, https://doi.org/10.1007/978-981-97-9140-8_1

and sometimes opposing ideas of a "better life." Whose lives are improved and uplifted? And at whose cost? Where does "development" exist? Where is it absent?

In his inaugural address, United States (US) President Harry S. Truman answered the latter two questions. There he claimed that more than half of the world's population live in "underdeveloped areas." According to Sachs (2019, xii), this is the first time that the term "underdevelopment" was proclaimed by a prominent figure, laying down the conditions for the use of the term "development" as justification for national and international power. The pronouncement also signaled the beginning of an "era of development," which overlapped and closely followed the epoch of colonialism. For Rist (2019), such a claim casts development as a ladder which places countries in a hierarchy bookended by blanket notions of developed and un(der)developed.

Three years after Truman's speech, United Nations (UN) experts came together to outline a global project of development that they characterize as "rapid economic progress" that comes with "painful adjustments" (UNDEA 1951, 15). UN's claim that "very few communities are willing to pay the full price of economic progress" suggests that a "better life" requires "underdeveloped" peoples to leave progress in the hands of the "developed" few. Escobar (2018, 6) views this "utterly arrogant" vision as an elimination of already existing modes of economic, social, political, and cultural development: "almost overnight, a diverse range of rich and vibrant traditions were reduced to being worth, literally, nothing: nondescript manifestations of an allegedly indubitable fact, 'underdevelopment.'" As Tortosa (2019, 10) puts, this is not about "growth and technology," but about who/what are central and peripheral in terms of power. Indeed, development must be described as "uneven," but not because of a geopolitical hierarchy of competitive advantage or chronopolitical race to attain a "better life." Development is uneven because the agency to advance it is monopolized by the few.

Engaging the Southeast Asian context, this book offers a provisional answer to the same questions: development has existed across the globe as all societies are capable of advancing progress in their own terms. A "better life" is possible outside the developmental framework imposed by the same political forces that benefitted from colonialism. It is possible because, for many years and on their own, communities and grassroots organizations have engaged in alternative, heterodox, and non-mainstream practices that encompass economic, political, and socio-cultural dimensions of development (Tadem et al. 2020, 16).

The book at hand is a product of collective efforts from community organizers, activists, and committed scholars. It presents nineteen case studies of social movements, community-civil society partnerships, and intra- and international alliances across Southeast Asia as evidence that peoples on the ground have been building their own paths to development, which they define based on tradition marginalized by Western epistemologies, knowledge exchanges with other formations, and long histories of struggle. Questions about development are asked to communities and civil society organizations instead of states and financial institutions: how do grassroots actors address the inadequacies of the hegemonic models of development? How do they make sense of their difficult situation in relation to issues faced by other sectors? What types of strategies do they employ as a response? And most importantly, what lessons can we learn from these alternatives to mainstream development?

This collection contends that surfacing these narratives from Southeast Asian peoples can help realize alternative development paradigms, policies, and projects that challenge notions of "developed" and "underdeveloped." Privileging these alternatives contribute to the emergence of new and progressive models of development, which embrace cooperation, solidarity, mutual benefit, the commons principle, and joint development instead of cutthroat competition, the thirst for profits, and narrow patriotism that leads to chauvinism.

1.1 Colonialism and Capitalism Behind Modern "Development"

The dominant framework of development, which privileges economic progress at all costs, cannot be separated from histories of colonialism and capitalism justified by the discourse of modernity. In the broadest sense, "modernity" captures social, economic, political, and cultural dynamism, situating people in time (Ogborn 1999, 53). To define development as modernity, which underpins the work of international financial institutions such as the World Bank prods people to look at development in largely economic terms (Willis 2005). Kothari et al. (2019, xxii) see this framing as a "critical part of our problems." This is not to say that everything modern is pernicious to societies. Even human rights are modern inventions that have become liberatory for some groups. However, a key feature in modernity is the conception of "universalism," which asserts that human civilization develops in a single, globalizing world in a linear timeframe, with high-end technology as the only reliable harbinger of progress even in social and cultural spheres (Blaser 2019; Wills 2005).

There is little doubt that this discourse of modernity-as-development is rooted in colonialism. The European colonial enterprise and the systems it imposed across the globe mandates that all human societies must occupy the "same temporal, geographic and ideological space" (Rist 2019, 66). The "developed/underdeveloped" divide finds it roots in colonialism's "civilized/native" and "West/East" discourses, which for centuries have justified colonial subjugation of already peopled and organized territories. The colonial mandate sees this epoch as shining light on allegedly "dark continents" (Bassey 2019, 4). These systems have already ushered in several aspects of modernity, such as the bureaucratization and governmentalization of life by a centralized state and institutions linked to it (Escobar 2018). Deeply personal and communal practices such as child development and food production became the object of calculation, scientization, and designing for colonial use. For Harvey (2003, 46) these conquests among European nations to occupy other territories opened the floodgates to "the pillaging of much of the world's resources by the imperial powers; and the widespread deployment of virulent doctrines of racial superiority; all matched by a total and predictable failure to deal with the surplus capital problem within closed imperial domains, as seen in the great depression of the 1930s."

By then, the world changed along another axis, European colonial rule seemed to have reached its zenith on the eve of the First World War, followed by the emergence of national liberation movements first led by native elites educated in leading academies in the West, and later involving peasants, workers, and other marginalized sectors (Motta and Nilsen 2011). The Second World War turned everything upside down (Rise 2019, 69). The early post-war periods introduced the "discovery" of mass poverty, which has provided the anchor for the need to restructure global culture and political economy (Escobar 1999, 382). The victors took it upon themselves to lead in this "restructuring." Alongside the unraveling of colonial enterprises, the bloodshed and wreckage left by the World Wars opened formerly colonial territories and nations struggling for liberation to development assistance or cooperation with superpower countries (Gould 2019). The newly independent nation-states came to constitute the Third World geopolitical entity united by their heritage of colonialism (Motta and Nilsen 2011). Bursting with nationalist fervor, many of these countries have committed to geopolitical non-alignment as a strategy of internal capitalist development (Desai 2004).

These emergent "modernities," so to speak, were arrested in their early stages during the post-war years, which were waged in the context of the Cold War between two competing factions—one led by US, the other by the Soviet Union—and their ideological positions—capitalism and communism, respectively. The economic depression and decolonization projects in the wake of these geopolitical tensions, which includes regional proxy wars influenced by the Cold War, inspired the strategy for world order devised by the US-led faction (McMichael 2019). Mainstream theories of development have evolved since the mid-twentieth century, shaped by historical events and geopolitical tensions. In the aftermath of decolonization, post-World War II reconstruction, and the Cold War, modernization theory came to the forefront of development paradigms during the 1950s and 1960s. Initiated by major international organizations such as the World Bank, the International Monetary Fund, and the United Nations, this approach advocated for rapid modernization of newly independent nations following the paths of industrialized countries. Proponents stressed the importance of industrialization, free trade, and free markets, assuming that economic growth would occur linearly through predefined stages of progression (Rostow 1960). Factors considered crucial included capital accumulation, technology transfer, and integration into the global market.

As dissenting viewpoints gained traction, dependency theory surfaced as an answer to modernization theory toward the end of the 1960s and early 1970s. Dependency theorists contended that underdevelopment in the Global South stemmed from developing economies being integrated into the worldwide market structure, resulting in subjugation by developed economies. Intellectuals including Frank (1967) proposed that the prevailing global financial system established a dependent relationship between the periphery—comprised of developing countries—and the core—consisting of developed countries—thereby maintaining impoverishment in the former. Development scholar Satake (2003, 15) astutely links dependency to colonial relations: "accordingly, the present economic disparity between the two dates back to the colonization of the latter by the former in the sixteenth century, and the dependent relations will not change unless the centuries-old global division of labor is altered." As such, the population of underdeveloped nations in the Global South, such as many in Southeast Asia, was polarized into rich urban and poor rural spheres (Frank 1967).

The so-called demise of communism marked by the dissolution of the Soviet Union shrunk the arena of competing models of modernity. What is left is the model of neoliberal globalization which "retains the state in the 'nation-state,' but makes it difficult to sustain the nation" (D'Souza 2011, 230). Beginning in the 1980s, neoliberal policies became increasingly influential within the development discourse due to promotion by institutions like the IMF and World Bank. These policies mandated the adoption of free-market reforms through programs such as Structural Adjustment Programs (SAPs) implemented in debtor countries. Such initiatives necessitated trade liberalization, diminished government intervention, reduced regulations, and enforced austerity measures. The contradictory anti-statist thrust of neoliberalism "advocates a redeployment of state power to cement ruling-class domination, break down barrios to the penetration of capital and to further the commoditization of both society and nature" (Bolden 2011, 85). The myths of "scarcity and limits to growth" calcified by neoliberalism have then and again justified the ravaging of the environments of resource-rich countries for the sake of profit (Sison 2020, 43). Indeed, such a model pushes hitherto colonies into a dependent relation with the metropoles of capitalism, but not without the backing of governments ran by the local ruling class. Critics assert that these prescriptions exacerbated income disparities and societal strife without fostering enduring development (Bello 2023; Sachs 1992; Sen 1999).

Such a program maintains, albeit in a bureaucratic form, the imperialistic notion that the West must "arrogate for itself the right to be 'the world,' and to subject all other worlds to its rules" (Escobar 2018, 86). An imaginary geography divided into Global North—the centers of capitalism, imperialism past and present—and Global South—the Third World assimilated into the discourse of globalization—justifies the economic, political, and cultural intervention in what is framed as "maldeveloped" or "underdeveloped" nation-states (D'Souza 2011, 228). These are not merely adjectives that discriminate one country from another. As Satake writes (2003, 16), "underdeveloped countries in the South have unduly-grown export sectors such as plantation agriculture and mining, on the one hand, and meager manufacturing sector, on the other," stunting the possibility of heavy industrial manufacturing that could have been a foundation for a self-sustaining mode of production.

Modernity continues to describe economic and societal forms based on the experiences of Western Europe, which encompasses industrialization, urbanization, and the intensified use of technology to speed up mass production (Willis 2005). As Escobar (2018, xiii) proclaims, "today we would say (ontologically) that development policy and planning, as well as much of what goes on under the banner of design, are central political technologies of patriarchal capitalist modernity and key elements in modernity's constitution of a single globalized world." Administrative processes at the national level such as domestic migration and economic liberalization echo a process of "internal colonialism" that characterized "the formation of most modern

Western nation-states" (Scott 2009, 3). This proclivity casts development as "growth, expansion, entanglement, and spread" (Bassey 2019, 3). None of these terms espouse social or ecological justice.

The use of the term "development" in this context indicates an underlying biological metaphor: that living beings develop in a natural, gradual, and beneficial genetic code, and in case irregularities take place, expertise must intervene. Underdevelopment is not the opposite of development, but rather the latter's incomplete or—to use another biological term—embryonic form (Rist 2019, 74). This framework, which dovetails with Truman's 1949 address, envisages a development program based on the "democratic" fair-dealing among nation-states (Tortosa 2019, 9). Western notions of "civilized/native," which has evolved into the dynamics of "developed/underdeveloped" and "North/South," frame a model of development characterized by the concentration of surplus capital, the commercialization of basic social services, the propagation of modern values, and the creation and hegemony of international institutions tasked to develop the underdeveloped world. In this era, "poor people's ability to take care of their own lives was eroded in a deeper manner than perhaps ever before" (Escobar 1999, 382).

1.2 The ASEAN's Development Framework

Mainstream notions of development heavily involve the participation of governments, international institutions and intergovernmental bodies, mainstream economists and planners, and non-government organizations (Phillips and Pittman 2009; Sachs 2019; Willis 2005). In the Southeast Asian region, such a framing has been adopted and is advanced by the Association of Southeast Asian Nations (ASEAN). ASEAN was first convened in August 1967 in Bangkok by Indonesia, Malaysia, the Philippines, Singapore, and Thailand, represented by their foreign ministers who were hailed as its "Founding Fathers" (ASEAN n.d.). Often compared to that of the European Union, the ASEAN model of regional integration has been lauded to have benefitted economic progress and cooperation (Idris and Kamaruddin 2019).

However, ASEAN's framework of socio-economic "growth" has been criticized for prioritizing profits—in the form of foreign direct investments from multinational and local corporations—over the peoples of Southeast Asia. In fact, despite the geographical proximity among ASEAN member countries, the organization's early years were marked by disparities in economic status, limited integration, and ongoing reliance on capital inflows from highly industrialized nations (Nesadurai 2008; Reyes 2015). Moreover, ASEAN's principle of regional peace aligned with the US' agenda to end communist-led insurgencies in the region during the Cold War (Eccleston et al. 1998; Söderbaum 2012). In this high-tension political context, ASEAN, as the only functioning intergovernmental organization in the region, laid down its regional economic cooperation program through the ASEAN-Post Ministerial Conference (PMC). The PMC casts ASEAN as a "single corporate entity" that can engage with external parties the states see vital in regional economic progress (Nesadurai 2008, 235).

In 1992, the ASEAN Free Trade Area (AFTA) was created to attract foreign direct investments to a single, geographical economic space that is Southeast Asia (Stubbs 2012). AFTA transformed ASEAN as a "hub of free trade agreement (FTA) networks in East Asia, thus, taking the 'driver's seat' in economic integration in the region" (IBON 2015, 2). Maximizing the geo-economic potentials of the region, ASEAN opened its doors to Western transnational corporations which have extended and relocated their procurement, production, and sale processes across Southeast Asian countries. Cheap labor and rich natural resources that can be found in the region attracted these investors (Reyes 2015).

Facing critics and hurdles in economic integration, ASEAN announced its vision of an integrated Southeast Asian market and production base through the ASEAN Economic Community (AEC) in 2003. While it was instituted along with two other pillars, the Political and Security Community (APSC) and the Socio-cultural community (ASCC), AEC overrides the others as it links directly to the project of economic progress envisioned by ASEAN (146). Like AFTA, the AEC seeks to liberalize trade and investment rules, eliminate non-tariff barriers, and encourage the free movement of skilled and professional workers across Southeast Asia (Aban and Sy 2022; Nesadurai 2008; Reyes 2015). Western multinational corporations have expanded and shifted their sourcing, manufacturing, and distribution operations throughout ASEAN nations (IBON 2015; Reyes 2015). AEC privileges the rights of business through increased protection for investors and corporations based on intellectual property rights. However, the social and economic protection of Southeast Asian peoples who face discrimination, workplace abuse, and corporate crimes on the ground remain absent in the ASEAN's vision (Aban and Sy 2022; Langlois et al. 2017; Weiss 2021).

The ASEAN's norms of sovereignty and non-interference in the name of non-coercive, consultative, and consensus-seeking among nations exacerbate the marginalization of the grassroots in designing regional developmental framework. Nesadurai (2008, 228) denounces the intergovernmental body's state-to-state conduct, or the "ASEAN Way," as "governed by a set of informal, procedural norms emerged through repeated practice, consensus decision making, non-confrontational, and deliberative style of interaction." For Collins (2008, 315), these principles of non-interference and economic integration only serve to safeguard bureaucracy of "the elite," closing the door for the participation of civil society and grassroots communities. The ASEAN People's Assembly (APA) was formed by ASEAN in 2000 to sustain its veneer of a "people-centered agenda." However, as critics note, the APA and the ASEAN at large provided very limited avenues to civil society organizations (CSOs) and grassroots communities, still tending toward exclusivity and secrecy, and offering only token attention to CSO recommendations (Aban and Sy 2022; Clarke 2012; Gerard 2014; Weiss 2021). While consultative channels are present, peoples' issues are left to the care of their respective nations based on the ASEAN's principle of non-interference and privileging of national and religious contexts (Clarke 2012; Uy 2012). Tadem et al. (2020, 3) summarize these criticisms: "founded in 1967 as a mechanism to support the US-led Western faction of the Cold War," the ASEAN evolved into a platform for the proliferation of "the neoliberal market-led agenda of development promoted by global capitalism."

1.3 Globalization Versus Development

Some four decades ago, the world witnessed the surge of globalization—the increased interconnectedness of countries through trade, investments, and cultural relations. Functioning under a neoliberal paradigm, the phenomenon was marked by heightened liberalization, privatization, deregulation, and standardization where states gave up most of their economic prerogatives to the private sector. Prominent players in this process included global institutions such as the World Bank Group (WBG), the International Monetary Fund (IMF), and the World Trade Organization (WTO).

The economic changes that took place particularly in developing societies saw the influx of investments in big-ticket infrastructures such as mega dams, extractive industries such as mining, large-scale agri-business plantations that utilize monocropping modes and chemical inputs, special economic zones that were divorced from the rest of the national economy, and high-end property development, among others. Another feature was the rise of speculative portfolio investments whose short-term character often resulted in volatile national current accounts.

Joseph Stiglitz, in *Globalization and its Discontents* (2002), while acknowledging the resulting rise in economic growth and other conventional indicators, also recognized the problems accompanying this type of globalization and raised several issues of concern. First, he observed the rise in economic and social inequalities between and within nations where the needs of the poor and marginalized were sacrificed. Free trade policies were harmful to developing economies while privatization privileged conglomerate businesses to the disadvantage of workers, and micro, small, and medium enterprises. Stiglitz further criticized the austerity measures imposed by the IMF and WBG that constricted the state from increasing public spending for social services. He wrote that while the "IMF is often called upon … when the country is facing a crisis … (its) structural adjustment policies "led to hunger and riots in many countries" as "benefits went disproportionately to the better-off, with those at the bottom sometimes facing greater poverty" (xiv). For Stiglitz (2002, 235), the problem with the rules of the game set by the IMF, WBG, and WTO is that "all too often, (they) have served the interests of the more advanced industrialized countries—and particular interests within those countries—rather than those of the developing world."

Furthermore, the lack of transparency in decision making accompanying the making of these rules also undermined democratic institutions. Stiglitz (2002: 219) also remarked on the environmental costs of globalization as natural resources are degraded as "little attention is paid to environmental concerns" in trade negotiations under the WTO. Stiglitz (2002, 269) concludes that:

> For millions of people globalization has not worked. Many have actually been made worse off, as they have seen their jobs destroyed and their lives become more insecure. They have felt increasingly powerless against forces beyond their control. They have seen their democracies undermined, their cultures eroded.

As the first decade of the new millennium unfurled, economic growth started slowing down and recession affected even the most advanced nations. As the Economist (2009) pointed out, the "global movement of goods, capital and jobs, is now in trouble" and that "the downturn affected has been sharpest in countries that opened up the most to world trade." The normally conservative **Economist** grudgingly admitted "of the failure of globalization to deliver many of its advertised benefits, especially to the poor."

The Global Financial Crisis (GFC) of 2008–2009 (also known as The Great Recession) is considered the gravest to hit the world's economy since 1945 (Knowlton 2009). Originating in the United States, it spread worldwide and caused the global economy to contract for the first time since World War II (Knowlton 2009). Causal factors include the US housing bubble (subprime mortgage crisis), an unstable financial sector (risky banking practices, deregulation, securitization, and credit derivatives), inadequate state regulatory oversight, and global trade imbalances between countries.

The socio-economic results were deep recession, financial turmoil, widespread job losses, increased public debts, augmented inequality, social unrest, decline of trust in financial institutions and governments, and significant economic hardship for millions of people around the world. Politically, populist governments were ushered into power, ultra-nationalism came into vogue, and the viability of globalization to address poverty and inequality became questionable. Recovery from this crisis was slow, uneven, and fraught with uncertainties and lingering after-effects that continue to haunt the global economy.

Responses from governments to the global economic meltdown were varied and uncoordinated such as "higher taxes, lower public spending, devaluation, inflation, and investment in green growth" (Taylor 2009). Other responses were massive bailouts of failed banks and other financial institutions, easing of monetary policy, fiscal stimulus packages, regulatory reforms, and international gatherings of leaders. Corporations, on the other hand, instituted cost-cutting measures by downsizing, merging, and focusing on their core businesses—resulting in job losses in the tens of thousands. In any case, the Economist (2009) lamented that "the integration of the world economy is in retreat on almost every front."

The responses from civil society organizations and social movements were more pointed, rejecting the neoliberal model and calling for reforms and restructuring. A prominent case is the "degrowth" movement. The central message of this campaign is invoked by Kohei Sato in his bestseller ***Slow Down: The Degrowth Manifesto*** which states that "the quest for infinite growth in gross domestic products will crash into the limits of a finite planet" (Hoang 2024). Appalled and dismayed with the "corporate capture of agriculture" and its depletion of natural resources that "puts profits over people," non-mainstream groups are pushing for "people to reclaim autonomy" and buying "directly from farmers' collectives" (Hoang 2024). A milder and more accommodating version of "degrowth" is "doughnut economics." As propounded by economist Kate Raworth in her doughnut diagram (Hoang 2024),

> countries should grow within two rings. The inner ring represents basic development -- countries need growth to meet a minimum quality of life and not fall below this development level. The outer ring represents the Earth's confines and resources. Typically, industrialized nations have already grown enough that if they continue unchecked, they threaten to push past what the globe can handle.

By far the most radical response to globalization's failures is "deglobalization," an idea first propounded by Walden Bello in *Deglobalization: Ideas for a New World Economy* (2004). Asserting that "the current system of global economic governance is a systemic one" that cannot "be addressed by mere adjustments within the system," Bello called for "an alternative system of global governance" and "reorienting economies from the emphasis on production for export to production for the local market" (134, 139, 140). The essential features of this deglobalization paradigm are (140–141):

1. Drawing most of a country's financial resources for development from within rather than becoming dependent on foreign investment and foreign financial markets.
2. Carrying out the long-postponed measures of income redistribution and land redistribution to create a vibrant internal market that would be the anchor of the economy and create the financial resources for investment.
3. De-emphasizing growth and maximizing equity in order radically to reduce environmental disequilibrium.
4. Not leaving strategic economic decisions to the market but making them subject to democratic choice.
5. Subjecting the private sector and the state to constant monitoring by civil society.
6. Creating a new production and exchange complex that includes community co-operatives, private enterprises and state enterprises, and excludes TNCs.
7. Enshrining the principle of subsidiarity in economic life by encouraging production of goods to take place at the community and national level if it can be done at reasonable cost in order to preserve community.

In sum, deglobalization can be grasped as "a program, the main points of which were refocusing the economy back on production for the domestic market than for export markets, resubordinating the market to society, reasserting cooperation over competition, replacing the pursuit of narrow efficiency with that of social effectiveness, leaving what could be done locally at the local level, and allowing for diversity of ways of organizing and economy rather than fitting all economies into one mold, the neoliberal template."

Within the past decade, however, deglobalization (or at least some of its features) was incorporated into the agenda of rightwing and far right groups particularly in Europe. Bello (2023) notes how the far-right National Front in France of Marine Le Pen "cleverly mixed valid working-class anxieties about globalization with anti-European Union and anti-immigrant sentiments." What was taking place was the capture of progressive themes under the deglobalization paradigm within a completely divergent ideology espoused by a faction of the ruling class that sought to distinguish and divide peoples and communities on the basis of race, gender, ethnicity, geography, nationality, and cultural traits.

This development is eerily reminiscent with how, in the recent past, concepts of human development, environmental sustainability, social protection, agrarian reform, minimum wage, eight-hour workday, universal health, education, and suffrage and inclusive growth have been similarly adopted by mainstream institutions like governments of developed states, UN agencies, ASEAN, OECD, WBG, ADB, etc. For many decades, these advocacies have been fought for in the streets by progressive movements but have been institutionalized and made to conform and be subservient to mainstream and conservative development strategies. In the process, the policies and programs that arise accommodate the political and economic agendas of the ruling classes, thus diluting their original objectives.

1.4 Alternatives to What? Development for Whom?

The criticisms against mainstream (that is, Western-modern) development and the ASEAN outlined in brief above also inspire CSOs, politically engaged intellectuals, and social movement activists to think of alternatives. While this framework has improved literacy, health, and the number of people living in absolute poverty, it is not without multiple adverse consequences. Chambers (1997, 1) sums it up in saying that "linear and convergent development through stages of [...] unlimited growth" where industrialization accelerated by foreign intervention is considered key has been exposed as "misconceived and, with easy wisdom of hindsight, naïve." The question development scholars and practitioners must ultimately ask is, in a given framework of development, "whose reality counts?".

Alternative proposals, representing radical re-imaginings of how development may look like, arose alongside the growing disillusionment over mainstream development models (Hapal 2020). This dream and later movement toward alternatives gained more prominence in the late 1980s and 1990s, (4) bolstered in part by Korten and Klauss' (1984) clarion call for a people-centered development to usurp the hegemonic "production-centered" model. Their model of development forwards that peoples, especially in the Third World, are capable of self-organizing and directing their own paths toward development, now reframed as a participatory process. "People-centered" means that those in the grassroots and civil society design their own benchmarks of progress and employ their own methods for achieving it. This is development for the peoples, by the peoples (Hapal 2020, 4). The irony is that, since it has become a staple in development practice especially at the community level, "participation" has been co-opted to legitimize the same undemocratic processes shaped by the discourse of modernity. Pieterse (1998, 348) points out how these institutions see "alternative" as a way of being progressive without endorsing any "radical" ideologies that may challenge hegemony. Cooke and Kothari (2001) expounded on this critique in the provocatively titled *Participation: The New Tyranny*. They observe that the term has been used by states and international institutions tokenistically to acclimatize communities to mainstream notions of development, thus justifying the continued imposition of profit-centric models.

By the turn of the twenty-first century, critics had pinpointed additional deficiencies plaguing conventional development strategies. Notably, they targeted the Eurocentric orientation inherent in prevalent development models, arguing that such methods superimposed Western norms and financial blueprints onto various cultural landscapes devoid of proper consideration for native customs, traditions, and histories. Escobar (1995) maintained that development functioned as a manifestation of cultural hegemony, silencing non-Western modes of cognition and existence. Furthermore, emerging foci revolved around environmental sustainability given the destructive implications tied to resource-intensive expansion endeavors, along with mounting concern over unequal socio-economic results produced by neoliberal policy implementations. Demands then intensified for development approaches centered on social fairness, parity, and human rights, exemplified by Amartya Sen's work (1999) accentuating expanded personal liberties and competences as authentic determinants of improvement.

Traditional development narratives primarily emphasized economic growth and modernization. Escalating criticisms originating from both the Global South and alternative development thinkers paved the way for more accommodating, culturally aware, and environmentally conscious approaches. Responding to these developments, new alternative development paradigms took shape. Among them is the post-development paradigm that challenges the fundamental idea of development itself as a Eurocentric construction propagating inequality and neocolonialism. Ongoing debates seek to recast development in manners that elevate human welfare, environmental stability, and social equality. Advocates propose place-based, pluralistic methodologies rooted in communal control and attuned to cultural nuances, opposing universal templates (Sachs 1992; Esteva and Prakash 1998). Additionally, movements such as Buen Vivir ("good living" in Spanish), as described by Gudynas (2011), draw inspiration from Indigenous belief systems across Latin America, espousing harmony among humans and nature alongside community collaboration over individual prosperity. Gudynas (2011, 444) praises the plural and highly embedded orientation of Buen Vivir as a wide net of Indigenous and community-based principles:

> All these different ideas of Buen Vivir are specific to each culture, with its own language, history, specific social and political contexts, and placed in diverse environments. Thus, the Ecuadorian *sumak kawsay* is not identical to the guarani *ñandereko*, and these two are different from all the others. There is no room for an essentialist position. Furthermore, is not possible to identify one idea of the Buen Vivir as the best one that became a standard reference to be followed by all others indigenous groups in Latin America.

Such a paradigm is borne out of the actual practice of a multicultural and multilingual people disentangling their communities from the vestiges of colonial history and imperialist rule. The aforementioned deglobalization also emerges as another alternative strategy, stressing decentralization of economies toward domestic production, renouncing unfettered competition in favor of cooperative efforts, embracing varied avenues of organization distinct from the neoliberal prototype, and promoting local solutions tailored to unique circumstances (Bello 2023).

Despite this challenge of keeping alternative models of development as challenges to hegemony, community organizers, civil society leaders, and scholar-activists continue to explore and research on the possibilities of an emancipatory standpoint emerging from radical political projects practiced on the ground. Alongside the question of "whose reality counts," we must also ask of emergent development models, "alternative to what?" Serrano and Xhafa (2011, 289) provide a clear answer: "an ongoing, non-deterministic process of economic and political struggle of people trying to move beyond the capitalist logic, be it at the macro, meso or micro level, and simultaneously transforming themselves in the process." At the core of any alternative is the promise of the full development of human potential that necessitates principles absent from capitalism: equality, solidarity, sustainability, and participatory democracy. Alternatives prove insubstantial what Lessem and Schieffer (2010, 124) describe as the "axiom of the selfish nature of human beings" that laid out the blueprint of neoliberal capitalist and imperialist designs.

What about alternative development? For Satake (2003), a framework of development that serves as an alternative to the dominant model critiques and transcends the circumstances that render a people dependent and exploited by others. Synthesizing several socio-economic theories of self-reliance, political movement, and sustainability, he concludes that alternative development must "satisfy basic human needs such as clothing, food, shelter, and medical care and to guarantee a social condition where people can realize their plans in life" (22). Alternative development must provide a space for the nurturing of critical consciousness and transformative capacities of participants in pursuit of developing in full their human potential.

At the risk of instituting new totalitarian regimes, alternatives must not be defined by a coherent inventory of concepts and prescriptions. Rather, as Lang et al. (2018) forward, alternatives emerge from on-ground praxes of dismantling systems that impose power over others, hindering individuals and sectors from accessing their democratic rights and maximining their potential. The term "alternative development" itself conveys an active connotation of choice and underscores numerous "people's initiative[s] in the development process in a given society or community" (Satake 2003, 23). This particular development discourse disentangles itself from mainstream indicators such as gross domestic product, per capita income, and other metrics from the lexicon of capitalist thinking (Shiva 2019). Instead, it aligns itself with praxes of solidarity that help people live a life of dignity. Practitioners of alternatives nurture arenas—both material and symbolic—where even different social dynamics and productive capabilities can thrive in harmony. Many of these arenas show potential for facilitating and fostering the emergence of alternative pathways (Serrano and Xhafa 2012).

Various socio-economic arrangements can challenge and surpass the fundamental principles of neoliberal capitalism, i.e., shared ownership and administration of production as opposed to private ownership and capitalist-controlled production;

the market as a mechanism for redistribution rather than a tool for private accumulation; dignified and socially beneficial labor instead of fragmented and alienated work; inclusive and democratic representation and popular participation of peoples across class, gender, race, and ethnicity instead of selective elitist representation based on social status; the democratization of basic social services such as education instead of subsuming them under the charge of corporations; the conservation and free practice of Indigenous traditions and belief systems instead of a totalizing profit-centered culture; and the restoration of communal resources and environmental justice in opposition to industrial practices reliant on social, extractivist and ecological exploitation. The list is by no means exhaustive as new alternatives continue to crop up in defiance of dominant and unsustainable ways of actualizing and advancing a good life.

1.5 Visions from the Grassroots and Social Movements

Alternatives are nothing new. In every epoch of human and class-differentiated history, the dominated and exploited classes envision and commit to various practices that counter the logics of the dominating and exploiting classes. As Tadem (2012, 162) puts it, "the poor and their communities have (through the ages) been able to manage their own economic and political lives through mechanisms that lie outside the formal systems of governance and economics." Hence, turning to the collective wisdom of the grassroots and social movements that have attempted to pave their way toward their own vision of development can help concretize what alternative development could and should be.

The grassroots, that is, the ordinary folk of communities across the world, often have a deeper understanding of their community's needs and issues than an outsider, more so a bureaucrat. Encumbered by a myriad of economic, political, social, and even environmental problems dealt by those in power, many grassroots communities mobilize into collective action as a form of popular democracy (Kaufman 1997). Alternative practitioners in the grassroots, by appropriating spaces where they can manage their own, get "to control their lives and livelihoods" (162). As Esteva (2009, 20) provocatively states, marginalized peoples "are increasingly dedicated to the marginalizing of the economy." In other words, the more the grassroots are driven into destitution, the more they renege the dominant systems that marginalize them in the first place. One can even interpret the alternatives operationalized by and in the grassroots as prefigurations of what the world can be outside capital and empire.

Typically, the grassroots work with, overlap, and/or develop into social movements and people's organizations, formations intervene "in the political process through advocacies and campaigns on regime and system change or devising strategies and practices that engage the state" (Tadem 2012, 161). What distinguishes social movements from other types of organization, like protest gatherings and community mobilization, is their longevity and structure. While shared grievances often spark collective action initially, structural political openings, such as mutual education, ideological growth, and robust social connections, inspire participants to continue their involvement in the movement (Sy 2022). The (un)favorable conditions from which movements arise are often stirred further by their disillusionment with existing socio-economic and political structures, which they then seek to reform, overcome, or replace totally. Grassroots communities that are pushed further into the margins of society and social movements that struggle to change societal systems provide the configurations of new paradigms and strategies for alternative development (Tadem 2012).

Practices of decolonization, anti-capitalism, plurality, anti-racism, effacing patriarchy, and ecological sustainability designed and executed by grassroot communities and civil society organizations (CSOs) in pursuit of the upliftment of their lives are alternatives to the long history of colonization, imposed modernity, and capital accumulation operationalized through oppression and repression. In response to the shortcomings of the state, grassroots movements have emerged to address issues of development and resist state repression and various forms of exclusion and dispossession characteristic of neoliberalism (Bolden 2011). These popular initiatives operate independently of the state, concentrating their efforts within communities and civil society. So, rather than universalist solutions, Blaser (2019, 236) calls for "unleashing" heterogenous and diverse practices for the better life of communities and collectives. Therefore, the task of scholar-activists and politically engaged researchers is not to impose a new model of development but instead draw critical vocabularies and replicable designs from practices that already stand as alternatives to the dominant system. In this way, we can "transform development's order of expert knowledge and power" (Escobar 2018, 148).

While heterogenous in form and objective, alternative practices may be categorized based on common themes. Tadem et al. (2020, 16–18) provide a useful typology of alternatives covering four spheres of life. Firstly, economic alternative practices subscribe to a production that is environmental- and people-friendly. On the production side, social enterprises, producer cooperatives, agroecology, zero-waste production, Indigenous agronomic practices, and technology exchanges of sustainable food production systems. Consumer-producer linkages via alter-trade organizations and trading cooperatives also strengthen local markets and the cooperation between direct producers and consumers. Secondly, political alternative practices involve informal and formal networks of CSOs and social movements to advance together workers', peasants', women's, and Indigenous peoples' rights, among others. Concerns such as human rights and security framed by the grassroots also belong in this sphere. Thirdly, social alternative practices cover self-help groups and local networks that work toward social protection. Community-based and "barefoot" health practitioners and folk schools, which both critique and make up for the inadequacy of basic services are social alternatives that tap into marginalized and Indigenous knowledges. Finally, cultural alternative practices appear in artists and performers' participation in human emancipation. Such participation in political and economic issues goes beyond the mainstream spaces of art such as national museums and artsy studios.

Instead, cultural alternative practitioners engage in community organizing and mobilization, too. These spheres are not mutually exclusive and commonly intersect praxially. In fact, several of the alternative practices featured in this volume cut cross these four categories.

Thinking about alternatives through the concept of "pluriverse" forwarded by post-development scholars such as Escobar (2018) and Kothari et al. (2019) can be productive in this endeavor. The design of this decolonial project can be summed up as setting up "a world where many worlds fit" (Escobar 2018, 54). Such a claim does not aim to privilege cultural relativism. The pluriverse as a post-development framework seek to address the multiple contradictions cemented by capitalism through the cultivation of principles of natural repair and generation spearheaded by local peoples (McMichael 2019, 13). Pluriversality envisages human emancipation "within nature," a position that privileges the material realities and environments of those paving their way to development (Kothari et al. 2019, xxix). Alternatives that make up this pluriverse do not call for a new hegemony, "but for an end to the hegemony of any system, [which requires] a refounding of the societies of the continent based on other principles of sociability" (Escobar 2018, 181). This renewed vision of the planet must be oriented toward the alternatives to its imperialized past. Thus, values embraced by the "degrowth" principle such as sharing, simplicity, conviviality, care, and the commons take center stage in the pluriverse (D'Alisa et al. 2015, 5–6).

1.6 Weaving Together Alternatives from Southeast Asia

Learning from the above critiques of development and principles of post-development, it can be confidently argued that alternative development in the context of Southeast Asia is growing and evolving. For development to privilege the reality of the diverse peoples in the region, it must come from the peoples themselves. Rightfully so, the collection *Reimagining Development in Southeast Asia: Alternative Practices from the Grassroots and Social Movements* gathers case studies written by politically committed scholars, grassroots community leaders, and social movement activists alike from across the region. Some chapters are even collaborations among academics and alternative practitioners.

The repeated use of "Southeast Asia" instead of ASEAN is a conscious decision and at the same time a criticism of the ASEAN framework of development that limits its association with nation-states willing to participate in globalization. Various scholars have already argued that Southeast Asia should not be confined to the ten ASEAN member states. Several key points provide rationale for new approaches and strategies among Southeast Asian civil society organizations and movements. Firstly, it is crucial to differentiate Southeast Asia as a geographical area with diverse cultures, populations, and histories from ASEAN, a regional organization driven by a market-oriented and elite-controlled development philosophy. Established in 1967 during the height of the Cold War to bolster the US-backed Western bloc, ASEAN has morphed into a proponent of the neoliberal, free-market development agenda championed by global capitalism.

Scott (2009, 14) categorizes Zomia—a massive highland expanse spanning Southeast Asia—as comprising southern and western portions of Sichuan, Guizhou, and Yunnan in China; northeastern India; most of northern Myanmar; northern and eastern sections of Thailand; all of Laos excluding the Mekong valley; northern and central Vietnam; plus the northern and eastern edges of Cambodia. This zone differs significantly from lowland and coastal regions due to its persistent defiance toward state formation. Adhering rigidly to the traditional Westphalian state system focusing on unfettered sovereignty and exclusive territorial rights allows ASEAN members and ruling elites to evade responsibility regarding human rights abuses or environmental harm within their jurisdictions. Scholar-activist Mushakoji (2013, 133–137) argues that this outdated paradigm struggles to remain applicable in today's context characterized by fluid international borders and intricate population migrations fostering multifaceted social, ethnic, and cultural affiliations. Thus, "multi-ethnic, multi-identity, and multi-cultural societies" blur fixed definitions of distinctive nationalistic selfhood.

In light of these observations, Southeast Asian civil societies must foster unity and collective consciousness surpassing strict national boundaries, political divisions, and ethnic differences through harnessing shared past experiences, goals, and visions transcending restrictive concepts of territory, nationality, and individual identification. From this standpoint, this volume contends that the Southeast Asian region includes countries and territories whose peoples bear similar ethnic, cultural, and social characteristics to those in areas commonly referred to as Southeast Asia. This includes Papua New Guinea and Timor Leste, both of which continue to be denied full ASEAN membership, and areas such as Northeast India and provinces in Southwest China whose peoples share historical affinities to those labeled as Southeast Asians today (Scott 2019, 13–14). Our aspiration of a borderless region is not predicated upon specific citizenry or physical location.

The volume at hand consolidates several efforts from across Southeast Asia as a way of forwarding alternative development. The documentation of stories and practices collected in this book was primarily facilitated through the Program on Alternative Development (AltDev) of the Center for Integrative and Development Studies (CIDS), the main policy research unit of the University of the Philippines (UP). AltDev's primary goal is to popularize alternative paradigms, policies, grassroots practices, and projects to pose a critical challenge to mainstream or dominant models of development. While based in the Philippines, the documentation activities of AltDev expand to various actors within the Southeast Asia region who are undertaking various types of alternative practices. AltDev envisions that through its documentation efforts, it can contribute to the linking of peoples and communities in the Philippines and the rest of the Southeast Asia region to form the building blocks of a peoples' alternative regional integration—a form of integration that puts people first, rather than profits, politicking or vested interests.

AltDev and its orientation of scholarship-activism build on the project "Integrating Alternative Development Efforts in Asia" (IADEA) which was a workshop jointly organized by Kerala Sastra Sahitya Parishad (KSSP), Asian Regional Exchange for New Alternatives (ARENA), and PP21 Rural Urban Alternatives (PP21 RUA)/ and Japan Committee for Negros Campaign (JCNC) at Palakkad, Kerala, India, in March, 1996. An encounter between South, Southeast, and Northeast Asian practitioners, the workshop sought to conduct "actual constructive work" in linking diverse forms of alternative practices across Asia after the realization that mere "critical analysis of and reaction to the dominant path of development are insufficient" (Raina 1997, 14). IADEA was not just a moment; it evolved into a movement whose continuing task has been to challenge the dominant model of progress by reimagining development and concretely building networks of solidarity. As intimated earlier, these efforts among Southeast Asian scholars recognize what Choudry (2019, 6) elegantly posits: "the fact that it is often the ordinary people in struggles who create knowledge and ideas: much of the intellectual labor of organizing does not take place through panels of experts, media conferences, and statements; it is worked out in the course of action."

The first section of *Reimagining Development in Southeast Asia: Alternative Practices from the Grassroots and Social Movements* pulls together practices of community-based resource management. Chapters in "Community-based Resource Management" show how different communities and civil society groups in the region tend to local resources by tapping Indigenous knowledges, planning from the bottom-up, and critiquing state initiatives to shape resource management. Donald Takhell and Rafael Dimalanta's "Indigenous Lifeways as Grassroots Resistance and Environmental Preservation: The Experience of Fishers in Loktak Lake, Northeast India" explores the experiences of Indigenous, highlighting their socio-economic and ecological connection with the largest freshwater lake in the region. The study discusses their grassroots resistance strategies, including educational campaigns and community-led conservation initiatives, aimed at protecting their Indigenous territory and right to stay in the wetlands.

Following this is Honey Tabiola's "Local Energy Planning in Vietnam and 'the Local' in Building Alternatives in Energy and Environmental Policy." Focusing on Vietnam's energy challenges, this case study delves into Local Energy Planning (LEP) as a bottom-up approach to energy planning involving local communities and authorities. It examines LEP's mechanisms, benefits, and transformative potential, emphasizing the importance of participation and its role in constructing alternatives in energy and environmental policy at the local level.

The fourth chapter in this collection critiques Thailand's reliance on rice production for export raises concerns about its sustainability and impact on food security. Civil society organizations like the Sustainable Agriculture Foundation advocate for alternative agricultural models to promote food sovereignty. Nathaniel P. Candelaria's "Promoting Food Sovereignty in Thailand and the Task of Civil Society" documents their efforts and explores how they address challenges such as the use of chemicals and the degradation of soil.

"Promoting Citizens' Participation in the Budget Process: The Social Watch Philippines' Alternative Budget Initiative" by Ma. Victoria R. Raquiza and Roja Estrelita Salvador looks into civil society efforts to carve a democratic space in state budget management. It documents the case of Social Watch Philippines, which advocates for alternative budget proposals to increase funding for social services and promote citizens' participation in the process. Despite some successes in influencing national budget priorities, challenges remain, including ensuring budget allocations reach the intended beneficiaries and overcoming resistance from legislators to institutionalize citizen engagement in budget advocacy.

This section shows the different strategies of resource management from below. Chapters two and four feature Indigenous fishers and agricultural advocates who are engaged in community-led conservation efforts to protect their natural resources and promote food sovereignty. This reflects a bottom-up approach to addressing environmental and socio-economic challenges. Chapters three and five underscore the importance of local participation in policy development processes. In Vietnam, Local Energy Planning (LEP) involves collaboration between local communities and authorities to design energy plans tailored to local needs and resources. Similarly, Social Watch Philippines pushes for citizen engagement in budget advocacy to ensure that national priorities align with the needs of local communities, emphasizing the role of participation in shaping policy outcomes.

The second section of this collection revolves around "People-centered Agrarian Systems." These case studies demonstrate the empowerment of local communities in shaping their alternative agricultural projects. For instance, in Nusa Tenggara Timur, Indonesia, a grassroots movement led by the Perhimpunan Petani Sorgum untuk Kedaulatan Pangan is reviving sorghum cultivation to combat food insecurity and reclaim cultural identity. Despite challenges like low rainfall, farmers are reintroducing local sorghum varieties, sparked by local leader Maria Loretha's passion. Ananeza P. Aban's "Power from the Grain Bringing Back the Lost Heritage of Sorghum in Nusa Tenggara Timur, Indonesia" tracks their journey toward biodiversity-based farming, emphasizing resilience and cultural heritage.

Meanwhile, land conflicts in Thailand spurred by neoliberal development goals are challenged by grassroots movements like the Southern Peasants' Federation of Thailand (SPFT). Investigating alternative practices, Supatsak Pobsuk's study explores community rights as a basis for land rights movements, countering centralized land management. Despite violence, the SPFT advocates for collective ownership, challenging mainstream development narratives and demanding equitable distribution. Grassroots movements such as the Perhimpunan Petani Sorgum untuk Kedaulatan Pangan in Indonesia and the SPFT in Thailand advocate for collective ownership, community rights, and participatory decision-making in land management. This highlights the importance of decentralization and community-driven approaches that prioritize the needs and aspirations of local farmers and residents.

Also by Aban, A. "Rediscovering the Sacred among Agrarian Societies: The Tara Bandu Case in Timor Leste," the book's eighth chapter documents how Timorese communities have revived "tara bandu," a customary law system, to manage resources amidst political disruptions. Tara bandu fosters community decision-making and underscores the importance of preserving collective life amid market pressures. Synergy between peasant unions, village leadership, and cultural elders shapes a just agrarian society, highlighting the interplay between social, ecological, and sacred aspects.

Amidst corporate dominance in Asia, social movements advocate for food sovereignty to empower small-scale food producers. Chapter nine, "Agroecology trends and Initiatives on Seed Saving/Banking in the Southeast Asia Region: The La Via Campesina Experience" by Raphael Baladad investigates how La Via Campesina resists corporate control, promotes agroecology, and reclaims traditional farming practices, they aim to build resilient food systems. Movements advocating for agroecology and seed saving epitomize the principles of food sovereignty in action.

In the face of neoliberal development agendas and corporate dominance, social movements advocating for food sovereignty emerge as powerful agents of change. By resisting corporate control, promoting agroecology, and reclaiming traditional farming practices, these movements seek to empower small-scale food producers and build resilient food systems that prioritize local needs and sustainability over profit-driven interests. This highlights the importance of challenging dominant narratives and advocating for policies that prioritize the well-being of people and the planet over economic gains.

The third section covers alternative ways of transferring knowledge across generations, particularly those conducted in Indigenous-led and community-based learning platforms. The tenth chapter, "The Organizing Work of Ayta Mag-indi Women in Porac, Pampanga in Advancing Women's Rights and Community Governance" by Venarica Papa, Jette Roldan, Micah Magaro, Eugene Balbas, and Ayta Mag-indi women leaders and elders look into how an Indigenous community in the Philippines emphasizes harmony and environmental stewardship in their unique form of community governance. Empowering women, preserving Indigenous knowledge, and collaborating with external institutions, the Ayta Mag-indi challenge conventional notions of development. Their resilience and continuity offer insights into alternative pathways for inclusive and culturally sensitive development practices.

Chapter eleven, "Teaching 'Pangiyak Ki!': The Lumad School as a Struggle for Land, Life, and Liberation" by Jose Monfred C. Sy, written with the help of students of the Lumad Bakwit School, delves into the establishment and challenges faced by alternative tribal schools for the Lumad people in the Philippines, addressing educational disparities and integrating Indigenous knowledge. Despite their noble objectives, these schools confront government militarization efforts, leading to attacks, closures, and displacement of thousands of Lumad students. In response, Bakwit Schools ("bakwit" being a popular languaging of "evacuate" in the vernacular) in Metro Manila emerge as temporary learning spaces and platforms for protest, highlighting the broader struggle for land, life, and liberation among the Lumad.

Andrew Aeria's thoughtful piece on "The Rural Internship Training Program in Sarawak, Malaysia," which is chapter twelve of this collection, examines the emergence of the Rural Internship Training Program (RITP) amid concerns about the shift in Malaysian university education toward individualism and corporate culture. The program challenges this trend by providing hands-on experiences in rural communities, fostering self-awareness and social consciousness among participants. Through the RITP, graduates gain practical skills, deepen their understanding of societal issues, and commit to advocating for marginalized communities, highlighting the importance of community-oriented learning experiences in nurturing socially responsible citizens.

These chapters demonstrate a range of educational approaches beyond conventional schooling, including the Ayta Mag-indi community's emphasis on environmental stewardship, the struggle for Indigenous education among the Lumad, and the practical learning experiences offered by the Rural Internship Training Program. These diverse models challenge traditional notions of education and highlight the importance of culturally sensitive and community-oriented approaches. Despite facing adversity, the communities and programs discussed in these chapters demonstrate a commitment to empowerment and advocacy. Whether through the preservation of Indigenous knowledge, the establishment of temporary learning spaces for displaced students, or the promotion of social consciousness among program participants, there is a shared goal of fostering resilience and promoting social justice through education. This emphasis on empowerment highlights the transformative potential of education as a tool for individual and collective liberation.

The fourth section of this volume, "Women-led Development and Communities of Care" documents the experiences of women shaping their own development. Chapter thirteen, Aban's "Vietnam's Peace House Shelter as a Safe Space for Women to Reclaim Their Rights," describes how women pioneer support for other women affected by gender-based violence in Vietnam, offering refuge and assistance aligned with national laws and international commitments. Recognized by law enforcement and authorities, the Peace House Shelter stands as a beacon of the women's movement, advancing women's rights amidst ongoing challenges.

Angeli Fleur G. Nuque's "Women's Empowerment through Weaving: The Case of the Women Weavers Cooperative of Lepo Lorun," chapter fourteen, tracks the efforts of the Lepo Lorun Women Weavers Cooperative to preserve their weaving heritage while empowering over 1200 women in Flores Island, Indonesia, economically and culturally. Through meticulous craftsmanship and plans for an ikat museum, the cooperative embodies the enduring significance of textiles in everyday life and cultural identity. Chapter ffifteen, also situated in Indonesia, zero in on women managing their households. Candelaria's "Women Household Heads in Indonesia and their Role in Addressing Gender Inequality" looks into how the Pemberdayaan Perempuan Kepala Keluarga (PEKKA) tackles gender inequality by mobilizing women and providing essential services, highlighting the intertwined nature of women's rights with societal welfare. PEKKA's work underscores the necessity for sustained efforts to integrate women's rights into societal norms and practices.

Elsewhere in Southeast Asia, women's movements continue to question conservative views on gender and sex. The Association for Development of Women and Legal Education (ADWLE) in Laos empowers women through legal aid clinics, challenging patriarchal norms and promoting gender equality. By engaging community resources and advocating for institutional change, ADWLE seeks to combat violence against women and advance women's human rights in Laos. Nuque and Aban's "Women's Access to Justice through Legal Aid Clinics and Paralegal Formations" documents this alternative practice in Laos.

As shown in this section, women-led initiatives are pivotal in addressing gender-based violence and advancing women's rights. Vietnam's Peace House Shelter, Lepo Lorun Women Weavers Cooperative, PEKKA, and ADWLE are prime examples of women-led endeavors that provide critical support, preserve cultural heritage, and empower women economically and socially. These initiatives demonstrate the effectiveness of women taking the lead in driving development efforts tailored to address gender-specific challenges.

The fifth and final section in *Reimagining Development in Southeast Asia* highlights praxes of solidarity in economic production and labor. These case studies do not only emphasize fair trade, collective enterprise, and grassroots empowerment. The communities and movements documented here also advocate for gender equity, decent work opportunities, and community engagement. Chapter seventeen, "Building Sustainable Social and Solidarity Economy in Southeast Asia: Towards More and Better Jobs for Women in the Informal Economy" by Rosalinda Pineda Ofreneo and Josephine Parilla limns how grassroots women in Southeast Asia's informal economy face intersecting crises, highlighting the urgency of inclusive, sustainable development models. Organizations like HomeNet, which this chapter elucidates, advocate for fair trade and solidarity economy principles, empowering disadvantaged producers and advancing gender equity amidst systemic challenges.

The eighteenth chapter in this collection, Benjamin Quiñones' "Collective Enterprise as a Workplace: The Case of Payoga-Kapatagan Multipurpose Cooperative (PK-MPC)" demonstrates how collective enterprise can transform communities, promote sustainable agriculture, and contribute to economic and ecological sustainability. Through a social solidarity economy approach, PK-MPC prioritizes people, planet, and profit, contributing to economic and ecological sustainability while fostering community engagement.

By challenging systemic injustices and prioritizing women's leadership, initiatives such as what Bianca Franco documented in "Workers' Information Center in Cambodia: Transforming the Narrative of Garment Workers" transcend conventional labor activism, fostering solidarity and paving the way for long-term systemic change. The Workers' Information Center (WIC) empowers women garment workers in Cambodia, challenging labor injustices and fostering solidarity in the face of systemic challenges. By prioritizing women's leadership and providing essential support services, WIC offers a model of grassroots empowerment that transcends conventional labor activism, paving the way for long-term systemic change.

The twentieth chapter in this collection, "Collective Action to Attain a Better Life: Maigting na Samahan ng Panlipuang Negosyante ng Towerville" (Philippines) by Regie Placido and Karl Arvin Hapal" examines the challenges faced by women entrepreneurs in the Philippines and the role of social networks in addressing these challenges. Through qualitative research strategies, it highlights the importance of social capital in women's entrepreneurship, emphasizing the need for supportive networks and access to resources. By understanding the dynamics of social networks and leveraging them effectively, women entrepreneurs can overcome barriers and succeed in their ventures, contributing to economic growth and empowerment.

The final chapter, Benjamin B. Velasco's "Worker-Driven Social Responsibility as an Alternative: Interrogating Export Zone Workers' Struggles in the Philippines" investigates labor rights suppression and workers' resistance within a Luzon-based ecozone, focusing on five case studies. Despite the importance of unionization and freedom of association, these rights are systematically stifled through various tactics, leading to worker protests and strikes. The research underscores the conflict-ridden relationship between claim holders and duty bearers, arguing that without a strategy of struggle, rights and entitlements remain contested terrain.

By giving precedence to gender equity, ensuring access to decent work, and fostering community involvement, these endeavors expand the scope of the traditional labor activism we know. They offer empowerment frameworks that directly address the systemic challenges faced by marginalized communities, particularly women in informal economic sectors. Additionally, the examination of social networks' role in assisting women entrepreneurs highlights the importance of community-led methods in overcoming barriers to economic empowerment. Taken together, these case studies emphasize how social solidarity economy strategies can profoundly advance sustainable development and cultivate solidarity amidst systemic obstacles.

The grassroots and social movement paradigms highlighted in this collection seek to challenge and change the dominant model of development instituted by the ASEAN, which have largely failed in meaningfully addressing the issues faced by different sectors. However, the university faculty and researchers affiliated with AltDev do not own any of the stories told in this book. The following chapters were written by or with close coordination with the communities or CSOs themselves. Some cases were documented by activists organizing in their respective sites. Other chapters faithfully reflect interviews and knowledges learned from fieldwork. All of these narratives also went through a validation process where community members or organizers can correct or augment what has been written with them. By sharing their experiences and learning from others, AltDev intends to reinforce the solidarity between different grassroots organizations who, in many ways, share similar issues and challenges.

This current volume seeks to reposition the discourse on a different track from conventional and macro perspectives whether coming from state and corporate mindsets or progressive outlooks wedded to long-held orthodoxies to one that emanates from the communities, groups, and sectors who have been marginalized from the mainstream. It is hypothesized that long-running Indigenous and localized practices at the ground level and the underlying principles drawn from them could contribute to the search for a new development paradigm built on a platform of social change.

Ultimately, we hope that these case studies can serve as active or potential arenas for fostering critical consciousness in critiquing the systemic nature of oppression and injustice. Drawn from lived experiences, this book acknowledges the imperative of the grassroots and social movements to resist the root causes of oppression and exploitation. It testifies to the capacity of Southeast Asian peoples to effect change in their own circumstances. By highlighting the liberating and transformative possibilities cultivated in the examined struggles, we underscore their significance in reimagining development.

References

Aban, Ananeza P., and Jose Monfred C. Sy. 2022. Queering solidarity: Civil society at the fringes of ASEAN regionalism and alternatives for the LGTBQ+. *Philippine Journal of Public Policy* 2022: 139–172.

Association of Southeast Asian Nations [ASEAN]. n.d. About ASEAN. Association of Southeast Asian Nations. Accessed March 2022. https://asean.org/asean/about-asean/.

Bassey, Nnimmo. 2019. Breaking the chains of development. In *Pluriverse: A Post-Development Dictionary*, ed. Ashish Kothari, Ariel Salleh, Arturo Escobar, Federico Demaria, and Alberto Acosta, 3–6. New Delhi: Tulika Books.

Beeson, Mark, and Richard Stubbs. 2012. Introduction. In *Routledge Handbook of Asian Regionalism*, ed. Mark Beeson and Richard Stubbs, 1–8. London/New York: Routledge.

Bello, Walden. 2023. Deglobalization as an alternative economic paradigm. In *Speech at the 4th Southeast Asian Conference on Alternatives*. University of the Philippines. 30 October. Typescript.

Blaser, Mario. 2019. Life projects. In *Pluriverse: A Post-Development Dictionary*, ed. Ashish Kothari, Ariel Salleh, Arturo Escobar, Federico Demaria, and Alberto Acosta, 234–237. New Delhi: Tulika Books.

Bolden, Mark. 2011. Neoliberalism and counter-hegemony in the global south: Reimagining the state. In *Social Movements in the Global South: Dispossession, Development and Resistance,* ed. by Sara C. Motta, and Alf Gunvald Nilsen, 83–103. Hampshire/New York: Palgrave Macmillan.

Boothroyd, Peter, and Craig Davis. 1993. Community economic development: Three approaches. *Journal of Planning Education and Research* 12: 230–240.

Chambers, Robert. 1997. *Whose Reality Counts? Putting the First Last.* London: Intermediate Technology Publications.

Choudry, Aziz. 2019. Reflections on academic, activism, and the politics of knowledge and learning. *The International Journal of Human Rights*, 1–19. https://doi.org/10.1080/13642987.2019.1630382

Clarke, Gerard. 2012. The evolving ASEAN human rights system: The ASEAN human rights declaration of 2012. *Northwestern Journal of International Human Rights* 11 (1): 1–27.

Collins, Alan. 2008. A people-oriented ASEAN: A door ajar or closed for civil society organization? *Contemporary Southeast Asia* 30 (2): 313–331.

Cooke, Bill, and Uma Kothari, eds. 2001. *Participation: The New Tyranny?* London/New York: Zed Books.

D'Alisa, Giacomo, Federico Demaria, and Giorgos Kallis, eds. 2015. *Degrowth. A Vocabulary for a New Era.* New York: Routledge.

Desai, Radhika. 2004. From national bourgeoisie to rogues failures and bullies: 21st century imperialism and the unravelling of the Third World. *Third World Quarterly* 25(1), 169–185. https://doi.org/10.1080/0143659042000185390

D'Souza, Radha. 2011. Three actors, two geographies, one philosophy: The straightjacket of social movements. In *Social Movements in the Global South: Dispossession, Development and Resistance,* ed. by Sara C. Motta, and Alf Gunvald Nilsen, 227–249. Hampshire/New York: Palgrave Macmillan.

Eccleston, Bernard, Michael Dawson, and Deborah J. McNamara, eds. 1998. *The Asia-Pacific Profile*. London/New York: Routledge.

Escobar, Arturo. 1995. *Encountering Development: The Making and Unmaking of the Third World*. Princeton, New Jersey: Princeton University Press.
Escobar, Arturo. 1999. The invention of development. *Current History* 98 (631): 382–386.
Escobar, Arturo. 2018. *Designs for the Pluriverse: Radical Interdependence, Autonomy, and the Making of Worlds*. Durham/London: Duke University Press.
Esteva, Gustavo. 2009. "What Is Development?" Unpublished manuscript, Universidad de la Tierra, Oaxaca.
Esteva, Gustavo, and Madhu Suri Prakash. 1998. Beyond development, what? *Development in Practice* 8 (3): 280–296.
Frank, Andre Gunder. 1967. *Capitalism and Underdevelopment in Latin America*. New York City: Monthly Review Press.
Gerard, Kelly. 2014. ASEAN and civil society activities in 'created spaces': The limits of liberty. *The Pacific Review* 27 (2): 265–287.
Green, Gary Paul, and Anna Haines. 2002. *Asset Building and Community Development*. Thousand Oakes: Sage.
Gudynas, Eduardo. 2011. Buen Vivir: Today's tomorrow. *Development* 54 (4): 441–447.
Hapal, Karl. 2020. Introduction: Alternative development from the Philippine grassroots. In *Towards a Peoples' Alternative Regionalism: Cases of Alternative Practices in the Philippines,* ed. by Eduardo C. Tadem, Karl Hapal, Venarica B. Papa, Jose Monfred C. Sy, Ananeza Aban, Nathaniel P. Candelaria, and Honey B. Tabiola, 1–9. Quezon City: University of the Philippines Center for Integrative and Development Studies.
Harvey, David. 2003. *The New Imperialism*. Oxford: Oxford University Press.
Hoang, Lien. 2024. Asia develops an appetite for degrowth 'doughnut economics. *Nikkei Asia*. 11 February. https://asia.nikkei.com/Economy/Asia-develops-an-appetite-for-degrowth-doughnut-economics.
Kaufman, Michael. 1997. Community power, grassroots democracy, and the transformation of social life. In *Community Power and Grassroots Democracy: The Transformation of Social Life,* ed. by Michael Kaufman, and H.D. Alfonso, 1–26. London/New Jersey: Zed Books.
Knowlton, Brian. 2009. Global economy called worst since 1945. *The New York Times*. 22 April. https://www.nytimes.com/2009/04/23/business/economy/23outlook.html.
Korten, David C., and Rudi Klauss. 1984. *People-Centered Development: Contributions Toward Theory and Planning Frameworks*. Sterling: Kumarian Press.
Kothari, Ashish, Ariel Salleh, Arturo Escobar, Federico Demaria, and Alberto Acosta. 2019. Introduction: Finding Pluriversal Paths. In *Pluriverse: A Post-Development Dictionary,* ed. by Ashish Kothari, Ariel Salleh, Arturo Escobar, Federico Demaria, and Alberto Acosta, xxi–xl. New Delhi: Tulika Books.
Lang, Miriam, Claus-Dieter. König, and Ada-Charlotte. Regelmann. 2018. *Alternatives in a World of Crisis*. Brussels: Rosa Luxemburg Stiftung.
Langlois, Anthony, Cai Wilkinson, Paula Gerber, and Baden Offord. 2017. Community, identity, orientation: Sexuality, gender and rights in ASEAN. *The Pacific Review* 30 (5): 710–728. https://doi.org/10.1080/09512748.2017.1294613.
Lessem, Ronnie, and Alexander Schieffer. 2010. *Integral Economies*. Farnham, UK: Ashgate/Gower.
IBON International. 2015. ASEAN Community 2015: Integration for Whom? http://iboninternational.org/resources/15/04/21/asean-community-2015-integration.
Idris, Aida, and Nurliana Kamaruddin. 2019. Overview of ASEAN. In *ASEAN Post-50: Emerging Issues and Challenges*, ed. Aida Idris and Nurliana Kamaruddin, 1–12. Singapore: Palgrave Macmillan.
McMichael, Philip. 2019. The development project. In *Pluriverse: A Post-Development Dictionary*, ed. Ashish Kothari, Ariel Salleh, Arturo Escobar, Federico Demaria, and Alberto Acosta, 12–15. New Delhi: Tulika Books.
Motta, Sara C., and Alf Gunvald Nilsen. 2011. Social movements and/in the postcolonial: Dispossession, development and resistance in the global south. In *Social Movements in the Global*

South: Dispossession, Development and Resistance, ed. by Sara C. Motta, and Alf Gunvald Nilsen, 1–33. Hampshire/New York: Palgrave Macmillan.

Mushakoji, Kinhide. 2013. Identity politics in the developmentalist states of East Asia: The role of diaspora communities in the growth of civil societies. *Asian Studies: Journal of Critical Perspectives on Asia* 49 (1): 113–123.

Nesadurai, Helen E. S. 2008. The Association of Southeast Asian Nations (ASEAN). *New Political Economy* 13 (2): 225–239.

Ogborn, Miles. 1999. Modernity and modernization. In *Introducing Human Geographies*, ed. Paul Cloke, Phillip Crang, and Mark Goodwin, 153–161. London: Arnold.

Peet, Richard, and Elaine Hartwick. 2015. *Theories of Development: Contentions, Arguments, Alternatives*. New York: Guilford Publications.

Phillips, Rhonda and Pittman, Robert H. 2009. A framework for community and economic development. In *An Introduction to Community Development*, ed. Rhonda Phillips and Robert H. Pittman, 3–19. London and New York: Routledge.

Pieterse, Jan Nederveen. 1998. My Paradigm or yours? Alternative development, post-development, reflexive development. *Development and Change* 29 (2): 343–373. https://doi.org/10.1111/1467-7660.00081.

Raina, Vinod. 1997. The idea behind IADEA. In *Integrating Alternative Development Efforts in Asia: Report on the Workshop at Palakkad, Kerala. India, March 1–4, 1996,* ed. by Dessa Quesada, 13–18. Tokyo, Japan: PP21 RUA, JCNC, Asian Regional Exchange for New Alternatives and Kerala Sastra Sahitya Parishad.

Reyes, Raniel SM.. 2015. Neoliberal capitalism, ASEAN integration and commodified education: A Deleuzian critique. *Budhi* 19 (2–3): 135–168.

Rist, Gilbert. 2019. The history of development: From western origins to global faith. London/New York: Zed Books.

Rostow, Walt W. 1960. *The Stages of Economic Growth: A Non-Communist Manifesto*. New York: Cambridge University Press.

Sachs, Wolfgang, ed. 1992. *The Development Dictionary: A Guide to Knowledge as Power*. London: Zed Books.

Sachs, Wolfang. 2019. Foreword: The development dictionary revisited. In *Pluriverse: A Post-Development Dictionary*, ed. Ashish Kothari, Ariel Salleh, Arturo Escobar, Federico Demaria, and Alberto Acosta, xi–xvi. New Delhi: Tulika Books.

Satake, Masaaki. 2003. *People's Economy: Philippine Community-based Industries and Alternative Development.* Manila/Kagawa: Solidaridad Publishing House, Inc. and Literary Society (Bunka Gakkai), Shikoku Gakun University.

Scott, James C. 2009. The art of not being governed: An anarchist history of upland Southeast Asia. New Haven and London: Yale University Press.

Sen, Amartya. 1999. *Development as Freedom*. New York: Knopf.

Serrano, Melisa, and Edlira Xhafa. 2011. Beyond the capitalist logic: Theoretical debates and social experiences. In *Trade Unions and The Global Crisis: Labour's Visions, Strategies and Responses*, ed. by Melissa Serrano, Edlira Xhafa, Edlira and Michael Fichter, 19–34. Geneva: International Labour Organization.

Serrano, Melisa, and Edlira Xhafa. 2012. Introduction: The 'process of becoming' alternatives. In *The Pursuit of Alternatives: Stories of Peoples' Economic and Political Struggles Around the World,* ed. by Melisa R. Serrano and Edlira Xhafa, 1–6. Munich: Rainer Hampp Verlag.

Shiva, Vandana. 2019. Development—For the 1 percent. In *Pluriverse: A Post-Development Dictionary*, ed. Ashish Kothari, Ariel Salleh, Arturo Escobar, Federico Demaria, and Alberto Acosta, 6–9. New Delhi: Tulika Books.

Sison, Jose Maria. 2020. *Basic Principles of Marxism-Leninism: A Primer*. Paris: Foreign Languages Press.

Söderbaum, Fredrik. 2012. Theories of regionalism. In *Routledge Handbook of Asian Regionalism*, ed. Mark Beeson and Richard Stubbs, 11–21. London/New York: Routledge.

Stiglitz, Joseph E. 2002. *Globalization and its discontents*. New York and London: W.W. Norton and Company.

Sy, Jose Monfred C. 2022. 'It takes a movement to build schools': A historical and pedagogical sketch of the Lumad Schools vis-à-vis the Lumad social movement. *Philippine Studies: Historical and Ethnographic Viewpoints* 70 (2): 423–464.

Tadem, Eduardo Climaco. 2012. Grassroots democracy, non-state approaches, and popular empowerment in rural Philippines. *Philippine Political Science Journal* 33 (2): 161–177.

Tadem, Eduardo C., Karl Arvin F. Hapal, Venarica B. Papa, Ananeza P. Aban, Nathaniel P. Candelaria, Honey B. Tabiola, Jose Monfred C. Sy, and Angeli Fleur G. Nuque. 2020. *Deepening Solidarities beyond Borders among Southeast Asian Peoples: A Vision for a Peoples' Alternative Regional Integration*. UP CIDS Discussion Paper Series 2020–04. Quezon City: University of the Philippines Center for Integrative and Development Studies.

Taylor, Paul. 2009. In Europe, differences over crisis exit strategy. *The New York Times*. 29 June. https://www.nytimes.com/2009/06/30/business/global/30inside.html

The Economist. 2009. The integration of the world economy is in retreat on almost every front. 19 February. https://www.economist.com/international/2009/02/19/turning-their-backs-on-the-world.

Tortosa, José María. 2019. Maldevelopment. In *Pluriverse: A Post-Development Dictionary*, ed. Ashish Kothari, Ariel Salleh, Arturo Escobar, Federico Demaria, and Alberto Acosta, 9–12. New Delhi: Tulika Books.

United Nations Department of Economic Affairs (UNDEA). 1951. Measures for the economic development of underdeveloped countries. Report. United Nations.

Uy, Veronica. 2012. 5 Reasons NGOs reject current version of ASEAN human rights declaration. *ASEAN Trade Union Council*. 13 September, 2012. http://aseantuc.org/2012/09/5-reasons-ngos-reject-current-version-of-asean-human-rights-declaration/.

Weiss, Meredith. 2021. Building solidarity on the margins: Seeking SOGIE rights in ASEAN. *Journal of Human Rights* 20 (2): 193–210.

Willis, Kate. 2005. *Theories and Practices of Development*. London/New York: Routledge.

Jose Monfred C. Sy is an assistant professor of Philippine studies at the Department of Filipino and Philippine Literature, University of the Philippines (UP) Diliman. He received his Master of Arts (Philippine Studies) and Bachelor of Arts (Comparative Literature), summa cum laude, from the same university. He is currently a research fellow of the Program on Alternative Development of the UP Center for Integrative and Development Studies, where he is a co-editor of the multi-volume monograph series *Alternative Practices across Southeast Asia* (2020, 2023). He is interested in social movements, anticolonialism and decoloniality, environmental humanities, and children's literature. His research appears in journals such as *Kritika Kultura, Humanities Diliman, Social Science Diliman, Daluyan, Suvannabhumi, AlterNative*, and *Philippine Studies*. He is a member of the Congress of Teachers/Educators for Nationalism and Democracy (CONTEND) and the Alliance of Concerned Teachers (ACT). He is also the author of several children's books.

Eduardo C. Tadem, Ph.D. is Professor Emeritus and Professorial Lecturer of Asian Studies, University of the Philippines Diliman and Convenor of the Program on Alternative Development, University of the Philippines Center for Integrative and Development Studies (UP CIDS AltDev). He has a Ph.D. in Southeast Asian Studies from the National University of Singapore. Until 2015, he was Professor of Asian Studies at UP Diliman where he served as Editor-in-chief of *Asian Studies (Journal of Critical Perspectives on Asia)*. He is a founding Governing Board Member of the Consortium on Southeast Asian Studies in Asia (SEASIA). As a civil society activist, he was President of the Freedom from Debt Coalition (FDC) from 2014 to 2018, Co-Convenor of the Asean Civil Society Conference/Asean Peoples' Forum (ACSC/APF) Philippine National Organizing Committee (2017), International Committee member of the NGO Forum on the ADB (2014–2018), and current Co-chair of the Asian Regional Exchange for New Alternatives (ARENA). He has researched and published extensively on inequality and tax justice, political economy of foreign aid, Southeast Asian regional integration, peasant society, agrarian reform and rural development, industry studies, Mindanao regional development, rural unrest and social movements, Philippine Japan relations, foreign investments, special economic zones, and contemporary politics.

Part II
Community-Based Resource Management

Chapter 2
Indigenous Lifeways as Grassroots Resistance and Environmental Preservation: The Experience of Fishers in Loktak Lake, Northeast India

Donald Takhell and Rafael V. Dimalanta

Abstract This chapter explores Loktak Lake, the largest freshwater lake in Northeast India, focusing on its ecological, socio-cultural, and political significance. Loktak is a biodiversity hotspot, home to numerous species of fish, birds, and the endangered Sangai deer. The indigenous Meitei community relies on the lake for food and livelihood. It is also the material basis for their cultural practices. However, state-led development projects, particularly the Loktak Hydroelectric Project and Ithai Barrage, have disrupted the lake's hydrology, causing permanent flooding, loss of biodiversity, and the degradation of Phumdis. These changes have severely impacted the traditional livelihood patterns of the Meitei people. In response, indigenous communities have mobilized against state policies, forming organizations like the All Loktak Lake Areas Fishermen's Union Manipur (ALLAFUM) to advocate for conservation and indigenous rights.

This chapter examines the intersection of environmental degradation, state interventions, and indigenous resistance in Loktak. The indigenous Meitei community asserts their right to life, livelihood, and resources in the lake, emphasizing their deep entanglement with the ecosystem and their traditional fishing practices. They demand government recognition of their rights, including the preservation of floating huts and community-based fishing traditions, which they see as part of their heritage. In response to ecological and developmental threats, they resist through conservation efforts, educational campaigns, and solidarity networks to protect both their way of life and the wetlands.

Keywords Environmental degradation · Fisherfolk · Indigenous resistance · Loktak Lake · Meitei · Northeast India

D. Takhell (✉)
Indigenous Perspectives, Manipur, India
e-mail: donaldtakhell@gmail.com

R. V. Dimalanta
University of the Philippines, Diliman, Quezon City, Philippines

J. M. C. Sy and A. P. Aban (eds.), *Reimagining Development in Southeast Asia*, Asia in Transition 21, https://doi.org/10.1007/978-981-97-9140-8_2

2.1 Introduction

Situated in the valley districts of Bishnupur and Imphal West of Manipur, Loktak Lake is the largest freshwater lake in the northeast region of India, presently covering an area of 246.72 km^2 (National Wetland Atlas 2009), and a direct catchment area of 980 km^2 and an indirect catchment area of 7157 km^2 (Thakur 2020). Loktak's catchment area comprises nine sub-catchments—Imphal, Iril, Thoubal, Khuga, Nambul, Kongba, Heirok, Sekmai, and Western (Khwairakpam 2021). There are three zones in Loktak: the central zone, northern zone, and southern zone. The open water areas are mainly located in the central zone; there are also a few patches of water bodies in the northern zone. Meanwhile, the southern zone is the site of the Keibul Lamjao National Park. 14 hillocks that arise as islands are also located in the central and southern parts of Loktak (Tombi 1994).

Floating vegetative mats locally known as 'Phumdis' are one of the most prominent features of Loktak Lake. Phumdi is a heterogenous biomass composed of living and decaying aquatic and terrestrial plants, soil, and other organic matter. It is created through the accumulation of water plants with suspended silt and grasses. The high proportion of vegetable matter in Phumdi results in high buoyancy that allows it to stay afloat. It has been observed to have an average thickness of 8 feet, and float on the lake with one-fifth of their thickness above the water, and four-fifth under the water. However, Phumdi regularly changes shape, size/thickness, and location, responding to the changes in season (Singh 2011).

This floating biomatter plays a key role in the ecological processes of the lake such as nutrient absorption, and carbon sequestration. It also serves as a mechanism for governing the water and nutrient dynamics, as well as functions as a source of biofertilizer in Loktak (Kangabam et al. 2018).

The Loktak Lake is a biodiversity hotspot with 425 species of animals (249 vertebrates and 176 invertebrates) (Trishtal and Manihar 2004) inhabiting the area. These include 186 fish species, 106 bird species, 42 mammal species, 117 butterfly species, 6 amphibian species, 150 arthropod species, 10 mollusc species, 16 annelid species, 186 zooplankton species, 43 aquatic and semi-aquatic plants, 25 amphibian flora, and 120 rotifers (Kangabam et al. 2015).

The adjoining rivers bring nutrient-rich sedimentation to the Loktak Lake and the Phumdis. Phumdis in the Loktak Lake contribute to the localized biodiversity as many species of aquatic, semi-aquatic and semi-terrestrial plants inhabit the Phumdis. Due to its rich biodiversity, it was declared as a wetland of international importance by the Ramsar Convention in 1990 and nationally by the National Wetland Conservation Program of the Government of India in 2011 (Khoiyangbam 2021).

2.2 Lifeline

As many as 220,017 persons or 9% of the total population of the state of Manipur are located in 65 villages inside and around the area (Thaoroijam 2017). Loktak Lake provides food, livelihood, and shelter to the indigenous Meitei community who depend on the vast resources present in the area. Forty-seven Phumdi plants are utilized by the indigenous Meitei community for various purposes: food, medicine, fodder and fuel, handicrafts, and habitation (Meitei and Prasad 2015).

Food: There are 27 Phumdi plants used in the indigenous Meitei community's food preparations. Furthermore, the edible parts of the various plant species growing on the Phumdi (whole plant, shoots, roots, leaves, and fruits) are also a component of the staple diet of the people of Manipur (Meitei and Prasad 2015).

Medicine: 25 Phumdi plants are used by the Maiba (traditional healers) for different medicinal purposes. The Maiba concoct extracts, powders, and pastes from these plants and uses them to treat wounds, cough, asthma, fevers, boils, and even rheumatism, hemorrhoids, epilepsy, stomach disorders, blood diseases, skin diseases, liver troubles, and nervous disorders (Meitei and Prasad 2015).

Fodder and Fuel: Phumdi plants are also used by the indigenous Meitei community as feed for their domestic animals such as cows, pigs, buffaloes, and goats. There are also 4 plants commonly used as fuel (Meitei and Prasad 2015).

Loktak was the largest fishery resource of Manipur prior to the 1950s, contributing 60% of the total fish production of the state (Trisal and Manihar 2004). Loktak Lake is referred to as the 'lifeline' of the people due to its contributions to the ecological, social, economic, and cultural life in Manipur (Kangabam et al. 2015). Despite its importance to the life of both Loktak indigenous fishing community and the people of Manipur, it has suffered severe degradation in the past decades.

2.3 Impact of Development Project

The Ministry of Irrigation and Power, Government of India initiated the Loktak Multipurpose Hydroelectric Power Project in 1971. The project was then transferred to the National Hydroelectric Power Corporation Limited (NHPC) and was commissioned in 1983. The objectives of the LHP was to (1) generate uninterrupted hydroelectric power supply, (2) control floods, (3) reclaim land, (4) provide lift irrigation, and (5) ensure water supply for various purposes (Khoiyangbam 2021). The LHP involved the construction of the Ithai Barrage at the convergence of Manipur River and Khuga River, and the diversion of water from the Loktak Lake through an open channel (Singh 2015). The Ithai Barrage converted Loktak into an artificial water reservoir, creating catastrophic shifts in the area's hydrology and ecosystem, and displacing human populations.

Alterations in Hydrological Regime: Loktak used to comprise more than 20 smaller lakes, experienced regular water level changes, and had varying extent and depth depending on the season. During the dry season (when water level is low), the separation of these lakes was evident. These smaller lakes would only merge into one water body during the rainy season (when water level is high). Loktak Lake's inflow comes from various rivers connected hydrologically to it (Moirangleima and Singh 2009). However, the Barrage blocked the flow of the Manipur River, changing its normal flow, and effectively making the Barrage function as a water gate (Singh 2015). Since then, Loktak Lake has become a stagnant single water body, permanently flooded with only small water level changes because of the water regulation by the Barrage (Moirangleima and Singh 2009; Thakur 2020).

Flooding and Submergence of Land: With the water level of Loktak Lake being kept at 768.5 m above the mean sea level throughout the year (higher than the altitude of normal water level of Loktak), there has been frequent flooding in the upstream areas, especially during the rainy season. The refusal to open the gates of Ithai Barrage to maintain the water level has resulted in the inundation of vast productive agricultural land and settlement areas. About 83,450 hectares of arable and habitable land have been submerged. (Singh 2018).

Phumdi Degradation and Proliferation: Seasonal shifts in Loktak Lake's water level trigger the natural life cycle of Phumdis. In the rainy season, Phumdi float on the water surface and in the dry season, they sink to the lake bed and absorb nutrients (Kangabam et al. 2018). Through this cycle, the Phumdis take in nutrients brought to the Lake by the adjoining rivers. After the Ithai Barrage was constructed, there was an increase in water level which kept the Phumdi permanently suspended, depriving them of the essential nutrients for growth (Thakur 2020) and resulting in reduction in thickness (Khoiyangbam 2021).

While the Phumdis experienced thinning due to lack of nutrients, their number in Loktak increased massively as they have been unable to float away due to the blockade in the outlet of the lake created by the Ithai Barrage (Singh and Khundrakpam 2009).

Threat to the Survival of the Sangai Deer: In the southern part of the Loktak lies the Keibul Lamjao National Park, the last remaining habitat of the Sangai deer. The Phumdis provide food and shelter to the Sangai deer but continuous thinning of Phumdis is dangerous for the Sangai deer as thin Phumdis are not able to support the deers' weight (Tuboi and Hussain 2016). The Sangai deers' hooves are also more prone to get stuck in thinner Phumdi (Singh 2014). Both typically result in their drowning. Mani Mohan, a forest guard from the KLNP, explained that with the Phumdi cycle being disrupted by the Ithai Barrage, this has not only caused the thinning of the Phumdis but also left it without much nutritional value. He says, "The [Sangai] deers' habitat has become affected. The thinning of phumdis [also] means less food for the deer…their number has been dwindling".

Loss of Biodiversity and Fish Resources: After the construction of the Ithai Barrage, the diversity and population of migratory and resident animal species of the Loktak Lake has severely declined. 5 mammals, 3 birds, 9 reptiles, 3 amphibians, 2 molluscs, and 1 annelid which were previously in abundance in the area have disappeared (Trisal and Manihar 2004) due to the damages to Loktak's ecosystem. Adverse changes in the lake have caused a decline in the production of edible aquatic plants that are served as main vegetable by the indigenous Meitei community in and around Loktak (Singh 2015). About 30 species of these have become extinct in Loktak (Singh 2018).

Migratory fishes from the Chindwin-Irrawaddy river system in western Myanmar, which constitute 40% of the capture fishery in Loktak Lake have disappeared; 12 species recorded in the past have not been seen again (LDA and WISA 2003). With the Ithai Barrage blocking the fishes' migratory pathways, fish resources in Loktak have sharply declined both in diversity and quantity (Singh 2018; Khoiyangbam 2021). A fisher from Loktak laments on the sharp decline of fish resources: "We used to have plenty of fish in Loktak. Fishing was always a good [livelihood] option. We could catch fish at will. But after the Ithai Barrage was constructed, it became more difficult". The lake's contribution to the total fish production of Manipur has gone down from 60 to 11% (Trisal and Manihar 2004).

Changes in the Traditional Pattern of Indigenous Meitei Communities' Livelihood and Habitation: Before the Ithai Barrage was commissioned, indigenous Meitei communities in the area were farmers who also engaged in fishing. They engaged in rice farming during the dry season and harvested their produce before the start of the monsoon. During the monsoon, they would shift to fishing activities and gathering of edible aquatic plants. Fishing huts constructed on the Phumdis then, were largely temporary.

As Oinam Thoiba, a fisher from Loktak narrates: "Since my childhood, I have been living in the lake. I have farms in the lake periphery…but the Ithai Barrage submerged them. And so I shifted to fishing in the lake. Again the Ithai Barrage has caused the tremendous increase of Phumdis which hindered our boats' movement…and fish catch has [also] come down".

About 83,450 hectares of agricultural land is said to have been affected by the Ithai Barrage (Singh 2015). When the indigenous Meitei community in the area lost their agricultural land, their traditional settlements, and livelihood following the barrage construction, they were forced to become increasingly dependent on the Lake and solely on fishing (Khoiyangbam 2021). This has become evident in the increase in the population of lake dwellers. In 1985, there were only 28 fishing huts in Loktak Lake but this has since increased to 733 in 2001 (Census of India 2001). The barrage distorted their traditional pattern of livelihood and habitation, and decreased their earning capability (Vikalp Sangam 2020).

2.4 Indigenous Communities

The Indigenous Meitei community residing in Loktak are socio-economically, culturally, and ecologically connected with it (Singh 2014). Manipur-based environmental activist Salam Rajesh explains: These communities who are living inside and at the periphery of the lake have been living here for hundreds of years. They have been utilizing the resources of the lake: the fish and vegetation for their food and livelihood. This entire water body and peripheral areas can be considered as their indigenous territory.

Records from the Gazetteer of Manipur that date back to 1886 show that the indigenous Meitei community have been inhabiting Loktak for fishing (Wangkheirakpam 2014). These communities have taken the role of custodians of Loktak, caring for the waterbody and surrounding forests and hills. They also play a critical role in the cycle of Phumdis. When the water level is low (during dry season), they gather the Phumdi in heaps and burn them. During this time, the lake is also utilized by the indigenous Meitei community as grazing ground for their livestock. The movement from domestic animals help in further disintegrating the Phumdis which make them easier to be flushed down the Khordak channel which connects Loktak to the Manipur River system, when the water level rises. Additionally, when the water level is high (during the rainy season), indigenous Meitei community cut the Phumdi into strips and transport them to where the water channels can deliver them to the larger body of water and out of the lake. Through this, they help in maintaining the good water quality of Loktak.

The Ithai Barrage gravely harmed the peaceful co-existence of the communities with Loktak. As a result, the communities were compelled to re-configure their livelihood and habitation in the area to adapt to the barrage-induced changes.

Habitation: At present, the indigenous Meitei community residing at Loktak can be found in three locations: (1) inside the lake, (2) on the islands, and (3) at the lakeshore (Singh 2019). Those who dwell inside the lake build huts locally called 'Phumsang' on the Phumdi. These huts are able to transfer locations and respond to changes in water level. A Loktak fisherman explains their process of building on Phumdi: "We first check the Phumdi by shaking it…we choose the sturdiest one. When the thickness of Phumdi is uniform, we can build a hut over it. To make it more buoyant, we can put wood beneath it [Phumdi]".

A typical Phumshang has a floor area of 0.5 m^2. (Takhell 2021) and can accommodate 4–5 household members. The flooring should be about 2–3 m and is made by layering wooden planks or bamboo with phumdi layers. Frames made of light wooden pillars or bamboo are bored into the Phumdi for stability. To ensure that the huts would be light, only straw or reeds are used for its walls and roof. Bamboo poles help position the hut to the desired location for fishing and secure it from being moved by strong winds (Meitei and Prasad 2015). In recent times, due to the thinning of Phumdis, the indigenous Meitei community have been made to adjust and use lighter materials such as tin sheets and tarpaulins to further reduce the load. He adds, "If we don't use the appropriate materials, the huts will be blown away on a stormy day".

Communities living in Phumshangs inside the lake can be further classified into three: permanent, temporary, or migrant. However, a great majority of those living inside the lake are permanent dwellers who do not have a house elsewhere (Trisal and Manihar 2002). They navigate Loktak and the adjacent areas using plank built canoes, replacing the traditional dugout canoes. Plank built canoes are more economical and easier to build for the indigenous Meitei community. These canoes are made by connecting wood planks together using a special mixture of adhesive made from kerosene, sawdust, and Mekruk (*Vatica lanceaefolia*) powder.

Fishing: Traditional fishing has been practiced by the indigenous Meitei community in Loktak since they inhabited the area (Bharati et al. 2017). Fisherfolks display their local knowledge accrued through the years in their fishing activities: (1) they observe wind direction to predict fish accumulation; (2) they monitor color and temperature of water to foretell fish availability; (3) they formulate various traditional feeds to catch fish; and (4) they devise a wide array of fishing tools and methods appropriate to the season (Nongmaithem 2012; Devi et al. 2013).

According to the Manipur Science and Technology Council, the state has 5 distinct seasons: summer (May–June), rainy/monsoon (July–September), autumn (October–November), winter (December-February), and spring (March–April). While the fishers have basic fishing tools which they use throughout the year such as the Nup-il (lift net), Longthrai (scoop net), Long (spear fishing), and Taijep (box trap). They also utilize seasonal fishing tools only used on particular seasons. During the summer season, fisherfolks mainly use the Il-jao (dip net), Khoisang thakpa (long line), Moonamba (drag net), Khoi Choppa (pole line), Moirang lang (encircling net), and Long-oop (plunge cover-basket). During the rainy/monsoon season, they use the Sora-lu (conical trap), Kabo-lu (tubular trap) but still keep the Il-jao (dip net), and Khoisang thakpa (long line). During the autumn and winter seasons, they use the Moirang lang (encircling net), Long-oop (plunge cover-basket) (Devi et al. 2013).

Athaphum: Indigenous Meitei community in Loktak have invented a distinct fishing method that utilizes Phumdi known as 'Athaphum'. "This system developed after the Ithai Barrage came in…when people who lost their agricultural lands thought of [other] techniques to fish", Salam Rajesh explains.

Athaphum fishing involves two phases: (1) 'Phum thaba' or the preparation of circular enclosures and (2) 'Phum namba' or the harvesting phase (Tombi and Shyamananda 1994). Phumdis are cut into narrow strips (that are 1.5–2.3 m wide and 5.0 m long) (Khoiyangbam 2021), and then placed on the water in a circular formation (with circumference of 200-250 m) (Brajeshwori et al. 2002). This is used to lure in fish by dropping feeds inside the circle (Khoiyangbam 2021).

When fishers observe that enough fish have been trapped, the dip nets (Il-jao) are cast around the enclosure. A team of 30–40 fishers is required to drag up the nets. Collected fish are kept half submerged using the box traps (Taijep) until they are to be transported to the market. The harvesting of fish typically happens twice a year between October and February. Fishers' income from Athaphum is significantly higher than any other fishing method. They are able to harvest 100–180 kg of fish per yield (Singh and Khundrakpam 2009).

Utilizing 'Unjas' or middle persons has long been a practice for the indigenous Meitei community in Loktak alongside their traditional fishing. Fishing households would typically have a partnership with an Unja that usually starts through loan provisions.[1] They have a unique role in the trading system of fish resources caught

[1] Loaning through financial institutions is difficult for fishing households as these necessitate formal documents and other related paperwork, which make the requirements improbable and the process tedious for fishing households in Loktak. Unjas make loan services accessible to fishing households, albeit with (a) very high-interest rate and/or (b) conditions that borrowers can only sell to their lender.

in the lake: typically in charge of procuring fish resources from the lake to be sold in the markets (Nongmaithem 2012).

The indigenous Meitei community's local knowledge of building floating huts and fishing exemplifies their deep and intrinsic relationship with Loktak Lake. These traditional practices distinct only to the indigenous Meitei community of Loktak are born out of their age-old co-existence with the lake.

2.5 Loktak Development Authority

The Loktak Development Authority (LDA) was constituted by the Government of Manipur in 1986 for the 'management' and 'improvement' of 'Loktak', particularly it aims to "check the deteriorating condition of Loktak Lake" and "bring about improvements on the lake ecosystem along with development in the field of fisheries, agriculture, and tourism while conserving the catchment area involving concerned Departments of the State [Manipur] Government". LDA envisions the "conservation and sustainable utilization of wetlands for ecological security and livelihood improvement of local communities" (Loktak Development Authority 2012).

In 2006, the Manipur Loktak Lake (Protection) Act was enacted by the Government, empowering it to constitute an executive body, the LDA, that will exercise the following functions: (1) administer the affairs of the Loktak Lake, and to protect and improve the natural environment of [the] lake; (2) regulate by granting of approvals or otherwise requests for commercial utilization or bio-survey and bio-utilization of any resources; and (3) to do such other things as may be incidental or conducive to the efficient administration for protection and improvement of the lake (Government of Manipur 2006).

The Act, however, was hurriedly passed without consultation from the indigenous Meitei community residing in the area, independent scientists, and environmental experts (All Loktak Lake Area Fishermen Union 2016). Likewise, the members of the LDA are heavily dominated by government officials with no representation from local stakeholders (People's Resource Development Association 2016). 37 out of the 40 positions in the committee of LDA are reserved for government officials. The remaining 3 positions are non-official members which are to be nominated by the Government of Manipur.

The following provisions ratified in the Act put the LDA as well as the Government of Manipur's moral principles and integrity further into question:

> Section 2 defines the indigenous Meitei community residing in Phumshangs (floating huts) as 'occupiers'.
> Section 5 prohibits research endeavors related to lake resources of Loktak without the approval of LDA.
> Section 20 prohibits activities in the core zone of the lake (where fishing activities occur) such as the Athaphum method of fishing and building huts on Phumdis.

Section 29 allows the Government of Manipur to take measures it deems necessary to protect and improve the natural environment of Loktak.
Section 33 removes the right of local stakeholders to seek judicial remedy (in the case of state-perpetuated human rights violations).

Moreover, the Government of Manipur through the LDA and the Act has maliciously framed the indigenous Meitei community who have been custodians of Loktak as encroachers and polluters of their own indigenous territory. The degradation of Loktak was primarily because of the Ithai Barrage. Such framing has adversely affected the indigenous Meitei community's livelihood options and earning capability as well as their habitation.

However, under the guise of 'conservation', the LDA has escalated their efforts to dispossess and displace indigenous Meitei community of Loktak. Salam Rajesh cautioned against the conservation framework employed by LDA in Loktak. He explains: "Sometimes in the name of conservation, the local people are being evicted from their indigenous territory".

According to Babloo Loitongbam, Executive Director of Human Rights Alert, a Manipur-based human rights organization, the LDA is making a distinction between the people who are living in Loktak and Loktak as an ecosystem. Underscoring this point, Babloo Loitongbam added: "This [approach to conservation] is snatching away Loktak from the people who have been guarding it—those who have been the guardians [of Loktak] for thousands of years".

2.6 Heightened Attacks

"The plans and initiatives of LDA in Loktak Wetlands are not for the people", said Heisnam Chaoba, a community leader in Loktak. This orientation of the LDA has become more evident for the indigenous Meitei community of Loktak in the years that followed.

In November 2011, officers from the LDA and the Manipur State Police carried out deliberate burning of 777 Phumsangs (floating huts) inside the Loktak Lake. Fishing tools of the indigenous Meitei community were burned together with all their belongings, leaving them no means for food and shelter. A measly 40,000 Indian Rupees (524 US Dollars) was offered—not as a direct compensation to the arson victims but as a 'financial assistance for livelihood improvement'.

Speaking against this atrocity of the Government of Manipur and LDA, Heisnam Ashangbi, another Loktak community leader, says: "What is LDA? Why is the government forcefully bringing down people's homes and driving us to misery and death? Why is the government destroying peoples' floating huts in Loktak? We can't sleep peacefully. We can't work freely anymore. We feel unsafe. How can we live in dignity when there is merciless harassment from the government?"

When the indigenous Meitei community of Loktak mounted a women-led mobilization in December 2011 in response to the incident, they were mercilessly dispersed by the Manipur State Police. Arson victims who took part in the protest were subjected to brutality: physical assaults, baton charges, and firing of guns wherein at least 10 women were seriously injured (Schertow 2011).

Violent demolitions in Loktak continued to be carried out by the LDA despite the prohibitions of the Gauhati High Court through (1) a court order issued on January 25, 2012, prohibiting the clearance of floating huts; and a subsequent (2) contempt notice issued on February 16, 2013, ordering the Government of Manipur to cease their eviction drive in the area (Asian Human Rights Commission 2013).

From January 2012 to March 2013, the Authority demolished 70 Phumshangs (floating huts) and destroyed Athaphum fishing ponds in the lake (People's Resource Development Association 2016). Similar attempts to evict indigenous Meitei community in Loktak have persisted in 2015 and 2016 (Seema 2016).

However, new threats to Loktak and its indigenous Meitei community came in the form of policies and 'development' projects advanced by the Government of Manipur with the private sector through public–private partnerships. These push to 'unlock' the potentials of Manipur, and particularly, the Loktak Lake for profit-making.

In 2014, the Government of Manipur unveiled its Tourism Policy that aims to "promote sustainable tourism as a means of economic growth, social integration, and to promote the image of Manipur as a State with a glorious past, a vibrant present, and a bright future". It intends to "harness the tourism potential of Manipur, attract tourists and investments" through public–private partnership (PPP). In this policy, Loktak Lake and its vicinity was declared a Special Tourism Area for integrated development of tourist circuits and destinations, and was included in the Government of Manipur's first phase (2014–2017) of creating tourism infrastructure. The vision for the area was to be developed and beautified into the 'Loktak Lake Wetland Complex', a major ecotourism attraction with time-share resorts, restaurants, spas, and yoga centers; facilities for recreational fishing, cruise tourism and water sports (e.g., canoeing, kayaking, rowing, boat rides); and infrastructure for a cable-car and ropeway project (Manipur Government 2014).

With the objective to continue the "successful implementation of the previous projects and to push the Inland Waterways Transport sector to its full potential", the Loktak Inland Waterways Improvement Project was proposed by the Transport Department, Government of Manipur and the LDA to the Ministry of Shipping and Inland Waterways Authority of India, Government of India; the proposal was approved in November 2019. The project fundamentally aims to "augment the existing machinery and equipment to deal with the growing menace by clearing the waterways" as "all the machinery of LDA are no longer capable of dealing with the growing problems". It aims to achieve this through the (1) augmentation of machinery, (2) construction of infrastructure for waterway clearing, and (3) improvement of designated routes (through removal of Phumdis and other aquatic vegetation, and deepening of shallow portion of waterways) (Loktak Development Authority and Transport Department 2019).

Building on the vision for Loktak in the Manipur Tourism Policy 2014, the Government of Manipur in November 2020 invited tenders for the Loktak Lake Ecotourism Project. The projects aims to (1) develop Loktak as a world class tourist destination in

Manipur; (2) undertake catchment treatment of the area; (3) construct embankments or footpath around the lake; and (4) build tourism infrastructure (e.g., golf course, amusement park, artificial beach, integrated ropeways, water sports, adventure sports, home-stays, and accommodations) (Government of Manipur 2020).

According to Salam Rajesh, there is erasure of the symbiotic relationship between Loktak and the indigenous Meitei community in the policies and 'development' projects of the State. He elaborates: "These [policies and projects] are not addressing issues which are part and parcel of the entire lake system: the biomass (Phumdis), the people, and the fishing activities". Policies and 'development' projects that aim to 'uplift' the condition of Loktak has been fraught with misinformation and failure to address the root cause of the lake's deterioration; and has only increased the precarity of the indigenous Meitei community and disturbance of the ecological balance in Loktak.

Deterioration of Water Quality: The heavy use of mechanized equipment in the removal of Phumdis, promotion of motorized boats, and construction of tourism infrastructure run the risk of spillage of oil and other pollutants into the waterbody.

Eradication of Usable Aquatic Plant Species: The indiscriminate removal of Phumdis in Loktak in order to 'develop' the area may result in the eradication of both edible and non-edible aquatic vegetation which are utilized by the indigenous Meitei community for everyday life.

Disruption in the Ecosystem and Decline of Biodiversity: The speed and noise of motorized boats may drive away the remaining fish and bird populations in Loktak. Fish movement, in particular, may be greatly impeded when traditional dugout canoes that are amicable to the species in Loktak are phased-out in favor of motorized boats. These motorized boats also cut aquatic vegetation as well as fishing nets casted by indigenous Meitei community for their livelihood. Moreover, degraded water quality along with elimination of Phumdis may worsen the imbalance in the ecological processes of the lake. Vast changes in the landscape of Loktak may result in the accelerated ecological ruin and further loss of biodiversity in the area.

Economic Displacement Livelihood and Dispossession of Indigenous Territory: Massive PPP projects such as these would not only cause adverse effects to Loktak's ecosystem and biodiversity but would also lead to the privatization of the lake, which has been for generations a common resource for the indigenous Meitei community in the area. Privatization of Loktak would mean limited access to the lake and its resources; the indigenous Meitei community who, until the present time, are the stewards of Loktak may be deprived of their livelihood and habitation. These would increasingly convert Loktak and its resources into commodities for corporate and government bodies to exploit and extract maximum profit (Yumnam 2014).

Avli Verma, an environmental researcher from the research center Manthan Adhyayan Kendra, asserts that these two 'development' projects are "in complete contradiction to each other". The distinctive Phumdis and lifeways of the indigenous Meitei community are what attracts tourists to Loktak but both projects are seeking to erase these, replacing them with tourist circuits and destinations.

Similarly, Salam Rajesh, emphasized on the necessity of the inclusion of indigenous Meitei community in the conservation strategy of Loktak. He explains further,

"Conservation [..] is not kicking out the people who are originally there, protecting the forests...Conservation should be defined in a 'new' perspective. Indigenous Meitei community should be part of the [genuine] conservation program".

2.7 Resistance

After the construction of the Ithai Barrage and the enactment of policies and ratification of 'development projects' that brought an assortment of difficulties to the indigenous Meitei community and ruin to the ecological balance of Loktak, resistance and conservation strategies has been inevitable for the indigenous Meitei community in Loktak.

Since then, there have been numerous efforts from the indigenous Meitei community and advocates to organize resistance against development aggression. In 1985, the Loktak Flood Control Demand Committee was formed by 15 indigenous Meitei communities affected by the inundation of agricultural land caused by the Ithai Barrage. Indigenous Meitei community together with social scientists and activists formed the Action Committee—Loktak Project Affected Areas, Manipur in 1991. The following year (in 1992), the Loktak Khangpok Fisherman Association was formed by the indigenous Meitei community in Thanga (one of the island villages inside Loktak); at the same time, various organizations and progressive government allies formed the All Manipur Ithai Barrage Peoples Organization (People's Resource Development Association 2016).

"The Government of Manipur has no proper policy or agenda for the conservation or rehabilitation of the local people. We feel that we have to come up with one ourselves", says Salam Rajesh. Further, the policies and 'development projects' imposed on Loktak ignore indigenous rights, territory, knowledge, and culture. In 2011, 27 villages around Loktak formed the All Loktak Lake Areas Fishermen's Union Manipur (ALLAFUM) as response to the escalating threats brought by the Manipur Loktak Lake (Protection) Act 2006. He explains that ALLAFUM's creation was motivated by the indigenous Meitei community's desire to save Loktak by bringing the people together (1) to create awareness and build capacity for the conservation of Loktak (e.g., conserve migratory birds and fishes, rejuvenate the aquatic vegetation, and improve the lake ecosystem); and (2) to fight back against eviction and dispossession of indigenous territory.

The indigenous Meitei community under the ALLAFUM banner have been at the forefront of conserving the natural resources in Loktak, defending their rights to the indigenous territory, and resisting the profit-oriented and exclusionary model of 'development' imposed by the government. ALLAFUM has since engaged in various actions:

Educational Discussions: To conscientize the indigenous Meitei community, ALLAFUM conducts regular grassroots-level learning sessions about the interlocking issues they face such as change in the lake ecosystem post-Ithai Barrage, depletion of fish resources and edible aquatic vegetation, loss of migratory birds and other species in Loktak, among others. Through these, union members are capacitated to discuss these issues to their fellows, and potential allies outside of Loktak.

Awareness Campaigns: Because of the smear campaign of the government to pin the deterioration of the lake unto the indigenous Meitei community, it has become an important undertaking for ALLAFUM to hold different activities to dispel these false statements, reach out to community members outside the union membership, and attract allies that can support their struggle. Public consultations along with clean-up activities, and dugout canoe rallies are regularly held. Significant days such as World Wetlands Day, World Water Day, International Rivers Day, and International Day for Ecological Diversity are observed by ALLAFUM. The Loktak Arson Commemoration is also annually organized by ALLAFUM during 15–23rd of November, together with allied civil society organizations, to remember the inhumane acts of the government against the Loktak's indigenous Meitei community and to further engage them into active political participation.

2.8 Community Conservation

The scarcity of natural resources and loss of biodiversity in Loktak created by the Ithai Barrage necessitated the efforts of the indigenous Meitei community to save and protect the remaining ones. The union, in response, undertook a variety of conservation strategies:

Creation of a Fish Sanctuary: ALLAFUM has designated a portion of the lake as a community-controlled and administered 'fish sanctuary' to control excessive fishing of immature fish, promote a 'closed season' for fishing from the months of May to July, and allow the spawning of fishes during this three-month period (Rajesh 2020).

Exploration of Livelihood Alternatives: To augment the income loss during the closed season, ALLAFUM is exploring other livelihood alternatives during this time (e.g., poultry, light crops, weaving). The union has also started to plant seeds of edible aquatic plants for their own consumption and for sale in the markets (Rajesh 2020).

Protection of Migratory Birds: Recognizing the importance of biodiversity to the Loktak ecosystem, ALLAFUM has created a team of union members that will guard the stopover sites of migratory birds, and protect the birds from poachers especially during the months of October to February.

Broadening of the Struggle: To build broader resistance and a larger platform, the indigenous Meitei community saw the need to extend the collectivization of efforts to those outside of Loktak but residing in interrelated wetlands that will also be negatively impacted by the Loktak Inland Waterways Improvement Project and the Loktak Mega Lake Ecotourism Project. ALLAFUM started this process through going around different villages located in the periphery of Loktak and at the adjacent wetlands. Through these efforts, Ngamee Lup was formed in 2019: a federation of fisher unions and other association of Loktak and surrounding wetland areas (e.g.,

Pumlen, Ikop, Kharung, Khoidum, Lamjao, Waithou, Ngakrapat, Biraharipat, and Ungamelpat). At present, it functions as a body formed by executive members of different fisher groups in Manipur.

The union has also started to tap advocates outside of the lake, bringing the educational discussions and awareness campaigns to spaces (e.g., civil society, academics) where potential allies can be organized to support the struggle of the indigenous communities in Manipur.

Assertion of Rights: Because of the policies and 'development projects' of the government, the indigenous Meitei community spearheaded by ALLAFUM and Ngamee Lup formed the following demands:

(1) Wetlands International, Switzerland must withdraw the "Loktak, Manipur: An Integrated Plan for Wise Use (2020–2025)" unconditionally and immediately.
(2) The Government of Manipur must withdraw the "Loktak Inland Waterways Improvement" and the "Loktak Lake Ecotourism Project" projects.
(3) The Manipur Loktak Lake (Protection) Act of 2006 must be repealed. In its place a new law must be enacted after intensive consultation with the Loktak region people to ensure their traditional rights and livelihoods are protected and they continue as custodians to protect and conserve this biodiversity rich wetland complex for posterity.
(4) The Loktak region must be declared a World Heritage Site by UNESCO in acknowledgment of the living heritage that fisher and farming communities here have shaped over centuries.
(5) The Loktak region must be declared a 'Community Conservation Area' as per the Forest Rights Act, 2006 and as a 'Biodiversity Heritage Site' in accordance with the Biological Diversity Act, 2002. As a first step toward this goal, statutory mandates to constitute Forest Rights Committees and Biodiversity Management Committees must follow.
(6) All fundamental and traditional rights of indigenous communities living in the Loktak region, particularly their right of domicile, right to housing and right to livelihoods, must be recognized and with due dispatch.

The collective actions of the indigenous Meitei community have been and are continuously informed by these demands. In 2013 and 2014, ALLAFUM submitted the memorandum to government leaders demanding for the repeal of the Manipur Loktak Lake (Protection) Act 2006. The union together with Ngamee Lup has been relentless in urging the government to repeal the Act.

In 2018, ALLAFUM urged for the implementation of a new Act governing Loktak that is community-centered and environmentally friendly; and urged for the formation of a Wetland Development Board. At the same time, the union appealed for the government to include the affected indigenous Meitei community in the process of creating these.

ALLAFUM has also been consistent in countering disinformation from LDA and other government agencies concerning the Loktak and the other wetlands, and in exercising opposition to the government's repressive and tyrannical strategies to evict the community residing in Loktak and adjoining wetlands. Their continuous collective actions have led to substantial wins for the indigenous Meitei community.

In 2019, the High Court of Manipur issued an order that prohibits the Government of Manipur to take up any new project or development programs concerning Loktak without the authorization of the Court. The order also required the Manipur State Wetlands Authority (MSWA), Forest and Environment Department and Water Resource Department of the Government of Manipur, and the LDA to file an affidavit mentioning all the actions they have taken and the progress of these. The MSWA and LDA were also ordered to reveal how they have utilized the government funds concerning projects in Loktak (The Imphal Free Press 2019).

When the High Court of Manipur gave LDA the permission to proceed with the Notice Inviting Tender (NIT) process for the Loktak Inland Waterways Improvement Project and the Loktak Lake Ecotourism Project in 2020 (IFP Bureau 2020), ALLAFUM swiftly filed a review petition against this court order.

Following this and the continuous urge of ALLAFUM and Ngamee Lup, in 2022, the High Court of Manipur, reaffirmed its 2019 court order that no 'development' in Loktak can proceed without the court's permission (Mathur 2022).

According to Salam Rajesh: "The indigenous [Meitei] community is trying to assert their right to life and resources [..] the area of Loktak is a traditional territory. Fishers have the right to go there, fish and live. [..] They are part of the ecosystem and that they have the right to live there".

"What we have been demanding of the Government of Manipur is to honor and recognize the traditional livelihood and fishing rights of communities in Loktak. People should be allowed to fish…floating huts, [which is] a timeless heritage of Loktak should be allowed", explains Oinam Rajen, Secretary of ALLAFUM. He further highlights that the future of the indigenous community of Loktak lies in the healthy regeneration of the water body's ecosystem—their lives are entangled with it. The indigenous communities thriving in Loktak have no other option but to resist and fight back for their right to life, right to housing, and right to livelihood (Rajesh 2021).

For the people of Loktak who have been gravely affected by the Ithai Barrage and subsequent development aggression, the ways in which they live and thrive are their developed grassroots alternatives to preserve the ecology of the wetlands. These together with their repertoire of resistance strategies from conducting educational discussions and awareness campaigns; executing community-led conservation initiatives, building broader networks of solidarity, and proposing of alternative community-centered policies—comprise the resistance of the indigenous Meitei community to protect their right to stay and the wetlands.

References

All Loktak Lake Area Fishermen Union. 2016. Loktak Lake status report 2016. In *Indigenous perspectives*. Manipur Nature Society.

Asian Human Rights Commission. 2013. "INDIA: Immediately stop evicting and threatening the floating huts-dwellers of Loktak Lake." Asian Human Rights Commission. February 27. http://www.humanrights.asia/news/urgent-appeals/AHRC-UAC-028-2013/.

Bharati, Huirem, Asha Landge, Arpita Sharma, Yumlembam Singh and Thangjam Nirupada Chanu. 2017. Insight into the Socio-economic Life of Fishers of Loktak Lake. *Manipur-A Ramsar Site 54*: 215–220.

Brajeshwori, T.D., et al. 2002. Phumdi management—the community perspective. In *Management of phumdis in Loktak Lake,* 70–73. Wetlands International South Asia.

Devi, Nightingale, Sumanta Mishra, Lipi Das, Nilesh Pawar, and Thongam Chanu. 2013. Traditional fishing methods in central valley region of Manipur-India. *Indian Journal of Traditional Knowledge.* 12: 137–143.

Government of Manipur. 2006. The Manipur Loktak Lake (Protection) Act. Manipur, India. https://www.indiacode.nic.in/bitstream/123456789/13415/1/the_manipur_loktak_lake_%28protection%29_act%2c_2006.pdf#search=Loktak%20Lake%20(Protection)%20Act%202006

Government of Manipur. 2020. Selection of project management consultant for Loktak Lake eco-tourism project. *Notice Inviting Tender,* 2 November. http://www.manipurtourism.gov.in/wp-content/uploads/2020/11/PMC-LOKTAK-LAKE_compressed-1.pdf.

IFP Bureau. 2020. HC permits NIT process to execute Loktak inland waterways, eco-tourism project. *The Imphal Free Press,* November 9. https://www.ifp.co.in/3288/hc-permits-nit-process-to-execute-loktak-inland-waterways-eco-tourism-project.

Kangabam, Rajiv and Sarojini Devi Boominathan and Govindaraju Munisamy. 2015. Ecology, disturbance and restoration of Loktak Lake in Indo-Burma biodiversity hotspot—an overview. *NeBIO* 6: 9–15.

Kangabam, R.D., M. Selvaraj, and G. Muniswamy. 2018. Spatio-temporal analysis of floating islands and their behavioural changes in Loktak lake with respect to biodiversity using remote sensing and GIS techniques. *Environmental Monitoring and Assessment* 190: 118.

Khoiyangbam, R. 2021. Wetlands in Loktak: issues and challenges of merging wildlife conservation and hydropower generation—an overview. *International Journal of Lakes and Rivers* 14(2): 223–236. ISSN 0973-4570.

Khwairakpam, E., R. Khosa, A. Gosain, et al. 2021. Water quality assessment of Loktak Lake, Northeast India using 2-D hydrodynamic modelling. *SN Applied Sciences* 3: 422. https://doi.org/10.1007/s42452-021-04440-8.

Loktak Development Authority. 2012. Annual Administrative Report 2011–2012. https://manipur.gov.in/wp-content/uploads/2012/10/annual-report-2011-12-_16-3-12-loktak-development-authority.pdf.

Loktak Development Authority and Transport Department. 2019. Revised project proposal: Loktak Inland waterways improvement project. *Government of Manipur.* September. https://drive.google.com/file/d/1RSeuleE-JyOq-OxzYStlbGy1VliOalyz/view.

Loktak Development Authority and Wetlands International—South Asia. 2003. Loktak Newsletter. March. Vol. 3.

Manipur Government. 2014. Manipur tourism policy. https://manipur.gov.in/wp-content/uploads/2014/06/tourism-policy-full-final-new.pdf.

Manipur Science and Technology Council. N.d. Modern climate. *Mastec.* https://www.mastec.nic.in/modern-climate.

Mathur, Aneesha. 2022. No development works in and around Loktak Lake without court's permission: Manipur HC. *India Today,* February 27. https://www.indiatoday.in/law/story/no-development-works-in-and-around-loktak-lake-without-court-s-permission-manipur-hc-orders-authorities-1918522-2022-02-27.

Meitei, M.D.I, M.N.V. Prasad. 2015. Phoomdi—a unique plant biosystem of Loktak lake, Manipur, North-East India. *Traditional and ecological knowledge, Plant Biosystems—An International Journal Dealing with all Aspects of Plant Biology,* 149(4): 777–787, https://doi.org/10.1080/11263504.2013.870250.

Moirangleima, K., and A.L. Singh. 2009. Shrinking water area in the wetlands of the Central Valley of Manipur. *The Open Renewable Energy Journal* 2 (1): 1–5. https://doi.org/10.2174/1876387100902010001.

National Wetland Atlas. 2009. SAC/RESA/AFEG/NWIA/ATLAS/03/2009.Space Applications Centre (ISRO), Manipur: Ahmedabad, India (96).

Nongmaithem, Bijayalakshmi. 2012. Study of fishers' socioeconomic and cultural profile around the Loktak lake of Manipur, India. *IOSR Journal of Agriculture and Veterinary Science.* 1: 48–56. https://doi.org/10.9790/2380-0154856.

People's Resource Development Association. 2016. Issues on Loktak. Np.

Rajesh, Salam. 2021. Alienating community in Loktak. *Heinrich Boll Stiftung*, August 31. https://in.boell.org/en/2021/08/31/alienating-community-loktak

Rajesh, Salam. 2020. Community participation in conservation of Loktak Lake of Manipur. *Vikalp Sangam,* June 22. https://vikalpsangam.org/article/community-participation-in-conservation-of-loktak-lake-of-manipur/.

Schertow, John Ahni. 2011. Urgent Appeal: Manipur police assault indigenous woman at Loktak Lake. *Intercontinental Cry*, December 19. https://intercontinentalcry.org/urgent-appeal-manipur-police-assault-indigenous-women-at-loktak-lake/.

Seema, Thokchom. 2016. Proposed tourism infrastructure in Manipur puts Loktak Lake Residents at Eviction Risk. *Land Conflict Watch,* September 13. https://www.landconflictwatch.org/conflicts/loktak-lake-area-tourism-plan-displaces-local-people.

Singh, A.L., and Khundrakpam, M.L. 2011. Phumdi proliferation: a case study of Loktak lake, Manipur. *Water and Environment Journal* 25: 99–105.

Singh, A.L., and M. Khundrakpam. 2009. Shrinking water area in the wetlands of central valley of Manipur. *The Open Renewable Energy Journal.* 2: 1–5.

Singh, H.T. 2014. Impact of Ithai Barrage on the environment of Manipur: an overview. In *Loktak lake and Manipuri lifeworld, putting the Ramsar sites to inconsequential Abyss*, eds. Hanjabam, S.S., Singh, A.K. and Singh, R.K, 128–134. New Delhi: Forward Books.

Singh, Kh. Jugindro. 2018. The Ithai barrage of Manipur: its emerging impacts on anthropogenic and ecosystem to the catchment areas. *IOSR Journal of Humanities And Social Science 23, no. 5, Ver. 6*: 67–72.

Singh, M. Damudor. 2015. Loktak multipurpose project and its environmental implications in Manipur. *Online International Interdisciplinary Research Journal* V, July. Special Issue (Bi-Monthly).

Takhell, Donald. 2021. Holding ground. Centre for financial accountability. Centre for Financial Accountability (CENFA), June 3. https://www.cenfa.org/holding-ground/.

Thakur, Jaya. 2020. The Ithai barrage of Manipur: to decommission or not. *ORF Issue Brief No. 364,* May. Observer Research Foundation.

Thaoroijam, Kalpana. 2017. Who's lake, who's tourism? corporatising tourism in Manipur. *Kangla Online,* November 21. https://kanglaonline.com/2017/11/loktak-whos-lake-whos-tourism/.

The Imphal Free Press. 2019. HC bans new project in Loktak Lake, directs stakeholders for progress reports on preservation. *The Imphal Free Press*, August.http://www.ifp.co.in/page/items/60628/hc-bans-new-project-in-loktak-lake-directs-stakeholders-for-progress-reports-on-preservation/.

Tombi, H.S. and R.K. Shyamananda. 1994. *Loktak Lake, Manipur*, 10–12. New Delhi: Worldwide Fund for Nature India.

Trisal, C.L and Th. Manihar. 2004. *The Atlas of Loktak, Loktak development authority and wetlands International South Asia, Delhi.*

Trisal, C.L. and Manihar, T. 2002. Management of Phumdis in Loktak Lake. *Wetlands International South Asia.*

Tuboi, C., and S.A. Hussain. 2016. Factors affecting forage selection by the endangered Eld's deer and hog deer in the floating meadows of Barak Chindwin basin of North-east India. *Mammalian Biology.* 81 (1): 53–60.

Wangkheirakpam, Ramananda. 2014. Ecological and social impacts of the Ithai Barrage. In *Loktak lake and Manipuri lifeworld, putting the Ramsar sites to inconsequential Abyss*, eds. Hanjabam, S.S., Singh, A.K. and Singh, R.K., 23–40. New Delhi: Forward Books.
Yumnam, Jiten. 2014. CRA Manipur Report: Loktak wetlands eviction: a case of development crisis in Manipur 2011–2013. *Centre for Research and Advocacy, Manipur,* February. https://drive.google.com/file/d/1FP_lAZ6m4dc9UrO_lJ38RRBkJXksPQWa/view.

Donald Takhell is an Research Associate at Indigenous Perspectives, an organization that provides space for action-research on local responses to globalization processes. His work currently focuses on the intersection of environment, development, climate change and indigenous people. He holds a postgraduate degree in Development Studies from Ambedkar University, Delhi.

Rafael V. Dimalanta is a researcher at the Program on Alternative Development of the UP Center for Integrative and Development Studies where he co-researches with civil society organizations to help surface alternative development perspectives and models from the grassroots. He graduated with a degree in Architecture from the University of the Philippines (UP) Diliman; has taken units under the graduate studies program of the UP Department of Anthropology; and is currently taking his masters in Community Development at the same university. He primarily works with people's organizations in urban poor communities, street vendor associations, and informal transport unions.

Chapter 3
Local Energy Planning in Vietnam and "the Local" in Building Alternatives in Energy and Environmental Policy

Honey B. Tabiola

Abstract At the intersection of energy justice and local governance, Local Energy Planning (LEP) in Vietnam demonstrates how grassroots participation can shape sustainable energy policies. This chapter examines the role of LEP in addressing Vietnam's growing energy demand while promoting environmental sustainability and social equity. It explores how community-led initiatives contribute to inclusive decision-making, integrating local knowledge to enhance energy resilience and efficiency. This study illustrates how decentralized energy planning fosters adaptive solutions to environmental and socio-economic challenges, particularly in rural and underserved areas.

Despite its potential, LEP faces significant institutional, financial, and technical barriers that hinder its full implementation. Additionally, the integration of renewable energy sources within existing policy frameworks remains a challenge due to regulatory inconsistencies and the dominance of centralized energy planning. Through an assessment of policy frameworks and stakeholder dynamics, this chapter highlights the need for stronger institutional support, capacity building, and policy coherence. Ultimately, it argues that LEP can serve as a crucial mechanism for advancing Vietnam's energy transition, ensuring equitable access to clean energy while empowering local communities in decision-making processes

Keywords Community participation · Decentralized energy · Energy justice · Local Energy Planning · Vietnam

H. B. Tabiola (✉)
University of the Philippines Center for Integrative and Development Studies, Diliman, Quezon City, Philippines
e-mail: hbtabiola@up.edu.ph

J. M. C. Sy and A. P. Aban (eds.), *Reimagining Development in Southeast Asia*, Asia in Transition 21, https://doi.org/10.1007/978-981-97-9140-8_3

3.1 Introduction

Over the past two decades, Vietnam has undergone rapid economic growth, transforming it into one of the strongest and fastest-growing economies in Southeast Asia (Urban et al. 2018). This has brought increasing industrialization as well as environmental pressures. CO_2 emissions have steeply increased, energy demand has been growing, and environmental pressures have been mounting in relation to the pollution of air, water, and soil; deforestation; and natural habitat destruction (World Bank 2017). As in most countries, Vietnam has been figuring out how to reconcile economic development with ecological sustainability.

Energy issues are top priorities on the agenda of Vietnam due to the need to address energy supplies, solve large pockets of energy poverty, and the heavy reliance on fossil fuels, particularly coal (Urban et al. 2018). Over the years, the government implemented a number of energy-related policies which aim at energy efficiency, diversified and effective use of domestic natural resources in combination with a reasonable import–export balance, reducing GHG emissions for environmental protection, and increasing the share of renewable energy in the total commercial primary energy supply (APEC 2013).

One key site where these national policies meet the sticky realities of the local community and of the efforts of civil society is the Local Energy Planning (LEP). It is a bottom-up approach to energy planning which entails data collection by and joint planning with the local community members and local authorities, in consultation with professionals, with the aim of achieving consensus on the contents of the energy plan. Addressing the weaknesses of top-down approaches to energy planning, LEP puts a premium on "participation" of the local community and local authority to design energy plans that customized their energy needs and resources. Participation here means (1) voluntary involvement of people; (2) the ability of the people to influence and control over the initiative, decisions, and resources; (3) involvement of people at different stages of the initiative; and (4) having the ultimate aim to improve the well-being of the people who are participating (Narayanasamy 2009, 3). As a bottom-up local initiative, it is cognizant of the insight that "the effectiveness, success, and sustainability of development initiatives largely depend on wholehearted participation of the stakeholders, particularly the primary ones" (Narayanasamy 2009, xvii). This essay documents LEP as conceived and implemented by Green Innovation and Development Centre (GreenID), a non-government organization (NGO) in Vietnam.

More generally, as LEP champions "the local" as a site where alternatives might emerge and flourish, an analysis of LEP also presents an occasion to assess the transformative role of "the local" in energy and environmental policy. In particular, the essay considers the power and limitations of localism for a potent politics of constructing alternatives.

In a work entitled *The Limits of Eco-Localism: Scale, Strategy, Socialism*, Gregory Albo (2007) articulates the promise and pitfalls of localism as a virtue and venue for political action and social change. Localism has been a project for both the right and the left. Especially prominent on the left, the local is seen as the space and scale for "tangible solidarities" for "alternate ways of life." As such, for Albo, it is important to interrogate "the local" and confront the "systematic obstacles" for progressive projects that count on "the local" to prosper (ibid., 339).

On the conservative right, the neoliberal case for localism has been widely supported. Albo offers the example of the World Bank which advocates for the idea that "the political objectives to increase political responsiveness and participation at the local level can coincide with the economic objectives of better decisions about the use of public resources and increased willingness to pay for local services" (World Bank 2001). From this perspective, local government units are seen as better positioned to respond to the needs of the populace and to make more informed decisions about the use of public funds. This devolution of powers to the local level is often knotted with the move to defund the local governments and leave them to manage single handedly the basic social services of the populace.

On the left, embracing the local is read as either "skepticism about universal projects and collective struggles for societal transformation" (Albo 2007, 337) or as the celebration of "local resistance and community alternatives" to the overwhelming imperatives of neoliberal globalization (ibid., 338). This view holds that the territorial scale of the local is decidedly the site for political action and transformation. This entails a shift from centralized and concentrated forms of wealth and power (often wielded by a privileged few) to emphasis on democratic, less bureaucratic systems of decision-making on the local and individual levels. Albo points out that one area where this attraction of the local is seductive is in the ecological movement.

The first section of the paper discusses the energy situation in Vietnam, particularly its challenges and the national programs and policies addressing these challenges. The second section provides an overview of the Local Energy Planning (LEP)—its purpose, features, mechanisms, benefits, and its vision for the transformation of the energy situation in Vietnam. Finally, given the controversy about the status and role of "the local" in anti-neoliberal politics, the paper highlights the necessarily "double-sidedness" of the local; that is, the local is both a product of large social structures and a *potential* for structural transformation. Further, the case of LEP shows that this double-sidedness may be overcome by continuing engagement and contestation with the large social structures or the "extra-local" institutions such as the national government and its agencies.

3.2 Energy Situation in Vietnam

LEP should be located within the energy situation in Vietnam. In 1986, Vietnam began a transition from an economy that is centrally planned to one that is market-oriented (Do and Sharma 2011, 5770). This transition led to a series of developments, namely Vietnam's obtaining access to concessional international finance in 1993, the lifting of US embargo in 1994, membership into the ASEAN (Association of Southeast Asian Nations), APEC (Asia–Pacific Economic Cooperation), and WTO (World Trade Organization) in the years 1995, 1998, and 2007, respectively. Arguably, these developments have been attributed as the stimulus for more than 7% growth of the economy over the past fifteen years (IEA 2008, cited in Do and Sharma 2011). In its desire to maintain high rates of economic growth, the energy sector is bound to face several challenges which prompted the Vietnamese government to carry out a set of energy policies to shape the evolution of the country's energy policy formulation.

The energy sector in Vietnam is beset with various challenges (Do and Sharma 2011, 5770–72). First is on energy import dependency. Given the increasing demand of the country driven by high economic growth and growing population, there has been an emerging consensus among Vietnamese policymakers and the planners that the indigenous energy resources are unlikely to meet the energy demand. As such, dependency on imported energy is a growing concern. It is projected that by 2025, Vietnam will have to import energy as the reserve of oil and gas will not be sufficient beyond a 25-year time horizon. Data on energy demand from 1995 to 2014 reveal that the GDP and the energy demand of Vietnam have increased consistently in the last years (IEA 2017) while projections also show that total final energy demand will be 2.5 times higher in 2035 than in 2015 (Danish Energy Agency 2017).

The primary energy supply of Vietnam comes from fossil fuels, mainly coal and oil (over 70%), as well as natural gas (IEA 2017). In terms of electricity generation, between 1995 and 2014, hydropower has been the major source, followed by gas and coal (IEA 2017), although hydropower is projected to decrease from 29.5% to 15.5% (GGGI 2017; GIZ 2017; Urban et al. 2017). It is also noted that the share of both renewable (particularly hydropower, wind, and solar energy) and non-renewable sources has steadily grown. "Vietnam has a good potential for the development of renewable energy, such as wind, solar, biomass, and hydropower (geothermal under consideration)" (Urban et al. 2018, 10), and the government is actively promoting the use and generation of renewable energy as enshrined in its National Energy Development Strategy (2007).

The second challenge is the low energy efficiency due to old technologies (e.g., for most existing coal and oil-fired plants) and poor energy management practices (from conversion to processing and end-use levels). Third and related is the mobilization of funds for developing the energy sector which is preliminarily estimated at nearly USD 137 billion of investment over the period from 2008 to 2025 (JICA 2008). Fourth is on environmental consequences of energy policies. While Vietnam is one of the lowest per capita emitters of carbon emissions in 2007, its CO_2 emissions per unit of GDP are about 2 times the world average (IEA 2009, cited in Do and Sharma 2011). In the year 2025, it is projected that nearly 93% of the total commercial, primary energy consumption of the country will be driven by fossil fuel, with coal occupying 47% share.

By way of addressing these challenges, the Vietnamese government developed a "unified energy market, both domestically and internationally" to "promote investment and secure financial resources for sustainable development of the energy system" (Do and Sharma 2011, 5772). Essentially, this entails a shift from direct market control via the state-owned firms to a system decided by the markets. Do and Sharma (2011) argued that a unified energy market or a "market-based energy system" would enhance security of energy supply by encouraging the participation of the private sector and improving flexibility and transparency of the energy system. In effect, they believe that this arrangement would reduce financial pressures on the government and increase effective allocation of resources through market competition.

The shift to a "market-based energy system" is accompanied by a suite of policy measures (ibid., 5774) concerning energy resource development (strengthening exploration and development of indigenous resources), domestic energy supply security (strengthening domestic energy supply capacity and expanding energy infrastructure), capital resources for energy development (raising investment capital), development of renewable energy resources (increasing the share of renewable energy in power generation), and international cooperation and energy trade (diversifying energy supply sources by exploring opportunities overseas). In particular, to reduce its GHG emissions and increase energy efficiency, the Vietnamese government issued a number of policies such as National Target Programme on Energy Efficiency, on carbon trading, and the Law on the Economical and Efficient Use of Energy (Urban et al. 2018, 9).

The Vietnamese government has played an active role in shifting the country's approach from climate change adaptation to mitigation (Urban et al. 2018); the government has been seen as working closely together with the private sector (both national and foreign) toward economic development (Schmitz et al. 2015). For example, import tax exemptions and land fee exemptions are granted by the government for those investing in renewable energy.

In their interviews with policymakers and representatives from firms, entrepreneurs, civil society, academia, and multilateral organizations, Urban et al. (2018, 14) found that the Vietnamese government has been successful in integrating top-down and bottom-up approaches to planning. An example given was the draft of GGGI Vietnam Country Planning Framework (CPF) for the period from 2016 to 2020 as the result of intensive consultations held over the course of 2015, including bilateral meetings, donor roundtable, and workshops among ministries, civil society, agencies, donors, and development partners. Interviewees also revealed that project consultants were held responsible by the local government which, in turn, was held responsible by the national government. This "high control" by the socialist government over all levels of the state is also manifested in "strong boundaries" and "tough stances on freedom of speech, freedom of press, and freedom to assemble" (ibid.).

3.3 Energy Justice Framework

The paper analyzes LEP from the perspective of "energy justice." It situates LEP within four key elements, namely "recognition" which is putting a premium on those who are affected by plans and policies concerning energy, a fair "distribution" of the costs and benefits that might be derived from energy services, a fair "procedure" in any forms of decision-making, and "restorative justice" which refers to efforts to compensate for possible risks as a result of energy decisions (Heffron and McCauley 2017). From the standpoint of energy justice, all energy decisions and policies do not take place in a vacuum (Sovacool et al. 2016); they have important material and moral consequences.

3.4 Analytical Approaches to Energy Planning

Energy planning is defined as a "program based on a rigorous research on issues related to energy supply and demand, energy prices, technology supply and demand, population growth, environmental, social, technological success in harnessing the energy and influence the political situation of the country" (Ravita et al. 2014, cited in Mougouei and Mortazavi 2017, 128).

Energy planning may be categorized according to purpose, model structure, analytical approach, underlying methodology, mathematical approach, geographical and sectoral coverage, time horizon, and data requirements (van Beeck 1999). For this essay, what is most useful for analyzing LEP is the two analytical approaches, namely top-down approach and bottom-up approach. Top-down energy planning is "based on macroeconomic modelling principles and techniques and is intended to include all important economic interactions of the society" (Jacobsen 1998, 445). Top-down models are "characterized by behavioral relations at an aggregated level with parameters estimated based on historical relationships" (ibid., 446). If top-down planning deals with concepts in macroeconomics, bottom-up energy planning is based on "disaggregation and the inclusion of a large number of technical parameters" (ibid., 443). Bottom- up models are usually detailed and include a description of energy technologies with technical and economic parameters (ibid., 445). For example, they might include an analysis of an energy technology which could be used as fuel substitution. Further, bottom-up models also include "energy demand" (ibid.) categorized into different uses such as heating, lighting, and ventilation rather than categorized according to energy types.

Each model has different properties which yield different results despite analyzing the same issues on energy. This is due to the fact that the two approaches have been conceptualized from different theoretical backgrounds with different purposes in mind (ibid.). In the contrast provided by van Beeck (1999), top-down approaches use an "economic" approach while bottom-up uses "engineering approach." The former tends to be pessimistic in its estimates on "best" performance while the latter gives optimistic estimates. The former cannot explicitly represent technologies while the latter allows for a detailed description of energy technologies. It is in their differing theoretical underpinnings and operations that the complementarity of the two approaches may be considered. This probably explains why GreenID does not abandon top-down approaches to energy planning but simply complements them with bottom-up approaches through LEP.

3.5 Localism

Greg Albo (2007) expresses his anxieties about localism as a project. Specifically, he shows the virtues and the vices of "the local" as a form of action toward social transformation. First, he describes the three major strands of thinking about localism within the context of ecological movement and critiques these localisms (especially the first two strands) for their implicit assumption that ecology and economic growth are compatible. Most importantly, he shows that exclusively depending on "the local" is fraught with complications and dangers.

As classified by Albo, there are three strands of thinking under eco-localism. The first is market ecology which is heavily influenced by "green economy" and "market solutions." Inspired by neoliberals such as Friedrich Hayek to Milton Friedman, this holds the view that the prices set by the market are the real, effective mechanism to compel individuals (construed as buyers and sellers) and firms to behave in an ecologically responsible way through green taxes and other kinds of incentives. Prices, it is argued, are "transmitted across space to equilibrate all markets" (ibid., 342) aggregating and incorporating all information markets. Often, market ecology means that local markets or smaller locally based capitals tend to be seen as more sustainable or more inclined to ecological conservation than big corporations and big governments. Under this view, the answer is decentralizing environmental regulation to the markets through pricing which makes possible the abolition of extensive enforcement authorities of the state and its agencies. This promotes the view that the markets are essentially decentralized and place-based (that is, local) regulators of the conduct of individuals and corporations.

The second form of eco-localism is called *ecological modernization* which champions technological and organizational transformations. Essentially, this involves the replacement of "resource-using and pollution-generating techniques" with "resource-saving and pollution-reducing" technology toward an "eco-efficient transformation of the whole built environment" (ibid., 343). This also entails the implementation of policies toward developing community-based green industry, often through corporatist sustainability partnerships. Projects under this also include urban density and urban green-space through "green city" projects for extending public transit and "green-belting" urban sprawl (ibid.).

The third strand is called *social ecology* which has strong traditional anarchist roots. In this thinking, "ecological balance is restrained with decentralized communities by the need to find local solutions, eliminating at once both externalities and resource overuse, as well as the disastrous effects of mass production industrialism" (ibid., 344). This means that the idea of extensively self-sufficient communities with little or less intervention from externalities is feasible and that the downscaling of production and exchange to the level of the community would counter internationalization of commodities and long- distance trade and supply chains which would prevent overuse of energy resources. This approach embraces local self-governance and self-sufficiency through "direct action aimed at establishing artisanal and other appropriate technologies, alternate markets, the redefinition of needs, the preservation of peasant economies, seed diversity, local currency systems, 'getting off the grid'" (ibid., 346), among others.

Local initiatives such as LEP are easily discredited and dismissed as conforming to these three strands of eco-localism. But Albo's thought is useful here because it becomes a rubric through which this essay delineates the distinct contribution of LEP, departing from the three strands.

3.6 Local Energy Planning: An Overview

Local Energy Planning (LEP) is a local initiative implemented by Green Innovation and Development Centre (GreenID) in cooperation with Vietnam Sustainable Energy Alliance (VSEA). LEP is a bottom- up energy planning in which the local community and local state authorities, in consultation with a team of experts, work together to craft a common energy plan on the consumption, future demand, and potential energy sources of the locality (i.e., village, commune, or district) (GreenID 2014a, 3). The LEP approach takes into account the availability, efficient use, and environmental impacts of local renewable resources and the use of renewable energy and sustainable solutions.

When it was just starting as an NGO, GreenID saw gaps in NGO work believing that no one is working on the energy sector yet, specifically on the policy and community level. At the time of the interview (2018), Ngụy Thị Khanh, Executive Director and founder of GreenID, claims that there are many NGOs who are already working on the energy sector, but "they are more in line with the government, not critic or provid[ing] any alternative." For Khanh, the work of GreenID on LEP is an attempt at a "win–win solution" between economic development and the environment:

> We stand for the sustainable development in Vietnam, but we identify our mission to focus on energy sector and improve environmental governance and public participation in this sector … We do not want to see that development kill the environment … We believe that the only way we are able to achieve this win-win solution is based on the green development and the full stakeholder participation in the decision-making process.

LEP is a collaboration of various key players. First, a local energy team (LET) composed of eight to ten people from the local community will be selected by GreenID and VSEA in consultation with the local authorities. They are usually those who have great influence and interest in participating in the LEP. In the interview, Khanh stressed that members of the LET are "really from the commune. They are villagers. They may be the head of the village and [or] they are women representatives [who] all work together." The second key player is the local partner who could be a local non-government organization (NGO) who works closely with GreenID and meets with the local people and with the authority at the province or district level. Khanh pointed out that GreenID tries to involve different types of agencies and institutions as local partners:

> We do not stick on one local partner on each location. We try to diversify our partner. We try with the Province Technology and Science Association in another province. We tried with the Department of Industry and Trade. That is a stage agency. But in another province, we worked with the Department of Agriculture. And in another province, we worked with women union. So, we see how it worked.

The third key player is the local community. For the pilot testing of LEP, some criteria for choosing the local community were considered such as the presence of an environmental problem, some possibilities for renewable energy, active participation of the local mass organization, and support from the local authority.

Fourth, the private sector also has a role to play. During a solution demonstration or exhibition organized by GreenID (which is a key stage in the LEP approach), companies go to the community and show to the people the available sustainable energy solutions to their needs such as household biogas, solar water heater, solar storm light, microbial fertilizer, water supply system Reverse Osmosis, LED lights, among others. For its part, GreenID bridges the energy needs of the community and the solutions offered by the companies. As Khanh puts it, "We negotiate with them [companies/suppliers] to 'lower the price for this community.' So, we have to create the demand and also inspire them to do the [corporate] social responsibility."

Fifth, the Vietnam State and Party is engaged through its local authority. When asked about the reception of LEP in the communes, Khanh thinks that the start was not easy. The local officials "do not pay attention to anything." But GreenID started to explain the process and the intentions of the program. Then, the local authority started to help GreenID set up the LET. The local authority is also invited to workshops where the results of the survey conducted by the LET are presented and discussions among the stakeholders can take place. Khanh explained the complexities when navigating the tight political system in Vietnam and the different responses by the local authorities:

For some commune, we got one member in the local energy team [who] is the vice chair of the People Committee. We ask him to sit in the room … information come[s] to him, and then, we can also propose [to] him to do something with the local authority. For some commune, they bring the plan to mainstream it to their official plan, but some they look at this [plan] and they support for the team and the project and implement in parallel with their energy economic development plan. For some, the local authorities see that "Oh, it is a good plan for some activity" they also bring their budgets from their local budget to support for the plan.

Despite the constraints of the political landscape of Vietnam, GreenID succeeded in securing the cooperation and support of some of the local authorities. So far, the responses have ranged from passivity to active participation, from parallel implementation of the local energy plan to mainstreaming it within the official energy plan.

Sixth, experts are also involved to train the LET to use the methods and tools to collect relevant data from the community. With regular exchanges with the LET, the experts conduct data analysis, formulate energy options, and suggest appropriate technology suitable to the community. GreenID has an expert from Thailand. She lives in Hanoi and worked at the New Capacity for Earth. Finally, for its part, Khanh understands the role of GreenID as "facilitator for the process" who also "build[s] capacity for local people." Below are the 10 steps of the LEP process (Nguyen 2017, 3):

1. Establishing and training Local Energy Team (LET) by GreenID, VSEA, and the local authority.
2. Preparing tools for data collection by both the expert and the community.
3. Surveying and collecting data by both the expert and the community.
4. Analyzing data by the expert.
5. Identifying prioritized problems and goals by the expert and the community.
6. Presenting and choosing solutions by the community in consultation with the expert.
7. Developing energy scenarios by the expert.
8. Building energy plans by the community and the expert.
9. Consulting and communicating with the community representatives to achieve consensus.
10. Implementing plans and replicating good results.

In 2012, LEP was pilot-tested in Nam Cuong and Bac Hai rural communes in the Thai Binh Province and was then replicated in many more communes. As of 2016, LEP has been implemented in six (6) provinces, six (6) districts, 11 communes with seven (7) local energy plans available and 15 types of energy models being applied (ibid., 1).

3.7 Purpose

GreenID highlights the challenges plaguing the top-down approaches to energy planning while stressing the unexplored opportunities that bottom-up energy planning offers. Specifically, LEP is offered as a corrective to top-down approaches in terms of the *content* and *process* of implementation.

First, LEP launches a critique of the *content* of top-down approaches. In GreenID's telling, as Vietnam energy development policies and plans are "mainly built at the central level," the Power Development Plan emphasizes "the increase of generation capacity from traditional energy [such] as hydropower and thermal coal" (GreenID 2014a, 2) is projected to cover 50% of electricity generation in 2020. With the centralized planning's emphasis on traditional energy, it is not surprising that the role of renewable energy is not maximized. In GreenID's words, there is "less discussion on alternative energy scenario of energy demand as well as sustainable energy development" (ibid.) and "the role and potential of renewable energy [and] energy efficiency is quite modest" (ibid.).

In contrast to top-down planning, the bottom-up approaches advocated by GreenID promote clean energy development and renewable energy industry development, which essentially entails a shift from concentration to distribution, decentralization, and diversification on the application of energy sources and technologies (ibid.). Instead of heavy reliance on fossil fuel, LEP explores the potential of renewable energy solutions and the contribution of community resources in generating energy. GreenID draws inspiration and lessons from the successful experience of bottom-up energy planning in Denmark and Thailand.

Second, it is in the level of *implementation* that bottom-up energy planning stands superior to top- down planning as GreenID points out. LEP sources its strength from the active participation of the local community and the local authority. The success of this bottom-up initiative primarily rests on the local people who are more strategically positioned to act and respond, who are directly affected by any energy-related decisions imposed from the top, and who are more knowledgeable about their energy needs. With top-down planning, the process of crafting the energy plan is centralized. Khanh shares that the framework and the process of energy planning only involved high levels of government, meeting, and working within a predetermined, planning cycle:

> [With] energy planning before 2011, [the team which is tasked to craft] the energy plan will meet every five years. But after that, the new law on electricity approved [that] the planning cycle is ten years and revived every five years … As you may know, the energy [sector] is very like centralized … because the planning [team] will meet only at the national and province level, not from the district and the commune [level].

The highly centralized execution of energy plans is also reflected in the choice of leading agencies which are tasked to implement the policies and the plan. One example is the Green Growth Action Plan (GGAP) which is the plan which concretizes the Green Growth aspiration of the Vietnam State of balancing economic growth and reducing its carbon emission per unit of GDP. In the roles and responsibilities of the stakeholders of the action plan, it is clear that the twelve (12) activities are to be led by major institutions at the top (GreenID 2014b, 3) such as government ministries (leading seven of the activities), the Provincial People's Committees (five activities), and the Vietnam Chamber for Commerce and Industry (two activities). The expectation is for the goals of the GGAP to start from the top levels of government and "trickle down to the lower levels" (ibid., 1).

For GreenID, while central planning has its uses especially in reviewing the financial framework and ensuring that the plan obtains similar outcomes for every province in the country, top-down planning has serious limitations (ibid., 3). Because of its heavy reliance on the stakeholders at the top, it runs the risk of hampering implementation by not actively engaging the local stakeholders who might not see the necessity and the benefits of the national policies and action plans imposed from above. Further, top-down planning often leads to suboptimal outcomes by discounting the contributions of other agencies and the local stakeholders on the ground such as the provincial, district, and communal agencies and officials who have hands-on knowledge about their respective local contexts.

Decentralized bottom-up energy planning puts the premium on the active participation of local authorities and the local people to craft and implement the local energy plan, with the support of energy experts, government, and international organizations. As such, it has better chances of implementing national policies on the ground with more efficiency by tapping the knowledge and capacities of the local communities and by deploying innovative, context-sensitive ideas and local solutions which might already be in place, thereby spawning a locally customized, real data-driven, multidisciplinary, and integrative approach to crafting and implementing an energy plan (ibid., 4).

3.8 Benefits of LEP

One year after the pilot implementation of LEP in two communes, a survey was conducted by GreenID to assess the impact of LEP and gather the feedback of the local people (GreenID 2014a, 6–8). The survey clustered the results in terms of economic, environmental, and social impacts.

Economic benefits are felt by the local people as 70% of the households responded that LEP was realistic, useful, and economically beneficial (ibid., 7). On the level of households, the respondents claimed that the application of sustainable energy models helped reduce the energy costs of their respective households. For example, it has been documented by the same survey that the gas produced from the installed biogas model could replace the use of Liquefied Petroleum Gas (LPG) for cooking, helping save 112 VND per month in a household. According to Nguyen (2017), since 2012, there have been 930 households using improved cook stoves, 776 households applying LED bulbs, and 215 households using biogas systems and solar water heaters (Fig. 3.1).

On the community level, the kindergarten schools of the two communes saved money as much as 450.000 VND/month owing to the solar water heater and biogas which were installed through the LEP. Still in the same survey, the drinking water system using reverse osmosis technology and solar photovoltaic (PV) has been providing free potable water to the kindergarten schools, primary schools, and medical stations. According to Nguyen (2017), there have been LED bulbs installed in 28 classrooms across eight schools, as well as six clean water supply systems, two solar PVs for water supply systems, and one community biogas system.

LEP has also brought a positive *environmental impact*. In the survey conducted by GreenID, 77% of the households believe that LEP is contributing to protecting and preserving the environment. As LEP champions the use of renewable energy, the respondents believe that their use of improved cook stoves, solar water heaters, and LED bulbs helped reduce emissions. For those using the biogas model installed to their piggery, the foul smell from animal wastes has been solved while additional source of energy has been created, not to mention decreasing air pollution and reducing the risks of eye and respiratory illnesses in the community.

Finally, there has been a significant *social impact* of LEP. As already pointed out above, the key to the success of LEP is the active participation of local stakeholders and the tapping of their local knowledge and capacities. As such, a necessary social

Fig. 3.1 LED bulb model for a kindergarten in Nam Dinh province, Vietnam (Photo courtesy of GreenID)

outcome of LEP is the increased awareness of the local authorities and the participating communes of issues related to energy and environment (GreenID 2014a, 7–8). After their participation in crafting the local energy plan and in various exhibits, seminars, study tours, the local stakeholders have gained a considerable measure of knowledge on energy saving, energy efficiency, and application of sustainable energy models. The Local Energy Team (LET) of the communes has become very proactive in their promotion of the LEP that they have been willing to articulate the benefits of LEP to higher authorities, including district and provincial agencies, and the National Assembly who are interested in LEP (ibid., 8). Within the LET, women members have been at the forefront in advocating LEP to other women as women have been proven to be the direct beneficiaries of LEP.

For future studies, the preliminary survey conducted by GreenID may be complemented with evaluation studies which use more rigorous statistical computations on the economic, environmental, and social impacts of LEP. They might also cover a wider scope, including the additional communes that have also decided to adopt LEP. Nevertheless, at this point, it is worthy to recognize that the initial success of LEP as manifested in the encouraging responses of the participants already signals an impetus to the possibility of upscaling an approach to energy planning which takes into account the needs and capacities of the local stakeholders.

3.9 Upscaling and Integration into Existing National Plans and Policies

As explicitly stated by Nguyen (2017), ultimately, GreenID sees two opportunities for LEP: *upscaling* and *integration into existing national policies and programs.*

First, GreenID hopes that LEP is replicated to higher levels such as the districts and the provinces. Given the economic, social, and environmental benefits brought about by LEP in the participating communities, the possibility for replication to more communes and upscaling from the level of communes to districts and to provinces, respectively, is part of the agenda. In fact, GreenID has started to explore the upscaling potential of LEP. After the pilot implementation in Thai Binh, LEP was replicated in two more communes in Nam Dinh and in another commune in Thua Thien-Hue. In this experience, they find that "the [LEP] approach is highly adaptable and would therefore be applicable throughout the country" (GreenID 2014b, 7). The objective to expand LEP has been inspired by the experience in Denmark, and then eventually adopted in Thailand and in Mozambique. Upscaling would also not be a problem. It is projected that once data on energy use and on the availability of renewable resources are collected from representative communes in a district, an energy plan can be drawn up accordingly (Nguyen 2017, 7).

Replication and upscaling entail costs. Executing LEP requires investments in manpower, technical expertise, and sustainable energy solutions. Raising the necessary financial resources to make LEP possible has been a challenge for GreenID and the participating communities. Nevertheless, there have been schemes designed to address this challenge. According to Khanh, GreenID offers seed-funding to the community which the community appropriates to achieve the goals. Khanh describes seed-funding this way:

It is seed-funding; we never subsidize 100%. It is seed funds from GreenID. Then, we provide seed fund for [sic] the community, but it depends on each community. They will decide on how they operate that seed-funding. For example, for the Women Union, they set up and make it become like credit and serving fund … they give for [sic] their member to borrow within 10 months. And they pay back monthly with very small amount … For some, it requires interest, but for the poor households, [just] return the money.

As for the private sector, GreenID sees its work as bridge-building between the community and the companies that offer sustainable energy solutions. GreenID organizes an exhibit of the available technologies and brings the suppliers to the community, so both the community and the suppliers can negotiate for lower prices. For Khanh, GreenID "create[s] the demand and also inspire[s] them [the suppliers] to do the social responsibility."

To overcome the financial constraints, GreenID recommends that the government "foster promising bottom-up initiatives" (GreenID 2014b, 8) by supporting seed-funding. From this perspective, the fund would be sourced from the state budgets and other resources from the local communities and civil society organizations. The creation of such a fund, GreenID proposes, would be awarded on a competitive basis based on certain criteria, but should be easily accessible. The criteria might include scalability and replicability of the project, as well as other goals such as gender mainstreaming, disaster resilience, among others. GreenID adds that the fund could be set-up in cooperation with a donor organization which has experience handling grants while the Ministry of Planning and Investment could serve as a national partner to make sure that the money is well-spent.

With this proposed funding scheme, it is clear that GreenID needs to reckon with the issue of financial constraints and that LEP counts on the support of the state, civil society, and the private sector for its endeavors to be possible. It is not explicitly stated whether the replication and upscaling of LEP as a bottom-up initiative is seen as a way to an eventual overhaul of the top-down approaches currently in place in Vietnam. What might hint at the overall vision for the LEP is the second possibility that GreenID sees, namely the integration of LEP into existing national policies and programs.

The ultimate vision for LEP is for it to be integrated into the existing national policies and programs. It is hoped that the insights gained from conducting LEP might inform the subsequent crafting of action plans and policies which are national in scope. As discussed in the earlier section, LEP seeks to use and save energy by tapping on the sustainable energy solutions and the available local energy sources of the community through a participatory approach. It is hoped that LEP can possibly contribute to the goals of the national policies and plans currently in place concerning the energy sector, the environment, rural development, and poverty (Nguyen 2017, 5).

In GreenID's account, *first*, LEP may be integrated into the National Target Program on Use of Energy Saving and Efficiency, 2012–2015 and into Renewable

Fig. 3.2 An innovative solar water heater (Photo courtesy of GreenID)

Energy Development Strategy to 2030–2050. Due to the generation of renewable energy resources through the use of sustainable energy solutions, LEP can contribute to the goals of energy saving, energy efficiency, and increasing the share of renewable energy in the national energy production and consumption. *Second*, LEP may also be integrated into the National Green Growth Strategy 2011–2020 due to LEP's drive for reducing greenhouse emissions and promoting the use of clean and renewable energy. *Third*, LEP can also make contributions toward achieving the National Target Program on New Rural Development, 2010–2020 due to LEP's capacity to organize a rural commune which effectively addresses concerns relating to energy access, water supply, and the environment (Fig. 3.2).

3.10 Engaging Extra-Local Institutions: On the Question of Scale and of the State

Given the ultimate goals of LEP namely replication and integration within the existing national policies and programs, two aspects of LEP need to be made explicit because they have important implications for any attempts at constructing alternatives concerning energy planning and environmental policy. Because of the promotion of the local, LEP has to contend with the question of scale and of the question of the state. Is "the local" romanticized as decidedly the scale in which political action and transformation can emerge? How does it address extra-local institutions such as the state?

3.10.1 On Scales

First, to speak of the promotion of bottom-up energy planning does not mean abandoning the top- down approaches. Indeed, it would be a misrepresentation of LEP to think this way. A closer look at the publications of GreenID suggests that LEP is hoped to complement, not supplant, the existing top-down approaches. The antagonism is not between top-down and bottom-up energy planning, but between *integrating* and *dichotomizing* the two models.

Integrating top-down energy planning with bottom-up energy planning, in many ways, reflects the current thinking on integration in many fields such as urban and city planning and in energy planning. As the Cambridge Forum for Sustainability and the Environment (CFSE) Report puts it, in the case of city planning for example, "the challenge is to *redefine the balance* between these two modes of governing cities for sustainability as we learn more about which problems can be addressed top down and which can only be resolved through bottom-up, collective action" (Linden et al. 2017, 28; emphasis added). Aiming for harmony of the two modes of governing recognizes the unique strengths of each and attempts to overcome and address the weaknesses of each through complementarity.

On one hand, still drawing on the CFSE Report, top-down planning overcomes the weaknesses of the bottom-up planning. Decentralized bottom-up planning has the tendency to create "a patchwork of solutions which is less than optimal" (ibid.) and will inevitably create a set of winners and losers in various local contexts when demands are not reconciled and balanced well. To address these weaknesses inherent in bottom-up planning, top-down approaches can bring local solutions together to reach national sustainability. Further, top-down approaches allow for "trade-offs" (ibid.) between different parts and elements of the wide territory covered by the planning process. With the centralized and wider scope of top-down planning, different demands may be better considered and balanced, least overall harm may be better achieved, and the greatest number of people who are served is better secured (Linden et al. 2017). This is also to make sure that one local community does not just pass its burdens to another community.

On the other hand, bottom-up planning can also overcome the limitations of top-down planning. The latter has the tendency to undermine the control of the local people over their community and exclude "informal processes" in decision-making (ibid., 30) that is, the role of civil society who has various viewpoints about the reasons for success and failure of certain policies and initiatives being implemented on the ground. Bottom-up planning can address these weaknesses by allowing local citizens to offer local solutions and pooling their highly contextualized knowledge to bring to bear to any decisions affecting the community.

This complementarity is modeled by LEP in which the shortcomings of top-down energy planning are addressed by the strengths of the bottom-up energy planning and vice versa. According to GreenID, as energy planning in Vietnam is highly centralized, national plans and policies are rarely reflective of and responsive to the energy demands and available renewable resources of the community. This little involvement with the local communities often leads to "delays" or "cost overruns" (Nguyen 2017, 3).

With LEP, the crafting and implementation of the energy plan is spearheaded by the local community and the local authorities. This way, the community has a sense of ownership and responsibility for the outcomes of the project. Nevertheless, LEP could not completely give up top-down approaches because of their inherent value in making sure that energy planning is not a piece-meal patchwork of thinly spread local solutions across the country which lacks uniformity and coherence in terms of policies and therefore risks obtaining disparate and differing outcomes across different parts of the country. At this point, there is no available solid evidence which supports the superiority of one mode of governance over the other (Linden et al. 2017, 31).

Given the complementarity of both top-down and bottom-up approaches, dichotomous orientation to energy planning (separating and pitting the two against each other) could be less optimal and even counterproductive. The way forward seems hybridization, balance, and parallelism (Linden et al. 2017), however conflicted and tensed these processes may be at times. What is known, at the very least, is that some areas of governance are better governed with top-down approaches and other areas are better catered by bottom-up collective action. One possibility for scaling is "inwards" in which governance starts with local communities and governing bodies, then, scaled up to the level of the region, nation, and international community (ibid., 2). Another possibility for scaling is "outwards" in which initiatives start at the level of the nation and penetrate into the regions, cities, and local communities. Another possibility is thinking hard about a network of governance which sits at the middle of the national and local levels which facilitates "the seamless and collaborative transition" (ibid.) between these two diametric opposite models of governance.

Scholars and practitioners have only just begun to unpack and imagine the paths to this seamless and collaborative transition and transmission between top-down and bottom-up approaches to governance. Nevertheless, following the lines of inquiry opened up by Linden et al. (2017) might be useful in reconstructing the hybridization of the two. Adapting the CFSE Report into the context of energy planning, the following questions might serve as a guide for further exploration: How can LEP as a bottom-up approach be incorporated into and/or complemented with top-down energy planning currently in place in Vietnam? What are the risks involved in this hybridization process? When is it appropriate? How can nation-states use top-down energy planning while allowing for local flexibility that bottom-up energy planning offers? How do we bring local solutions together to bring sustainability to the province, region, and the nation overall? How do we design governance systems that facilitate a seamless and collaborative transmission between bottom-up energy planning (e.g., LEP) and top-down energy planning that are effective at the level of the local and the large scale? Answering these questions satisfactorily on the basis of solid evidence will be the work of engineers, natural and social scientists, environmentalists, and policymakers. But at this point, GreenID through its work on LEP is carefully walking the tightrope of bottom-up/top-down planning, of conformity/contestation, of theory/practice.

3.10.2 On the State

LEP does not idealize "the local" as the exclusive scale for transformation, and this is ultimately shown by its continuing engagement with extra-local institutions, foremost of which is the Vietnamese government. It is this constant engagement that saves it from other pitfalls in ecological movements which idealize clean technology, market incentives, and "the local."

On the surface, it might be argued that LEP easily falls prey to the three strands of eco-localism (Albo 2007). First, owing to its emphasis on eco-responsible technologies, LEP can be read as propagating *ecological modernization*. LEP champions energy solution technologies which are to be used by the local community and supplied at a lower price by private firms that see this endeavor as their corporate social responsibility. These technologies are seen as generating renewable energy resources, enhancing energy efficiency, and reducing fossil fuel use in the households and in the community. Also, LEP might also be seen as subscribing to *market ecology*. On the part of the citizens and their communities, it would appear that they are just strategizing in making investments and calculating ways to bring down the energy costs by generating their own energy sources and by using energy-saving technologies. Each household privatizes the supposedly public concerns of energy security and environmental hazards. Finally, LEP might be engaging in *social ecology* by romanticizing "the local community" as the locus of direct action through the use of local solutions and decentralized governance while viewing large-scale, state planning as ineffective, external exercise to the untainted, local initiatives of the people on the ground.

Indeed, one should be mindful about how projects like LEP can have these kinds of tendencies. It is true that LEP pins its faith on clean energy technologies, induces the local people to invest and calculate their energy costs, encourages private firms to engage in ecology-friendly enterprises, and embraces localism as a venue for direct decentralized action. But any attempt toward a defossilized future should involve the use of clean energy technologies, should involve the active participation of the people on the ground, and should welcome local initiatives. But these efforts are carried out, alongside the constant engagement with extra-local institutions: *GreenID continues to engage the local state agencies and, by extension, the Vietnam State apparatus.* As mentioned above, reducing LEP as romanticizing localism and market solutions ignores the fact that LEP engages extra-local institutions by virtue of its constant contestation with the state's heavy use of coal. This is revealed in the exchange below:

> Ananeza (AltDev): I am just interested on how you complement the energy plan of the government here ... What about coal? What is the role of GreenID?
>
> Khanh: In 2011, the government issued Power Development Plan No. 7, and in that plan it intend[ed] to develop 75,000 MW of coal; this means that more than 30 of the initial capacity, but in terms of the power productions is more than 60% from coal. But we do advocacy and by March 2016, revise the planning and cut out 20,000 MW of coal.
>
> Ananeza: So, the government accepted that recommendation?
>
> Khanh: I think the government recalculated and accepted this ... We proposed it to cut down. We are the first one to propose to cut coal. That is the proposal. For today, after that planning approve[d], we still see that it have [sic] have a 43% of coal exist[s] in the power plant. And we thought that it need[s] to cut further. That is the reason why we conduct this new study that we launched today. And we proposed that we review coal power more then make it below 30%.
>
> Ananeza: And the government is open to that dialogue?
>
> Khanh: Today, we have one person from the government, from the ministry come to attend our event. And he said he highly appreciates our efforts and open to receive the recommendation from an organization like us. I think we had a very good discussion today.

GreenID's work on LEP and the research studies about the energy sector are primarily geared toward active lobbying with the Vietnamese government to cut down on coal in its energy plans. Through various efforts, the people involved in LEP struggle to transform the institutions beyond their local community that inevitably shape their community by seeking support and involvement of the local state authorities and state ministries to serve as partners in LEP, by increasing their awareness of the perils of coal-driven development, by persuading them to explore options such as clean energy, and by encouraging them to earmark state funds for the implementation of LEP through grants and seed- funding schemes. At the very least, this shows that GreenID does not dispense with the state as a key player in attempts at social transformation; it is a strategy of "engagement with" the existing extra-local institutions (i.e., national, regional, and international), not "withdrawal from" them (Mouffe 2005). On the level of extra-local institutions, there has been a push toward New Multilateralism for a Global Green New Deal which attempts to design global rules, regulations, and public institutions toward "the overarching goals of social and economic stability, shared prosperity, and environmental sustainability" which are "protected against capture by the most powerful players" (Gallagher and Kozul-Wright 2019, 27).

For sure, choosing to engage with extra-local institutions such as the state has its own traps. For one, the implementation of LEP is always bound up with monitoring, surveillance, and approval of the local authorities and state agencies. Every step of the way, GreenID has to operate within the socialist framework of the Vietnam Party and State, specifically in relation to energy and environmental policy planning that is set up by the state. Second, the explicit recognition that LEP requires enormous financial investment, not exclusively but primarily, from the state reveals the fragility, if not dependence, of LEP on state funds. But because of these very same ties within the socialist framework, GreenID is also forced to work with the state and its local authorities. It is, at once, constrained but also compelled to maneuver opportunities for negotiation, contestation and maybe even transformation, no matter how tense, tentative, but nevertheless tangible these attempts are.

One might think that it would be too easy to make the case for the continuing contestations of GreenID with the Vietnam State precisely because it is operating within a socialist framework, that is, LEP in Vietnam might only be valid within the context of a socialist political order. This is not untrue; the scope of action of GreenID for LEP is always contoured by the socialist framework within which it operates. But, at the very least, what NGOs that champion "the local" in other countries might take away from this case is the key notion that they must not lose sight of engaging extra-local institutions such as the state and other major institutions as a key player to negotiate, to contest, and to transform in their attempts at large-scale transformation. LEP attests that extra-local institutions are entities which have enormous resources and capacity for transforming the major political and economic arrangements concerning energy and the environment. Further, LEP shows that "the local," seen as actively engaging and transforming extra-local institutions, should be the genesis and destination of any attempt at bringing about a defossilized, sustainable, and habitable future.

References

Albo, Gregory. 2007. The limits of eco-localism: scale, strategy, socialism. *Socialist Register* 43 (2007): 337–363. https://socialistregister.com/index.php/srv/article/view/5871/2767.

APEC (Asia-Pacific Economic Cooperation). 2013. *APEC energy demand and supply outlook*, 5th ed. Tokyo: Asia Pacific Energy Research Centre.

Danish Energy Agency. 2017. *Vietnam energy outlook report 2017*. Copenhagen: Danish Energy Agency. https://ens.dk/sites/ens.dk/files/Globalcooperation/Official_docs/Vietnam/vietnam-energy-outlook-report-2017-eng.pdf.

Gallagher, Kevin P., and Richard Kozul-Wright. 2019. *A new multilateralism for shared prosperity. Geneva principles for a global green new deal*. Boston, MA: Boston University Global Development Policy Center and United Nations Conference on Trade and Development. https://www.bu.edu/gdp/files/2019/04/A-New-Multilateralism-GDPC_UNCTAD.pdf.

GGGI (Global Green Growth Institute). 2017. *Scaling up biomass waste-to-energy in Viet Nam*. https://gggi.org/project/scaling-up-biomass-waste-to-energy-in-viet-nam/. Accessed 15 Apr 2019

GIZ (Deutsche Gesellschaft für Internationale Zusammenarbeit). 2017. Vietnam Power Development Plan for the Period 2011–2020: Highlights of the PDP 7 Revised. http://gizenergy.org.vn/media/app/media/legal%20documents/GIZ_PDP%207%20rev_Mar%202016_Highlights_IS.pdf.

GreenID (Green Innovation and Development Centre). 2014a. *Energy planning in commune level. Promoting green growth and developing sustainable community*. Hanoi: Green Innovation and Development Centre. http://en.greenidvietnam.org.vn/view-document/5aec240c5cd7e8a201a8254d.

GreenID. 2014b. *Fostering bottom-up alternatives: a valuable means for green growth implementation on the ground*. Hanoi: Green Innovation and Development Centre. http://en.greenidvietnam.org.vn/viewdocument/5aec240c5cd7e8a201a8254d.

Heffron, Raphael J., and Darren McCauley. 2017. The Concept of Energy Justice across the Disciplines. *Energy Policy* 105 (June): 658–667. https://doi.org/10.1016/j.enpol.2017.03.018.

IEA (International Energy Agency). 2017. Viet Nam. https://www.iea.org/countries/Vietnam/. Accessed 25 Apr 2019

Jacobsen, Henrik Klinge. 1998. Integrating the bottom-up and top-down approach to energy-economy modelling: the case of Denmark. *Energy Economics* 20: 443–461. https://doi.org/10.1016/S0140-9883(98)00002-4.

JICA (Japan International Cooperation Agency). 2008. *a study on national energy master plan*. Tokyo: The Institute of Energy Economics.

Linden, Paul, Doug Crawford-Brown, Rosamunde Almond, Simon Patterson, and Elizabeth Tyler, eds. 2017. *Cities for the future report*. Cambridge: Cambridge Forum for Sustainability and the Environment.

Minh Do, Tien and Deepak Sharma. 2011. Vietnam's energy sector: a review of current energy policies and strategies. *Energy Policy* 39(10) (October): 5770–5777. https://doi.org/10.1016/jenpol.2011.08.010

Mouffe, Chantal. 2005. *On the political*. New York: Routledge.

Mougouei, Farzad Rahimi, and Mahdieh-Sadat. Mortazavi. 2017. Effective approaches to energy planning and classification of energy systems models. *International Journal of Energy Economics and Policy* 7 (2): 127–131.

Narayanasamy, N. 2009. *Participatory rural appraisal: principles, methods, and application*. New Delhi: Sage Publications.

Nguyen, T. H. 2017. *GreenID LEP approach in Vietnam*. Hanoi: Green Innovation and Development Centre.

Schmitz, Hubert, Dau Anh Tuan, Pham Thi Thu Hang, and Neil McCulloch. 2015. Drivers of Economic Reform in Vietnam's Provinces. *Development Policy Review* 33(2) (March): 175–193. https://doi.org/10.1111/dpr.12097.

Sovacool, Benjamin K., Raphael J. Heffron, Darren McCauley, and Andreas Goldthau. 2016. Energy decisions reframed as justice and ethical concerns. *Nature Energy* 1(16024): 1–6. https://doi.org/10.1038/nenergy.2016.24.

Urban, Frauke, Giuseppina Siciliano, Linda Wallbott, Markus Lederer, and Anh Dang Nguyen. 2018. Green transformations in Vietnam's energy sector. *Asia and the Pacific Policy Studies* 5(3) (September): 558–582. https://doi.org/10.1002/app5.251.

Urban, Frauke, Giuseppina Sicilliano, and Johan Nordensvard. 2017. China's dam builders: their role in transboundary river management in Southeast Asia. *International Journal of Water Resources Development 34*(5): 747–770. https://doi.org/10.1080/07900627.2017.1329138.

van Beeck, Nicole. 1999. Classification of energy models. *FEW research memorandum* vol. 777. Tilburg: Tilburg University and Eindhoven University of Technology. https://pure.uvt.nl/ws/portalfiles/portal/532108/777.pdf

Vietnam Law and Legal Forum. 2007. "Decision No. 1885/QD-TTg of December 27, 2007, Approving Vietnam's National Energy Development Strategy Up to 2020, with 2050 Vision." Official Gazette Issue Nos. 8–9, December 2007. http://extwprlegs1.fao.org/docs/pdf/vie78162.pdf.

World Bank. 2001. *What, why and where.* The World Bank Group. www.worldbank.org/publicsector/decentralization/what.htm. Accessed 13 Mar 2019

World Bank. 2017. *Vietnam | Data.* https://data.worldbank.org/country/vietnam. Accessed 8 Mar 2019

Honey B. Tabiola was Junior Research Analyst of the Program on Alternative Development (AltDev) of the University of the Philippines Center for Integrative and Development Studies (UP CIDS). His research interests lie at the intersection of political theory, critical theory, and political economy.

Chapter 4
Promoting Food Sovereignty in Thailand and the Task of Civil Society

Nathaniel Punongbayan Candelaria

Abstract Thailand is one of the largest rice exporters in the world, with rice production playing a crucial role in its agricultural sector. However, the current model of industrialized rice production, driven by the Green Revolution, has led to significant challenges, including soil degradation, economic burdens on farmers, and threats to food sovereignty. This chapter examines the limitations of the dominant food security paradigm, which prioritizes food availability and economic access but overlooks cultural and ecological concerns. It highlights how Thai civil society organizations, particularly the Sustainable Agriculture Foundation (SAF), the Alternative Agriculture Foundation (AAF), and BioThai, are advocating for food sovereignty as an alternative framework. Through focus group discussions and case studies, the chapter explores how these organizations build networks, engage in policy advocacy, and promote sustainable agricultural practices. Their efforts emphasize grassroots mobilization, research collaboration, and public education to challenge industrialized agriculture. By reclaiming traditional farming knowledge and fostering solidarity, these movements present a viable path toward an ecologically sustainable and socially just food system in Thailand.

Keywords BioThai, food sovereignty · Rice production · Sustainable Agriculture Foundation · Sustainable agriculture · Thailand · Alternative Agriculture Foundation

4.1 Introduction

Thailand is known to be one of the largest rice exporters in the global economy in the current times. As of 2015/2016, Thailand's rice production accounts for 24.5% of the total number of exported rice in the global market (Krungsri Research 2017). Rice occupies the largest share of Thailand's exported agricultural products, as it

N. P. Candelaria (✉)
Department of Political Science, College of Social Sciences and Philosophy, University of the Philippines Diliman, Quezon City, Philippines
e-mail: npcandelaria@up.edu.ph

J. M. C. Sy and A. P. Aban (eds.), *Reimagining Development in Southeast Asia*, Asia in Transition 21, https://doi.org/10.1007/978-981-97-9140-8_4

accounts for 15% (USD 5,817 million) of its total agricultural exports (Pongsrihadulchai 2018). The agricultural sector employs 32.80% of its working population (World Bank 2018). However, the share of the agriculture sector in terms of Thailand's gross domestic product (GDP) is only 8% (Indexmundi 2021).

Thailand's inclination towards rice production is a gambit in food availability. The focus of this particular model of production is on the increase of rice supply in the market. But this pattern of food production, as argued by scholars and practitioners, is not sustainable in the long run. It was argued that the use of chemicals in agricultural production exacerbates the problems in food production, as the use of chemicals made food production more expensive (Thanukid 2014), since farmers are not financially capable in terms of purchasing the technology needed to sustain the production of rice (Barichello and Widanage 2018). Thai farmers, in particular, are finding it difficult to maintain the current agricultural system; this contributes to their insecurity in terms of food (Isvilanonda and Bunyasuri 2009). However, the impact of this technology does not only affect farmers as it also affects the quality of soil. The use of green revolution technologies has also contributed to the degradation of soil (Thanukid 2014).

Given this trend, Thai civil society organizations are at the forefront of the advocacy for an alternative system of agriculture. This paper attempts to document the experiences of organizations such as the Sustainable Agriculture Foundation (SAF), the Alternative Agriculture Foundation (AAF), and BioThai, for their advocacy towards an alternative model on agriculture and food. The paper also explores how these organizations promote the idea of food sovereignty in Thailand. Lastly, it also investigates the nuances and similarities in their campaigns towards the promotion of food sovereignty. To fulfil the objectives set by this paper, a case study research design is used as such is apt in finding out the experiences of these groups, in more detail (Flick 2009). And to generate data, a focus group discussion (FGD) was conducted by the members of the research team, together with some of the members of the abovementioned organizations.

4.2 Food Security as the Dominant Paradigm on Food

The current thrust of Thailand on food production falls under the purview of food security. Food security, as a framework, discusses the idea of being food-secure as a condition when "all people, at all times, have access to sufficient, safe and nutritious food to meet their dietary needs and food preferences for an active and healthy life" (FAO 2006, 1). This entails that people should have increased access not only in terms of availability but also in terms of their economic means to access food (Barichello and Widanage 2018). The current definition of food security, according to them, is defined during the 1996 World Food Summit (WFS). The definition considers a plethora of factors, which is contrary to its previous thrust that solely focuses on the issue of food availability during the 1950s (Shaw 2007).

Food security, as a framework, is currently operationalized based on the lens of food self-sufficiency and through the increase in access at the domestic level (Barichello and Widanage 2018). Most of the actions undertaken by different governments are actually geared towards these interventions in order to secure food under the food security paradigm. In the Philippines for instance, its Congress is considering enacting a policy on rice tariffication, which lessens the restrictions on import quota for rice (Gonzales 2018). And at the regional level, particularly for the Association of Southeast Asian Nations (ASEAN), trade has been a recognized avenue for food security to be ensured (Candelaria 2019).

However, self-sufficiency and the budget a family spends on food in measuring food security have limitations (Barichello and Widanage 2018). Barichello and Widanage have argued that most state interventions on food insecurity are implemented at the national level, despite the fact that food security is noted to be occurring mostly at the local level. This is seen as a problem on how food security is actually discussed as a policy.

Aside from its operational definition, there are also issues about food security as a framework with respect to culture on food production. One of the concerns on food security is that it is not cognizant of the role of culture and identity in food production (Cote 2016). Another concern is the loss of traditional knowledge on agriculture. The case of the indigenous peoples in North Vietnam is an example of this. Because of the policy of the Vietnam government, indigenous peoples in Northern Vietnam are

Fig. 4.1 An example of a sustainable farm in Thailand (Photo courtesy of Nathaniel P. Candelaria)

forced to adapt certain measures in support of the current rice policy, even if it is disadvantageous on their part (Turner and Bonnin 2018).

Based on a closer reading of the definition, the Food and Agriculture Organization (FAO) of the United Nations' definition of the concept of food security fails to acknowledge the role of culture and tradition on food production. As a way of promoting an alternative viewpoint on food, international non-government organizations such as the *La Via Campesina* movement promote the idea of food sovereignty as an alternative framework on food (Fig. 4.1).

4.3 Food Sovereignty as an Alternative Framework

In 2008, the world experienced a food crisis where the price of basic commodities has increased as an effect of the global financial crisis. Cases of food insecurity were reported in different areas of Mexico, the Philippines, China, and Africa due to the changing dynamics in the global market (Bello 2010). The impact of the food crisis in 2008 made the poor, composed mostly of farmers, vulnerable (Holt-Giménez and Altieri 2013). Nevertheless, the call towards an alternative framework on food did not start recently. In fact, food sovereignty as an alternative framework was already discussed in 1996.

In the second congress of the *La Via Campesina* movement in Tlaxcala, Mexico, the movement called for "changing the existing, inequitable, social, political and economy structures and policies ... (which) are the very causes of the social and environmental destruction in the countryside in both the North and the South" (Wittman et al. 2010, 3). The concept of food sovereignty further evolved in succeeding congresses such as during the World Forum on Food Sovereignty in 2001 and the NGO/ CSO Forum on Food Sovereignty in 2002 (Wittman et al. 2010). Food sovereignty is currently defined as:

> The right of peoples to healthy and culturally appropriate food produced through ecologically sound and sustainable methods, and their right to define their own food and agriculture systems. It puts the aspirations and needs of those who produce, distribute and consume food at the heart of food systems and policies rather than the demands of markets and corporations. It defends the interests and inclusion of the next generation. It offers a strategy to resist and dismantle the current corporate trade and food regime, and directions for food, farming, pastoral and fisheries systems determined by local producers and users. Food sovereignty prioritizes local and national economies and markets and empowers peasant and family farmer-driven agriculture, artisanal fishing, pastoralist-led grazing, and food production, distribution and consumption based on environmental, social and economic sustainability. Food sovereignty promotes transparent trade that guarantees just incomes to all peoples as well as the rights of consumers to control their food and nutrition. It ensures that the rights to use and manage lands, territories, waters, seeds, livestock and biodiversity are in the hands of those of us who produce food. Food sovereignty implies new social relations free of oppression and inequality between men and women, peoples, racial groups, social and economic classes and generations. (Nyéléni Forum for Food Sovereignty 2007, cited in Wittman 2011, 88)

Food sovereignty encompasses the following tenets: (1) it entails the rights of a country to define its own food policies; (2) it focuses on the priority of the state to feed the people; (3) it recognizes the rights of farmers to produce their own products, and the right to be protected against cheap agricultural products; and lastly, (4) it recognizes the roles of the community and women in promoting food for the population (La Via Campesina 2003). The concept of food sovereignty also entails the need to rediscover the cultural practices that people have with respect to the cultivation of their own food (Cote 2016). Particularly for the last point, the respect on cultural practices on food is not visible under the dominant paradigm on food.

Food, as a basic human need, should not be considered as a commodity in a fluctuating global market.[1] Because of the treatment of agricultural products more as commodity than food supply, it led to the conversion of agricultural lands into the production of high-value crops such as palm oil, in order to increase income under the agriculture sector (Bello 2010). In the case of Malaysia, for instance, communities have been forced to migrate to urban centres to earn a living while their ancestral lands were occupied by agricultural companies cultivating high-value crops such as palm oil (Lee 2018).

The production of high-value crops contributes to food problems precisely because it does not see the need for food to be produced solely for consumption (Bello 2010). Alternative practices on agriculture promote a better chance of having a harvest than subscribing to monocropping (Altieri and Nicholls 2008). This is also observed in Vietnam where indigenous peoples have also utilized their indigenous knowledge on agriculture in order for them to be safe from the problems related to climate change (Turner and Bonnin 2018).

Both food security and its counterpart food sovereignty share similarities and differences. Both ideas aim to secure food for the survival of mankind. For food security, this is covered by its second pillar, i.e., food accessibility. However, they differ on the manner of achieving these goals. Food security, as discussed earlier, has pillars intended for people to be food secure not only in terms of its availability but also in terms of people's economic accessibility towards food. Discussion on food security also includes the idea on how people will receive proper nutrients needed by the human body and how all of these will be stable. Food sovereignty, on the other hand, aims to promote the production of food, independent from the forces of the market. Based on the proponents of the framework such as the La Via Campesina, food is considered as a basic right that everyone should be able to enjoy. In this regard, food sovereignty is geared towards democratizing food access throughout the population (Salzer and Fehlinger 2017).

[1] In his book *Food Wars*, Walden Bello discussed how the current economic paradigm, together with the programs implemented by international financial institutions, contributed to how food was made as a commodity under the international market. He cited several cases such as Mexico, the Philippines, China, and Africa to narrate their experiences on food under the current economic system.

4.4 Alternatives in Action: The Case of the Sustainable Agriculture Foundation, the Alternative Agriculture Foundation, and BioThai

The growth of the agricultural sector in Thailand was a product of the so-called green revolution in agriculture. Due to the issue of food insecurity during the 1960s, leading donor agencies such as the Rockefeller Foundation and Ford Foundation funded initiatives and research projects on how to increase food production using modern technologies such as artificial fertilizers, chemical products, and the use of irrigation systems for water sources (Hazell 2009). During the 1970s, the Thai government integrated the technologies from the green revolution by building irrigation and road systems and adopted technologies, which contributed to the current status of Thailand as one of the top exporters of rice in the world (Isvilanonda and Bunyasiri 2009).

While the use of green revolution technologies remains in the Thai agriculture system, there are organizations advocating for alternatives. For this paper, the discussion will focus on three organizations, i.e., the Sustainable Agriculture Foundation (SAF), the Alternative Agriculture Foundation (AAF), and the BioThai, on their role in promoting food sovereignty in Thailand. These organizations form a larger consortium seeking to promote alternative practices on food production in Thailand. The research team from the Program on Alternative Development of the UP Center for Integrative and Development Studies (UP CIDS) visited the members of the consortium in Nonthaburi, Thailand, and a focus group discussion (FGD) was conducted at the BioThai Office in order to find out the experiences of these organizations on their advocacy for food sovereignty.[2]

During the FGD, the participants were asked to discuss the activities that their organizations have undertaken, their objectives, the challenges they faced, and their future goals. Based on the discussion, these organizations have different sets of goals and objectives, but they are united in the goal of promoting an alternative way of agriculture, which is in contrast with the current mode of agriculture production in Thailand.

4.4.1 *Profiles*

These are the following organizations, represented by their members, that the research team was able to have a fruitful discussion with on food sovereignty:

1. ***The Sustainable Agriculture Foundation.*** As one of the organizations comprising the consortium, the Sustainable Agriculture Foundation (SAF) aims to promote sustainable agriculture practices in the context of urbanization

[2] The FGD participants were: Witoon Lianchamroon, Kingkorn Narintarakul, Dr. Naruemon Thabchumpon, Supa Yaimuang, and Ubon Yoowah. The research team for the Thailand fieldwork is composed of Asst. Prof. Karl Arvin Hapal and Nathaniel Candelaria.

(Ashoka 2018). Based on their profile as winner of the Ashoka award, the SAF aims to bridge the gap between the traditional and modern agricultural practices and help others to promote the initiative in order for them to have better nutrition. As discussed by one of the members of the SAF, the organization collaborates with other organizations such as the Alternative Agriculture Foundation and BioThai in terms of research and network building.

2. ***The Alternative Agriculture Foundation.*** The Alternative Agriculture Foundation (AAF) is a farmers' group that seeks to "develop sustainable agriculture techniques based on the local ecology and expand their positive impacts by training and educating other members of their communities" (Alternative Agriculture Foundation n.d.). Based on further information available from their official website, they also aim to promote the cultivation of organic agricultural products in order to make it an alternative and sustainable source of livelihood for farmers. During the FGD, one of their members mentioned that the goal of the AAF is to promote an alternative model against the government-sponsored way of agriculture, i.e., the use of fertilizers and chemicals on rice production. The member from AAF further elaborated that this is because they observed that those who do not subscribe to chemical farming have better quality of life and produce better agricultural products.
3. ***BioThai.*** The organization was established first in 1995 as a network called "Thai Network on Community Rights and Diversity" and was turned into an organization, called as the "Biodiversity and Community Rights Action Thailand" in 1999 (BioThai n.d.). As elaborated further in their website, the organization was officially registered as a foundation and changed its name to BioThai Foundation in 2006. The website also highlighted the change of their name to Biodiversity–Sustainable Agriculture–Food Sovereignty Action Thailand in 2014. One of the members of BioThai mentioned that the organization helps the other member organizations of their consortium by conducting research. As mentioned in their website, the organization deals with contemporary issues on agriculture and food production such as genetic engineering, seed freedom, etc. And as highlighted in their website, BioThai also serves as the coordinator of the network.

4.4.2 The Perceived Issues and Problems in the Current Agricultural Practices of Thailand

As discussed during the FGD, one of the reasons why the SAF, AAF, and BioThai mobilized as a consortium is because of the problems of the current agricultural practices in Thailand. As discussed in the context of the paper, the Thai government continues to prioritize rice production with the use of chemicals. One of the participants of the FGD from the AAF mentioned how the promotion of the green revolution in Thailand led to the loss of forest lands. The same member also pointed out how it contributed to the worsening situation of farmers in Thailand and how the use of these technologies affected community relations (Fig. 4.2).

Fig. 4.2 A basket of soy harvest, fresh from the farm (Photo courtesy of Nathaniel P. Candelaria)

While the government prioritizes green revolution technologies, the FGD participants acknowledged the existence of a policy on alternative agriculture. However, they lamented that there is little government support as far as the implementation of the policy is concerned. The FGD participants have also recognized the role of the Assembly of the Poor in terms of mobilization, as the social movement was able to make the Thai government listen to their demands, in terms of reforms in government policies. As much as they campaign for the promotion of alternative practices in agriculture, given the little support that they receive from the government, a participant from the AAF mentioned that only around 3% of the total areas for agriculture are using organic farming techniques at present.

Another concern for the SAF, AAF, and BioThai, as a consortium of alternative practitioners, is the issue of organic certification. The FGD participants were able to share that the process of certifying their products as organic is problematic, since some of the factors needed for organic agriculture have already been contaminated with the use of technologies from the green revolution. One of the representatives of BioThai during the FGD shared their organization's experience on the problem. The member from BioThai mentioned that the results of the examinations conducted both by the government and the private sector differ from each other, thus making the certification of organic products an issue for their organization. Moreover, the same member also lamented the fact that people do not really understand the difference between organic and non-organic agriculture.

4.4.3 The Push Towards Food Sovereignty and Alternative Agricultural Practices in Thailand

To promote food sovereignty in Thailand, it was brought up during the FGD that they built networks among the farmers and built plans before mobilizing. One of the members of the SAF emphasized the role of network building and mobilization at the grassroots level. The same participant recalled that it was in 1997 when they mobilized the network, which then contributed to the change in policy. In the case of the AAF, one of its members has recognized the need to build solidarity on other issues such as trade, collective land management, and collective bargaining, bolstered by mobilizations.

Aside from network building, the FGD participants also recognized the need for collaboration in terms of research and policy analysis on different agriculture models and other issues affecting agriculture such as land, price, access to credit, and issues of empowerment. The FGD participants pointed out the division of roles within the consortium. For instance, the representative from BioThai mentioned that their organization deals with research and the dissemination of its results. The BioThai representative also mentioned that SAF is responsible for the managing of the research undertaken for the network. The AAF is in charge of establishing networks among farmers. For the purpose of disseminating information, the representative from AAF mentioned the critical role of the media as an important avenue for sharing knowledge on agriculture.

The FGD participants have recognized the importance of engaging their respective governments. The members of their respective organizations meet with government offices and lobby for pro- alternative agriculture policies. One example of this is the sustainable agriculture policy, as discussed both by the members of the SAF and AAF. Particularly for the SAF member, engaging the government is needed to ensure that policies promoting these initiatives will be enacted; they will continue to do so until these initiatives are enacted.

Another agenda that the FGD participants mentioned is to engage the public and to educate them about organic agriculture. The BioThai member in the FGD lamented that consumers do not really have an idea on how their food is produced to begin with. Moreover, the consumers are also not knowledgeable in distinguishing organic/inorganic agricultural products. The bottom line of this is that they want everyone to benefit from alternative agriculture practices. The FGD participant from BioThai highlighted the desire for farmers to benefit from agriculture, without sacrificing the environment. The same member envisioned a social market where people are free to access goods that they want to purchase, as an idea on how to operationalize this.

In this particular debate on food, the civil society groups discussed in this paper subscribe to the idea of food sovereignty. According to the FGD member from the AAF, there is a need for people to acknowledge the impact of the current system of agriculture in terms of its devastating effects on land. There is also a need to recover their forgotten culture on agriculture, as it was nearly wiped out by the government's insistence on the promotion of food security through the green revolution. The ideas mentioned in the FGD resonate with the arguments raised by food sovereignty organizations such as the La Via Campesina on the impact of the green revolution on indigenous agricultural knowledge and why there is a need to rediscover these practices once again (Cote 2016).

4.5 Building Solidarity for Food Sovereignty

Based on the experience of the Alternative Agriculture Foundation, the Sustainable Agriculture Foundation, and BioThai, there are several points that we can learn: the role of collective action, cooperation, the importance of research, and engagement with other actors in the society. These factors are important for these organizations to plan, to mobilize, and to build solidarity with others sharing the same ideas with respect to food sovereignty.

4.5.1 Collective Action and Defined Roles

While they are united towards a specific goal, i.e., the promotion of alternative practices on agriculture, their respective roles within the consortium are well-defined. Based on the discussion of the FGD participants, the Sustainable Agriculture Foundation oversees promoting ideas about agricultural sustainability such as urban agriculture techniques. The Alternative Agriculture Foundation, on the other hand, is in charge of promoting alternative agricultural practices, such as organic farming. And as for BioThai, it is involved with issues concerning contemporary technologies on agriculture.

It is noteworthy for these groups to coalesce and advocate change in terms of Thai agriculture as a collective, despite their different roles. Based on their experience, we can infer how important defining their roles are in avoiding a significant overlap of functions. The cases of these three organizations also highlighted how their advocacies are centred on the promotion of an alternative agriculture and food sovereignty. Moreover, the result of the FGD highlights that they have a united front in their advocacy against green revolution and its negative effects over organic agriculture. These organizations are united despite having different functions within the consortium.

4.5.2 Cooperation

The importance of cooperation among different organizations is manifested in the idea behind different social movements such as the degrowth movement, and how these organizations behind the movement can learn to recognize each other's strengths and weaknesses (Muller 2016; Kothari 2016). In the case of the SAF, AAF, and BioThai, it is noteworthy that these organizations value the importance of camaraderie and cooperation.

As mentioned by the FGD participants, they have established linkages with other organizations and farmers. Particularly for the AAF, they also deal with issues on agriculture and food trade and collective agriculture. The organization has recognized the role of cooperation with instrumentalities, such as the government as an arena for exchanging ideas. In this regard, we can infer that these organizations are cooperating with other organizations and stakeholders in order to promote the idea of alternatives in Thai society. And as a way of promoting cooperation, the FGD participants pointed out the importance of building networks, since it helps people mobilize towards a goal shared by everyone.

4.5.3 The Importance of Research and Development in the Promotion of Advocacies

While there is a need to have a unifying lens on how to view agriculture and issues on food, the FGD participants have also underscored the importance of research in their advocacy. Based on their experience, research is important in order for them to be able to share accurate information about the state of Thai agriculture. Research is also important for them in developing new ways and/or mechanisms on how to promote alternative agriculture without sacrificing nature. Whatever the outcome of their research is, they realized the need for their organizations to disseminate the findings towards a larger audience.

While they recognize its importance, these organizations also differ on what their researches are primarily focused on. For example, the SAF deals with concerns pertaining to sustainability and urban agriculture, as discussed by their representative in the FGD. BioThai deals with issues on contemporary agricultural techniques and their implications for Thai Agriculture, as mentioned in their website. And for the AAF, the organizations deal with issues pertaining to alternative agricultural practices and their impacts on the communities (Fig. 4.3).

Fig. 4.3 Aligned on the shelves are different eco-friendly products made from fresh materials from the farm (Photo courtesy of Nathaniel P. Candelaria)

4.5.4 Engagement with Other Actors

As discussed during the FGD, the participants have stressed the importance of engaging other actors in the society regarding their advocacy on alternative agricultural systems. Based on the comments of the participant from the SAF, one of the key actors that needs to be engaged is the government. The consortium keeps engaging the government, until the government sees the merit of setting policies supporting these kinds of alternative practices. Aside from the government, engagement with the public is also emphasized. As pointed out by the participant from BioThai, the consortium needs to engage the public in order to raise awareness about existing alternative agricultural practices and their benefits vis- à-vis mainstream practices. And for the AAF, a participant mentioned that their organization engages with other farmers by convincing them to abandon costly agricultural practices and to shift to alternative ones. While the AAF oversees the networking, as discussed during the FGD, other members of the consortium such as the SAF and BioThai also play a significant role in engaging other stakeholders of the society.

The abovementioned points are important as these groups can learn from each other's experiences and strengthen the solidarity among them. This realization is important given the differences in their respective tasks. For instance, each member-organization of the degrowth movement has its own objectives and priorities, but their experience has highlighted how they can build a much more integrated network of alternative practitioners, despite being different from each other in terms of goals (Muller 2016; Kothari 2016). The example shown by this movement demonstrates that, in spite of their differences, they can still work as long as they share the same advocacies. And this has been the case of the consortium on food and agriculture joined by the three distinct organizations.

4.6 Conclusion

While Thailand was able to maintain its status as one of the top exporters of rice in the global economy, a portion of its population, especially farmers, are food insecure as costs of food production increase with the use of modern technologies on agricultural production. And while the price of inputs is high, the green revolution made the price of rice low due to the increase in supply. Moreover, the green revolution technology had an impact on the long-term agricultural production as it contributed to the decrease of nutrients in the soil, and it also developed the resistance of pests to these chemicals. These factors contributed to the difficulty experienced by farmers, which resulted in hunger and poverty, as explained by the participants of the study.

In the case of the farmers, the prioritization of the Thai Government to pursue the green revolution to promote rice production, paved the way for the SAF, AAF, and BioThai to coalesce and promote the idea of food sovereignty as an alternative framework, with regard to food and economic issues felt by Thai farmers in general.

The case of the Sustainable Agriculture Foundation, the Alternative Agriculture Foundation, and BioThai is noteworthy in the sense that they are trying to address the gap in terms of food and livelihood concerns among the farmers brought about by the green revolution policy of the Thai Government. As mentioned during the FGD, these groups are trying to promote food sovereignty by providing information to farmers on how to cultivate land for organic and sustainable agriculture. These groups are also at the forefront of challenging existing policies on organic certification and promoting alternative practices on agriculture. This is due to the concern that these groups have regarding the differences on how the government and the private sector evaluate organic products. Given how difficult the challenges are, these groups have also recognized the importance of network building, cooperation, and pursuing research to support their respective advocacies.

Despite their position vis-à-vis the current agriculture policy of the government, they are still engaging with the government in order to change these policies. While they are still far from attaining their goal, their attempts on promoting food sovereignty among their farmers have nevertheless contributed to the improvement of their lives. As mentioned earlier, organic agriculture helped people earn more income than conventional agricultural practices. These practices are sensible efforts that can be investigated both by policymakers and other non-governmental organizations as well, as an alternative to the current system of food production.

References

Altieri, Miguel A., and Clara I. Nicholls. 2008. Scaling up agroecological approaches for food sovereignty in Latin America. *Development* 51 (4): 472–480. https://doi.org/10.1057/dev.2008.68.

Alternative Agriculture Network–Esan. N.d. About the AAN. *AAN-Esan.* https://aanesan.wordpress.com/about/.

Ashoka Fellows. n.d. Sustainable Agriculture Foundation. https://www.changemakers.com/ashoka-fellows/entries/sustainable-agriculture-foundation.

Barichello, Richard, and Rupananda Widanage. 2018. Food security in rural areas: alternate definitions, measurement, and gender issues. In *Ensuring a square meal: women and food security in Southeast Asia*, ed. Theresa W. Devasahayam, 23–42. Singapore: World Scientific.

Bello, Walden. 2010. *Food wars.* Mandaluyong: Anvil Publishing.

BioThai. n.d. Our Work. *BioThai.* https://www.biothai.org/node/129.

Candelaria, Nathaniel P. 2019. Price volatility and trade: implications for food security in ASEAN. *Social Science Research Network.* https://ssrn.com/abstract=3420745.

Coté, Charlotte. 2016. 'Indigenizing' food sovereignty: revitalizing indigenous food practices and ecological knowledges in Canada and the United States. *Humanities* 5(3): 1–14. https://doi.org/10.3390/h5030057.

Flick, Uwe. 2009. *An introduction to qualitative research*, 4th ed. London/Thousand Oaks, CA: SAGE Publications.

FAO (Food and Agriculture Organization). 2006. Food Security. In *FAO policy brief, issue 2, June.* Rome: Food and Agriculture Organization. http://www.fao.org/fileadmin/templates/faoitaly/documents/pdf/pdf_Food_Security_Cocept_Note.pdf.

Gonzales, Catherine. 2018. Bicam Panel Okays Rice Tariffication Bill. *Philippine Daily Inquirer* (November 22). https://newsinfo.inquirer.net/1056342/bicam-panel-okays-rice-tariffication-bill.

Hazell, Peter B.R. 2009. The Asian green revolution. *IFPRI Discussion Paper 00911*, November. Washington, D.C.: International Food Policy Research Institute. http://ebrary.ifpri.org/utils/getfile/collection/p15738coll2/id/29462/filename/29463.pdf.

Holt-Giménez, Eric, and Miguel A, Altieri. 2013. Agroecology, food sovereignty, and the new green revolution. *Agroecology and Sustainable Food Systems* 37 (1): 90–102. https://doi.org/10.1080/10440046.2012.716388.

Indexmundi. 2021. Thailand GDP—Composition by Sector—Economy. *Indexmundi,* Last updated September 18, 2021. https://www.indexmundi.com/thailand/gdp_composition_by_sector.html.

Isvilanonda, S., and I. Bunyasiri. 2009. Food Security in Thailand: Status, Rural Poor Vulnerability, and Some Policy Options. *ARE working paper no. 2552/1* (September). Bangkok: Department of Agricultural and Resource Economics, Faculty of Economics, Kasetsart University. http://agri.eco.ku.ac.th/RePEc/kau/wpaper/are200901.pdf.

Kothari, Ashish. 2016. Radical ecological democracy: reflections from the south on degrowth. *Degrowth,* August 30. https://www.degrowth.info/wp-content/uploads/2016/07/DIM_RED.pdf.

Krungsri Research. 2017. Thailand industry outlook 2017–19: Rice Industry. *Krungsri,* May https://www.krungsri.com/bank/getmedia/83a146ea-a14f-41c7-9e80-9214a5d9b963/IO_Rice_201705_EN.aspx.

La Via Campesina. 2003. Food sovereignty. *La Via Campesina,* January 15. https:// viacampesina.org/en/food-sovereignty/.

Lee, Poh Onn. 2018. Property rights, food security, and gender: the impact of the expansion of palm oil plantations on native communities in Sarawak, Malaysia. In *Ensuring a square meal: women and food security in Southeast Asia*, ed. Theresa W. Devasahayam, 147–170. Singapore: World Scientific.

Muller, Tadzio. 2016. Climate justice: global resistance to fossil-fuelled capitalism. *Degrowth*, December 13. https://www.degrowth.info/wp-content/uploads/2016/12/DIM_Climate-Justice.pdf.

Nyéléni Forum for Food Sovereignty. 2007. Declaration of the forum for food sovereignty. Selingué, Mali, 23-27 February 2007. https://nyeleni.org/en/declaration-of-nyeleni/

Pongsrihadulchai, Apichart. 2018. Thailand's rice industry and current policies towards high value rice products. *FFTC Agricultural Policy Platform*, July 02, 2018. https://ap.fftc.org.tw/article/1297#:~:text=The%20market%20share%20is%20about,exporting%20value%20of%20agricultural%20products

Salzer, Irmi, and Julianna Fehlinger. 2017. Food sovereignty: neither growing nor yielding, but fighting for good food for all! *Degrowth*, January 17. https://www.degrowth.info/wp-content/uploads/2016/12/DIM_Food-sovereignty.pdf.

Shaw, D. John. 2007. *World food security: a history since 1954*. Hampshire/New York: Palgrave Macmillan.

Thanukid, Supon. 2014. The green revolution that did more harm than good. *The Nation*, November 21. https://www.nationthailand.com/national/30248272.

Turner, Sarah, and Christine Bonnin. 2018. We women know more about the seeds: ethnic minority food security and the gendered implications of hybrid rice in upland Northern Vietnam. In *Ensuring a square meal: women and food security in Southeast Asia*, ed. Theresa W. Devasahayam, 103–122. Singapore: World Scientific.

Wittman, Hannah. 2011. Food sovereignty: a new rights framework for food and nature? *Environment and Society: Advances in Research* 2 (1): 87–105. https://doi.org/10.3167/ares.2011.020106.

Wittman, Hannah, Annette Desmarais, and Nettie Wiebe. 2010. The origins and potential of food sovereignty. In *Food sovereignty: reconnecting food, nature and community*, eds. Hannah Wittman, Annette Desmarais, and Nettie Wiebe, 1–14. Halifax/Winnipeg: Fernwood.

World Bank. 2018. Employment in agriculture. *The World Bank*. https://data.worldbank.org/indicator/SL.AGR.EMPL.ZS.

Nathaniel Punongbayan Candelaria is currently an Assistant Professor at the Department of Political Science, UP Diliman, where he also earned his Bachelor of Arts (Political Science) and Master of Arts (Political Science) degrees. He also previously worked as Teaching Associate at the UP Department of Political Science, as Senior Project Assistant at the UP Center for Integrative and Development Studies (UP CIDS), and as Research Assistant at the Institute for Strategic and Development Studies (ISDS Philippines) and at the Foundation for Integrative and Development Studies (FIDS). His research interests are non-traditional security, International Relations, religion and politics, and Southeast Asian politics.

Chapter 5
Promoting Citizens' Participation in the Budget Process: The Social Watch Philippines' Alternative Budget Initiative

Ma. Victoria R. Raquiza and Roja Estrelita Salvador

Abstract Citizens' participation in budget processes is crucial for ensuring that public funds are allocated toward social services that benefit marginalized groups. The Social Watch Philippines–Alternative Budget Initiative (SWP–ABI) exemplifies how civil society can engage in national budget-making to push for increased spending on social, agricultural, and environmental programs. While budget processes in the Philippines are often opaque and dominated by technocrats, SWP–ABI has systematically intervened, particularly in budget preparation and legislation phases, advocating for realignments that prioritize social development. Drawing from the model of participatory budgeting, SWP–ABI has collaborated with legislators to present alternative budget proposals, leading to policy victories in securing funding for targeted programs. However, challenges persist in budget execution and accountability, as funds allocated in the General Appropriations Act do not always reach intended beneficiaries. The experience of SWP–ABI highlights the inherently political nature of budget allocation and the need for sustained citizen engagement, strategic alliances, and institutional reforms to ensure transparency, equity, and responsiveness in public finance. Strengthening participatory governance remains key to advancing social justice and inclusive development in the Philippines.

Keywords Budget accountability · Citizen participation · Participatory budgeting · Social justice · Transparent governance

---

Ma. V. R. Raquiza (✉)
National College of Public Administration and Governance, University of the Philippines, 1101 Diliman, Quezon City, Philippines
e-mail: mrraquiza@up.edu.ph

R. E. Salvador
Social Watch Philippines, Diliman, Quezon City, Philippines

J. M. C. Sy and A. P. Aban (eds.), *Reimagining Development in Southeast Asia*, Asia in Transition 21, https://doi.org/10.1007/978-981-97-9140-8_5

5.1 Introduction: Principles on Financing Social Policy[1]

Citizens generally "aspire to live a life of dignity and well-being" (Raquiza 2021, 19) throughout their lives. This requires, at the minimum, access to "quality social services that are adequate, affordable and available as well as decent work." This aspiration is within reach if there is a strong developmental and welfare-oriented government, working hand in hand with a mature polity that includes vibrant engagement of citizens' movements, subaltern groups, the private sector, and individuals toward a shared vision for transformative change.

The provision of social services is embedded in a social contract between the State and its citizens. From a rights-based perspective, the State has the duty to provide these services, and the citizens have the right to claim these. Financing the provision of services comes from a variety of sources, such as taxes, private sector schemes, or out-of-pocket spending by individuals, and some sources will be utilized more than others depending on several factors such as a country's fiscal space, administrative capabilities and as importantly, the policy regime (e.g., welfare-oriented or neoliberal). The United Nations Research Institute for Social Development (UNRISD) puts it succinctly: "the amount of resources invested in social policy—and how they are spent—is largely determined by a country's policy regime" (UNRISD 2010, 209). However, the UNRISD also contends that mobilizing resources is, however, only part of the battle. Decisions about revenue policies and the allocation of public funds are the result of political processes, often dominated by elite groups. Consequently, such policies may not lead to the best outcomes in terms of providing public goods and reducing poverty. Furthermore, institutional capacity, including the quality and efficiency of public administration and service providers, influences how successfully resources are translated into social outcomes (2010, 207–08).

The type of policy regime as a primary determinant of a country's level of social spending does not negate the importance of fiscal space. As is widely believed, fiscal space is highly dependent "on a country's economic performance, including its capacity to produce income and savings and to produce government revenues" (UNRISD 2010, 207). Many developing countries find sustaining fiscal space difficult. For instance, the Philippines, whose economy is services-led, suffers from low agricultural output and underdeveloped industries, including its manufacturing sector. The country continues to be plagued with a long-standing balance of trade deficit. In the context of declining and worsening agricultural production and marked deindustrialization in the last few decades, the role and impact of overseas workers' remittances and the business process outsourcing (BPO) sector have served to keep its economy afloat.

The use of financing instruments is based on "normative principles of individualism or solidarity" (UNRISD 2010, 211) as these may be "distributionally progressive" (e.g., "redistributing from rich to poor" including favoring disadvantaged and socially excluded groups) , "neutral or regressive." "Progressivity" implies "that

[1] The authors acknowledge the contributions of Alce Quitalig, Jessica R. Cantos, Rene Raya and Janet R. Carandang, all from SWP, in writing this paper.

those who have more will give more in society"; hence, "the rich must pay a larger proportion of their income to taxes," which is contrary to the imposition of indirect taxes (e.g., value-added tax), user fees, and increasing levels of out-of-pocket spending for basic social services (UNRISD 2010, 211; Briones 2008, 12–13; Raquiza 2021, 19–20). In particular, the "most regressive and individualistic" financing schemes "are those in which people provide for themselves (self-provisioning) or that require out-of-pocket payment such as user fees" particularly in a context where the majority of a people are poor (UNRISD 2010, 211). Conversely, the most progressive source of financing public goods is via "direct taxation of personal and corporate income, including property" (212). Meanwhile, the principle of solidarity "emphasizes collective responsibility for individual well-being" and involves fostering "social cohesion and coalition-building among classes, groups and generations, working against different types of divisions in society," (1370) an enduring feature of universal social policies.

Power relations within and among different groups in society significantly shape the content and process of public decision-making on policies and programs. Thus, they largely determine social, economic and environmental outcomes. Power relations are, therefore, "at the core of development" (UNRISD 2010, 284). Many governments, including the Philippines, rely on technocrats to run the economic and more technical aspects of governance including the area of public finance. While they may have an arsenal of policy tools to address myriad development problems, oftentimes, its efficacy remains the subject of intense debate in some policy, academic, and activist circles. What is incontrovertible is their tendency to "[insulate] policy making from public scrutiny" and their aversion to messy but democratic processes of citizens actively participating in the various stages of public decision-making (284). This is however necessary because studies show that development outcomes oftentimes result in a significant reduction of poverty and inequality if subaltern groups—those who are poor and socially excluded—are able to influence public policymaking (286). This means that subaltern groups participating in the process of shaping policies—especially on matters that affect their lives—is not only rights-based but also makes economic sense.

This however necessitates a deliberate shift in the balance of power so that access to public decision-making processes is not concentrated to those with economic, political, and technical/bureaucratic clout. Active citizenship towards more participatory and democratic governance, especially of subaltern groups, makes for better policies and developmental outcomes.

Furthermore, increased and sustained spending for social and sustainable development occurs when policymakers view these types of expenditures as a "social investment that can generate multiple positive impacts, from reducing poverty and income insecurity in the short-term to enhancing the country's productivity and capacity for innovation in the long term" rather than as mere expenses (Raquiza 2021, 21; UNRISD 2010, 162).

To underscore the point that social spending is not only a function of fiscal space, there are in fact many countries with lower levels of GDP compared to the Philippines that run more universal social programs like in Bolivia, Lesotho, and Namibia (Raquiza 2021, 21; UNRISD 2010, 210).

5.2 A Policy Climate not Conducive for Participatory Budgeting

During the Martial Law period, the Philippines' public finance structures and practices were governed by a law that did not provide avenues for citizens' participation, as Presidential Decree (PD) 1177 concentrated all budgetary powers to the President. When liberal democracy was restored in the country after the 1986 People's Power Revolution, the government that was installed under the leadership of President Corazon C. Aquino adopted PD1177 as part of the country's Administrative Code.[2]

It is noteworthy that the 1987 Philippine Constitution gave importance to citizens' participation in public policymaking. In particular, it contained provisions such as Article II, Section 23, which stipulates that "[t]he State shall encourage non-governmental, community-based, or sectoral organizations that promote the welfare of the nation." Moreover, Article XIII, Section 16 provides that

> [t]he right of the people and their organizations to effective and reasonable participation at all levels of social, political, and economic decision-making shall not be abridged. The State shall, by law, facilitate the establishment of adequate consultation mechanisms.

Under Philippine laws, the Executive Branch prepares the budget, and the President formally proposes it to Congress; the President's budget is referred to as the National Expenditure Plan (NEP). Congress is not allowed to increase the total budget amount proposed by the President unless it identifies additional revenue measures. Since legislators oftentimes do not do so, they usually operate within the President's budget ceiling. Thus, the budget adjustment in the legislative phase is essentially a zero-sum game, where increased funding for one program means budgetary decrease in another. In this light, it is not enough for (budget advocates to identify what program budgets they want to increase and by how much; they also have to identify programs for budget cuts so that freed-up funds could be realigned to programs which they perceived as needing additional funding.

As is sometimes practiced, the previous year's budget is reenacted when the House of Representatives and the Senate are unable to agree on a final version of the budget before the preceding year ends. Furthermore, as stipulated in Article VI, Section 27 of the 1987 Constitution, and ratified in *Araullo v. Aquino* (2014), the President has a line-item veto power on the budget. This means that s/he can subject certain budget

[2] Presidential Decree No. 1177 (1977), "Revising the Budget Process in Order to Institutionalize the Budgetary Innovations of the New Society".

items to direct veto and conditional implementation, without affecting the entire budget.

The Philippine budget process consists of four phases: (1) budget preparation, (2) budget legislation, (3) budget execution, and (4) budget accountability. Each phase is distinctly separate from the other, with certain stages overlapping from one year to the next. In theory, civil society organizations (CSOs) can intervene in the budget process in all phases— for example, by engaging with the national agencies in the budget preparation and execution stages, lobbying with legislators in the budget legislation phase, and participating as citizen auditors in certain audit engagements.

Official budget-making is generally an opaque and non-transparent process exacerbated by a public that is generally unaware of the mechanisms, rules, and activities involved and insufficient mechanisms for citizens' engagement. Adding to the opaqueness of the process is the use of technical language and protocols mostly accessible only to technocrats and policymakers. The public is usually made aware of the uses and abuses of the national budget during the budget legislation phase as media, both traditional and social, banner stories of wastage and corruption during the annual budget deliberations in Congress. Toward the end of the year, the public is then presented with the General Appropriations Act (GAA), signed by the President. The GAA contains the country's spending plan, including budget priorities, for the following year.[3]

The Social Watch Philippine–Alternative Budget Initiative (SWP–ABI) was a pioneering effort where ordinary citizens engaged the national budget-making process in a systematic and sustained manner. These advocates pushed for increased spending for social, economic, and environmental programs that would lead to better social services, livelihoods, and a healthier environment. SWP–ABI sought changes to the proposed budgets of some national government agencies so that public funds would be realigned to specific programs that would benefit the poor and the socially marginalized more, as well as the environment. This initiative enhanced the capacities of citizens to autonomously influence national budget-making, challenging the government to practice their oft-repeated principles of transparency, accountability, and participatory governance. Reception to SWP–ABI program and budget proposals varied across government agencies and legislators, with some more open to integrating elements of their proposals into department budgets compared to others. SWP–ABI actively engaged the budget preparation and budget legislation phases, with limited experience in the last two phases of the budget process.

In numerous cases internationally, social movements and citizens' networks were able to sufficiently influence the orientation of politicians and policies to benefit the

[3] The Development Budget Coordination Committee (DBCC) is an interagency body composed of the Department of Budget and Management (DBM), the Department of Finance (DOF), Bangko Sentral ng Pilipinas (BSP), the National Economic and Development Authority (NEDA), and the Office of the President, with the DBM as Chair. According to Executive Order No. 292 (1987), "[t]he role of the DBCC is primarily to review and approve the macroeconomic targets, revenue projections, borrowing level, aggregate budget level and expenditure priorities and recommend to the Cabinet and the President the consolidated public sector financial position and the national government fiscal program".

poor. They catalyzed actions between government actors and citizens for the democratic distribution of resources that do not only favor the market and the conditionalities of donors (UNRISD 2010, 261–99). UNRISD cited some best practice cases, such as in Costa Rica, where farmers founded the National Association of Coffee Producers to "defend smallholders' interests on prices, taxes and credit" (297). In Mauritius, the peasantry and agricultural labor force converged to "[wage] active campaigns for labor rights, wage increases and improved working conditions" (297–98). Finally, in the State of Kerala, India, the institutionalization of networks of CSOs, unionization that includes the informal economic sector, and social pacts among the government, the labor sector, and the business sector led to better conditions and wages for workers (298).

5.3 Institutionalizing Citizens' Participation

SWP–ABI Background: How It Began. SWP was established in 1997 as part of the international Social Watch network. Social Watch focuses on addressing the causes of poverty and inequality and promotes equity, including gender justice and the fulfillment of human rights. It monitors governments' progress around the world on the delivery of international and national commitments for a rights-based social, economic and environmental development, including the Sustainable Development Goals (SDGs). It also creates spaces for participatory dialogue on financing for development and people's engagement in governance. SWP–ABI has been very active among Philippine civil society groups which calls for higher spending for equitable economic, social and environmental development, including the SDGs. SWP works amid a political environment where governments invoke fiscal and administrative constraints, including low absorptive capacities of certain government agencies, as reasons to justify low budgetary allocations for social development.

SWP advocates realized that while many politicians and bureaucrats spoke of the importance of social development such as in the areas of health, education, and social protection, oftentimes they did not match their passionate rhetoric with allocating the corresponding funds from the national budget. As is often the case, social spending is often viewed from the lens of fiscal constraints (except when it came to debt repayment, where technocrats would even engage in additional borrowings in order to service the country's debt). This is because social spending was seen more as an expense rather than as a "right" and a "social investment" that contribute to improvements in peoples' well-being as well as increasing the country's productivity and capacity for innovation in the medium to long term (UNRISD 2010, 162). SWP actively participated in the series of preparatory meetings that led to the 2001 United Nations Financing for Development (FfD) Conference, which provided a rights-based policy framework to promote funding for the Millennium, and then later, the Sustainable Development Goals.[4]

From then on, it was just a matter of time before SWP took their financing for development advocacies to the very heart of Philippine development—that is, to the engagement of the national budget. In 2006, SWP initiated the formation of the Alternative Budget Initiative (ABI), which began as an alliance of 20 CSOs but has since then ballooned to, at its peak, 200 organizations.

SWP–ABI was conceived in 2006 during a meeting among some civil society leaders and a number of legislators who were part of the political opposition at that time. The meeting was focused on the underfunding of social development programs, and the need to ensure that the national budget would be more responsive to the needs of the poor and socially excluded. The legislators of the minority bloc and civil society groups proposed the formulation of what they termed as an "alternative budget" to be presented to both the House and Senate members and various policymakers during Congressional budget deliberations.

This approach is akin to the participatory budgeting (PB) model developed in the 1990s in the City of Porto Alegre in Brazil and replicated in various localities globally. In Brazil, PB was characterized by communities taking center stage and actively working with government officials in identifying budget priorities and designing implementation plans. The communities are given the power to vote for or endorse plans for implementation by government authorities.

SWP–ABI Engagement of the National Budget-Making Process: How It Works. SWP–ABI usually kicks off its budget advocacy work by analyzing macroeconomic data and government plans and targets, interrogating budget assumptions, as well as development trends. It does so using social equity lens, figuring out how various budget items and its increase or decrease may hurt or enhance the State's capacity to address the needs and interests of subaltern groups.

SWP–ABI analyses of macro-economic trends, government targets, budget assumptions, and alternative budget proposals for select departments are compiled in what is referred to by them as the "Orange Book,"[5] which is submitted to the Philippine Congress for each fiscal year. It contains relevant data and information, as well as research done by SWP–ABI, to provide evidence justifying the need for increased spending on critical social, agricultural, and environmental services. The Orange Book has also served as a reference for decision-makers, executive officials, legislators, and mass media in assessing the President's budget (the NEP), which is deliberated on and revised during the budget legislation phase.

The SWP–ABI composition largely determines the thematic or sectoral clustering—education-focused CSOs organize themselves into the Education Cluster, social protection-related CSOs comprise the Social Protection Cluster, and so on. As

[4] The FfD principles outlined at that time were the following: 1. "Mobilization of domestic financial resources for development;" and 6. "Addressing systemic issues on the coherence and consistency of the international monetary, financial and trading systems in support of development" (UN 2002, 3–15).

[5] It is referred to as such because its front and back cover are colored orange.

such, the SWP–ABI had the following thematic or sectoral clusters—health, education, social protection, agriculture, and environment. These thematic or sectoral clusters formulated budget proposals, which included, among others, increased financing for more classrooms and teachers; alternative learning systems for out-of-school youth and adults; increased access to health services for the poor, including more reproductive health services; more farm-to-market roads and support for farmers through irrigation, post-harvest facilities and seeds; livelihoods and economic opportunities for victims of typhoons such as Yolanda ("Haiyan") in 2013 and Ondoy ("Ketsana") in 2009; support for biodiversity and conservation efforts; increased support for children's protection, persons with disabilities, older persons and disadvantaged women and girls, as well as increased funding for social protection programs to be implemented at local levels (SWP 2016, 5–6). It was a different case for NGOs which focused on promoting the interests of persons with disabilities and older people. They organized themselves through a two-pronged manner: they joined the various thematic clusters given the intersectionality of their issues with education, social protection, health, and the like, but they also organized themselves into distinct groups referred to as the Persons with Disabilities Cluster and Senior Citizens address issues unique to each of them. The Clusters on Persons with Disabilities and Senior Citizens actively promote its advocacies across different departments. Working with the various SWP–ABI thematic clusters, they seek to ensure that their disability and senior citizens' rights perspectives are integrated in the SWP–ABI proposals when engaging the various relevant departments. Finally, there are other themes, such as gender equality, childrens' rights, and climate change, which are often also taken into account by Cluster members when drawing up the alternative budget proposals.

The Budget Preparation Phase. Here, the various national government agencies draw up their proposed programs and budgets for the following year. During this period, each thematic cluster analyzes the proposed programs and budgets of the relevant agencies; for example, the Health Cluster engages the Department of Health, the Education Cluster engages the Department of Education, and so on. Each SWP–ABI thematic cluster generally advocates revisions to the proposed budgets of the agency in which they are engaging. They push for increased allocations of certain items which they think should be prioritized. Likewise, they propose reducing the budgets of certain programs or items which they think are either unnecessarily bloated, redundant or are too vague, thereby making these vulnerable to corruption.

These "alternative budget proposals" are advocated to departmental executives during this phase. Their cluster proposals are primarily based on data, feedback, and analyses generated from local communities and national CSOs groups whom SWP–ABI works closely with, as well as from data from other sources like the academe and policy think tanks. It is in this context that SWP–ABI budget proposals can be viewed as more attuned to social equity considerations and therefore are more responsive to peoples' needs. Many national organizations with a grassroots membership base comprised the Clusters and through democratic consultations with their members and other relevant partners, generate the SWP– ABI proposals. These are then presented to various national government agencies.

As previously mentioned, the SWP secretariat analyzes the entire budget and looks for items therein that seem redundant, superfluous, or just vulnerable to plain corruption. The secretariat envisions that the budgets for these can be "freed up" and realigned so they can be utilized for the proposed budgetary increases of the different SWP–ABI clusters. As such, when cluster advocates are asked by legislators during public hearings where will they source the money for the proposed increases in the budget of select departments, the SWP–ABI is ready to provide a list of possible sources of financing identified in the NEP where budgetary realignments can be made. A favorite source of budgetary realignments identified by the SWP–ABI advocates are the Special Purpose Funds (SPFs), which are a staple feature in the budgets across administrations.

According to the Department of Budget and Management (DBM) (2013, 1), SPFs are "budgetary allocations in the General Appropriations Act (GAA)" that "are usually lump sum in nature, as the recipient departments or agencies and/or the specific programs and projects have not yet been identified during budget preparation and legislation." However, precisely because of the vagueness in the description of many of these lump sum allocations, a number of analysts believe that these can sometimes be highly vulnerable to corruption. As such, SWP–ABI advocates would oftentimes propose reallocating budgets of some of these vaguely described SPFs to go to funding instead their proposals like in the areas of education and health, or for the needs of children, persons with disabilities and senior citizens. As such, these advocates are proud of the fact that they not only propose increased funding for certain budget items but that they also identify possible sources of funding from within the NEP.

The members of the various SWP–ABI clusters lobby the relevant departments to incorporate their budget proposals—or what is simply referred to as "alternative budget proposals"—into the agency budget proposals which are submitted to the DBM. Some agencies are more receptive than others, where the former will incorporate certain ABI budget proposals into the agency budget. The DBM, in consultation with other departments, further refines and consolidates these departmental budgets and submits these to the President for scrutiny and approval. After approval, the President then officially submits what is now referred to as the National Expenditure Plan (NEP) to Congress during the ceremonial State of the Nation Address (SONA). As soon as the President submits the Executive's proposed budget to Congress, budget-making is now in the hands of legislators, which signals the next phase of the budget cycle.

The Budget Legislation Phase. Here, SWP–ABI members continue their lobbying and advocacy work, particularly for the many proposed budget items that were not incorporated during the budget preparation phase. This time, lobbying efforts are targeted at legislators, both at the House and Senate, for them to incorporate the SWP–ABI budget proposals into the various agency budget which they will eventually approve at their level—now referred to as the General Appropriations Budget (GAB)—and recommend to the President.

During this phase, SWP–ABI advocates usually partner with specific legislators who either have their own advocacies which are consistent with SWP–ABI policy and budget agendas or simply express support for some of the proposals. These legislators then become latter legislative champions of some of the SWP–ABI proposals in Congress.

Furthermore, SWP–ABI members realized that to make the budget more responsive to peoples' needs and to promote citizens' participation in the budget process, specific legislative reforms were necessary. Therefore, SWP–ABI decided to work with legislators in the crafting of certain bills and lobbied for their enactment. A case in point was House Resolution No. 1376, entitled an "Urgent Resolution Allowing the Active Participation of Bona Fide People's Organization in Public Hearings in Congress Annual Budget Deliberations." Filed in 2006, this bill was championed by Congressman Lorenzo "Erin" Reyes Tañada III and co-authored by Congressman Teofisto Guingona. It was then adopted by the entire House of Representatives in 2007. Afterward, Senator Panfilo Lacson filed a similar bill in the Senate.

Unfortunately, the passage of this bill has consistently seen rough waters. Despite its continuous refiling in Congress across various administrations, it has not prospered. Indeed, the legislative attempt to institutionalize CSO participation in budget-making has not been adopted due to fierce opposition from many legislators. They viewed the CSOs' involvement as encroachments on their policy domain. Based on informal interviews with congressional staff, SWP–ABI advocates were informed that the long-running opposition to institutionalizing CSO involvement in the budget-making process was fueled by fears among certain politicians who thought that such a bill, if enacted into law, would make many of them more vulnerable to getting caught and exposed on the possible misuse of public funds.

Some Political Gains, Victories

In 2007, the year after the formation of the SWP–ABI network, alternative budget proposals from civil society, highlighting proposed increases in social development, agriculture, and environment expenditures, were officially presented by CSOs for the first time and discussed during the budget deliberations in both the House of Representatives and Senate. Former House Minority Floor Leader Teofisto Guingona III stated,

> Today, we think it is time for change. A change for a more dynamic Philippines; a change that will usher in genuine people's participation in the budget process. This will bring in a new perspective, and will make our system a more participatory democracy. This alternative budget has undergone a series of consultations among various groups. This is a clear indication that the opposition does not only confine itself to criticism but also proposes meaningful alternatives. (SWP 2010, 27)

Representative Douglas Cagas similarly observed,

> [W]hat the minority has done is something that is necessary in a democracy. This is the beginning of a constructive way of doing things. In fact, in all major issues that confront Congress and the coming Congresses, any opposition group should present an alternative proposal, especially the case of an alternative budget. (SWP 2010, 27)

Meanwhile, then-Senate President and Finance Committee Chair, Franklin Drilon, immediately met with the SWP–ABI consortium in order to review the alternative budget proposals, covering every item. During his sponsorship speech during the budget deliberations at the Senate, he physically held up the Orange Book for all to see and stated,

> Let it be said that we are crafting a national budget that does not focus solely on the imperatives of economic growth, but a national budget that pursues growth with equity. We are proposing a national budget which has as its core the welfare and development of every Filipino man, woman, and child. (SWP 2010, 28)

In 2008, the House Committee on Appropriations once again invited civil society groups to participate in the budget deliberations in Congress and included the SWP–ABI alternative budget proposals in the national budget. The then-chair of the committee, Representative Edcel Lagman, said in his opening statement,

> This morning, we are going to break traditions. For the first time, we are giving non-government and people's organizations that opportunity to present their views, comments and recommendations on the General Appropriations Bill which they call the alternative budget. (SWP 2010, 28)

Since its inception in 2006 to the present, some of the SWP–ABI proposals were either adopted at the agency level during the budget preparation phase or championed by certain legislators during the budget legislation phase and eventually integrated into the General Appropriations Act resulting in a cumulative increase of about PHP 66.9 billion in the national budgets for social services. At one point, the sector on health received the largest cumulative budget increase of at least PHP 24.906 billion. It was followed by the other sectors: education (PHP 11.666 billion); agriculture (PHP 11.07 billion); environment and climate change (PHP 9.078 billion); PWDs (PHP 2.285 billion); and social protection (PHP 178 million). This does not include the reported aggregate increase for health and agriculture in 2007 amounting to PHP 5.2 billion, as well as the aggregate increase for the sectors of education and agriculture of at least PHP 7.7 billion in the 2009 budget (SWP 2018, 3–6). Meanwhile, lobbying continued to include other strategic issues, like campaigning for the rationalization of special purpose funds (SPFs) and calling for the enactment of the bill to institutionalize CSO involvement in budget making. Budget gains from 2017 and onwards were not as far-reaching as in the previous years.

For fiscal years (FYs) 2019 to 2021, the SWP–ABI was able to incorporate increases for social protection for children as a result of their strong public advocacy campaign in partnership with UNICEF, Save the Children, and the Child Rights Network, to name a few. For example, the Child Protection Program under the Department of Education (DepEd) was given a dedicated budget item since 2018. Furthermore, notable budget increases were granted to the Last Mile Schools Program in geographically isolated and disadvantaged areas (GIDAs), the School-Based Feeding Program, textbooks and other instructional materials for children with disabilities, training on positive discipline for teachers, and the Data Interconnectivity Program.

The budget increases on child labor profiling; compensation for community facilitators, encoders, and enumerators; and livelihood assistance to beneficiary parents/guardians of child laborers contributed to the implementation of the Child Labor Prevention and Elimination Program (CLPEP) under the Department of Labor and

Employment (DOLE). However, the advent of the COVID-19 pandemic caused many budgets meant to benefit children to be reallocated to COVID-19 response initiatives. Examples of suspended and reallocated projects to respond to the COVID-19 pandemic are: (1) the building of 361 schools in GIDAs under the Last Mile Schools Program of DepEd, at around PHP 5 billion, and (2) the allocation of PHP 30 million allocated for profiling of child labor to be undertaken by the DOLE (SWP and Save the Children 2020, 1–14; GAA 2018, 2019, 2020, 2021).

As earlier mentioned, the SWP–ABI members worked with national government agency personnel, as well as legislators. The collaboration has yielded sustained partnerships. For example, in 2012, the Department of Health (DOH) institutionalized civil society participation in its budget formulation process through a Department Order requiring CSO consultation in its proposed budget. Since then, the agency has been annually presenting its proposed budget to the SWP–ABI Health Cluster and other CSOs for comments and inputs. Through this process, many of the proposals of the SWP–ABI Health Cluster's People's Health Agenda were considered and incorporated in the DOH budget prioritization process (SWP 2017, 11). Throughout many years,SWP–ABI actively participated in the budgeting processes of the Department of Agriculture (DA), the Department of Social Welfare and Development (DSWD), the Department of Health (DOH), the Department of Environment and Natural Resources (DENR), the Department of Education (DepEd), and other government agencies and institutions.

Moreover, the SWP–ABI members' common resolve to work for a budget that responds to the needs of the people and to promote participatory budgeting contributed to the institutionalization of certain citizen–government partnerships at the local level through ordinances and Memoranda of Agreement in selected local government units (LGUs). SWP–ABI forged partnerships with local executives and organizations of diverse marginalized communities in influencing local budgeting and public finance processes within the framework of the Millennium, and later, the Sustainable Development Goals. Examples of these communities include the upland and lowland farming communities in Guimba, Nueva Ecija; coconut farming communities in Baao, Camarines Sur; fishing communities in marine protected areas in Dauin, Negros Oriental; small fishing communities in Hinigaran, Negros Occidental; urban poor communities in Tondo, Manila and Quezon City; indigenous peoples communities in Magpet and Makilala in North Cotabato; Muslims in displaced communities in Tulunan, North Cotabato; and selected disadvantaged municipalities in Pangasinan.

Even prior to the formation of SWP–ABI, the 1991 Local Government Code (LGC) institutionalized the participation of CSOs in Local Development Councils from the barangay all the way to the provincial level.[6] This means that theoretically, CSOs should be in a position to participate in development planning and budgeting of LGUs. Thirty years after the LGC was implemented, however, studies show that this provision is honored more in the breach than in the observance (Guce 2020, 72–75, 81–83, 86–92).

[6] Republic Act No. 7160, Sections 2, 25, 107 (2), 111.

There are some good executive measures implemented, however, in more recent years. For example, transparency, accountability, and participation in budgeting were incentivized through the Department of Interior and Local Government (DILG) Seal of Good Housekeeping in 2010, and later, the Seal of Good Financial Housekeeping (SGFH) awards. This program, among other things, recognizes LGUs that spent their public funds along the principles of transparency, accountability, performance, and people's participation. This includes the areas of local legislation, development planning, resource generation, resource allocation and utilization, customer service, and human resource management and development, and good governance. Good Financial Housekeeping is a requirement for an LGU to receive the Seal of Good Local Governance (SGLG), which aims to push for local investments, employment, environmental safeguards, social protection, and disaster preparedness.

However, a 2020 assessment on the implementation of the SGLG revealed that this program is generally out of reach to many of the poorest LGUs, i.e., those belonging to the fifth- and sixth-income classes, as they were unable to pass the requirements for the SGLG. For example, more than 30 percent of LGUs in the Bicol Region, Eastern Visayas, and Central Visayas did not pass. The study also showed that it takes at least two years for many of the lower-income class LGUs to comply with the SGLG requirements. Furthermore, the study revealed that some local government officials were unaware of the SGLG. Also, a significant number of projects were delayed, and development funds were poorly utilized. Hence, the challenge is to design the SGLG by income class or regional location, undertake capacity building for improved governance, and conduct a thorough evaluation of development plans (Diokno-Sicat et al. 2020, 23–24). Equally important is the need to promote equitable access of the SGLG, especially for the poorest LGUs.

Further, with the citizens' clamor for transparency in the budget process, the DILG issued the Full Disclosure Policy, which requires local governments of barangays, municipalities, cities, and provinces to fully disclose the financial transactions of the LGUs to their constituents. This includes informing the public of how the LGU managed, disbursed, and used public funds. To what extent, and how this program is implemented, however, is a subject of another study.

In addition to influencing policy reforms, the SWP–ABI was also able to influence the public narrative on the budget process through its partnership with journalists. For example, according to Malou Mangahas, Executive Director of the Philippine Center for Investigative Journalism (PCIJ), in *Retake the People's Purse: The Role of Media and Civil Society,*

Thanks to Social Watch Philippines and its associates in the Alternative Budget Initiative (ABI)—the network of civil society groups that have demonstrated a curious passion for fleshing out, breaking down, scrutinizing and reordering the GAA—journalist[s] have started to get firmer and sharper handles for reporting better on the national budget. Or at the very least, for doing stories that are more accessible, or less Greek to the taxpayers from whose sweat and brow the national budget draws major sustenance(quoted in SWP 2010, 126).

Aside from the news releases, feature articles, and in-depth studies to aid investigative journalism, SWP–ABI conducted capacity-building activities for journalists to deepen their understanding of the workings of budget-making. SWP's regular media briefings enhanced the media's appreciation and understanding of ABI's work and the results it wants to achieve. This influenced the narrative on burning issues on the budget and promoted CSO-mass media partnerships for a transparent and accountable budget process (SWP 2010, 128–30). Furthermore, Mangahas observed that before SWP–ABI, most journalists found the budget to be an "inscrutable mystery" and a "colossal riddle." She said, "More than just a numbers game, the ABI has sufficiently impressed on journalists the policy and development content and implications of the GAA" (125–27).

SWP–ABI and Building Capacities

Building the capacities of SWP partners meant deepening their understanding of the process, substance, protocols, and procedures in formulating the national budget. Here, it is imperative to note that the then-Lead Convenor of SWP, Leonor M. Briones, is a trained and nationally recognized fiscal expert with the distinction of having served important roles in government, civil society, social movements and academe in the areas of economic and social development and public finance. As such, Briones played a decisive role in training SWP–ABI advocates to undertake budget advocacy work. Through workshops, lectures, and on-the-job training, SWP–ABI members learned how to advocate for an inclusive budget in a new terrain, Congress. These training processes included how to read and assess voluminous, technical budget documents using social equity lens and gauge its effect and impact on poor farmers, women, children, persons with disabilities, senior citizens, and the environment. As importantly, the SWP–ABI advocates had to learn how to draw up proposals to increase allocations for certain departmental budget items based on certain budget assumptions and analysis that would measure up in the face of departmental and congressional scrutiny. Part of the CSOs' capacity-building included learning protocols on how to dress, what language to employ in parliament, and many other procedural rules. The former SWP Convenor and former DepEd Secretary recalled,

> From the start, I had no apprehension at all because we thought that it was a very simple thing to do. It turned out to be very difficult because we needed to do capacity building for the members and briefings for the legislators. ABI demanded more time than I really anticipated. (SWP 2010, 35)

Alongside honing their capacity to understand the finer and more detailed aspects of the budget and its attendant processes and procedures, SWP–ABI members also developed their networking skills with policymakers and legislators who championed their advocacies. This partnership also helped deepen the understanding and appreciation of many legislators and government executives on the role of citizens' groups in public finance. Furthermore, Briones' reputation in this CSO initiative helped "open doors" to many government officials, as her expertise in fiscal policy was publicly known, and some of the policymakers, including their staff, were her former students during their university days. Apart from being an authority on the topic, Briones had the ability to popularize a technical subject like public finance to a broad audience. Therefore, she became a favorite interviewee of many journalists; this, in turn, provided a public spotlight to SWP–ABI's budget advocacy work.

As can be gleaned, the SWP–ABI clusters became a channel to reach out to many CSOs and help train them to push for increased public spending for their existing advocacies, whether in education, health, or the environment. The process also amplified the perspectives of those in vulnerable situations, such as children, senior citizens, those living with disabilities, disadvantaged women and girls, those in conflict-afflicted areas, in disaster-prone communities, in informal settlements, and in urban poor communities.

As previously mentioned, enabling citizens' engagement in the budget process extended to some local community processes. For example, Luz Anigan of Action for Economic Reforms and an active SWP researcher, narrated,

> In the beginning of ABI, the education cluster conducted several workshops in the National Capital Region, Central Luzon and Western Visayas focusing on budget monitoring. The training activities proved effective in enhancing the participation of more groups in formulating the alternative budget for education and in generating interest among the local constituencies. The education cluster also prepared a set of research modules which served as guidelines for conducting budget research and analysis focusing specifically on basic education. The modules were effective in encouraging local partners in conducting local budget analysis. (SWP 2010, 74)

Adding to this perspective is Patrick San Juan, an erstwhile member of the Philippine Rural Reconstruction Movement (PRRM), one of the organizations leading the SWP–ABI environment cluster. He stated,

> Budget advocacy for the environment and the Millennium Development Goals (MDGs) is a pioneering work and something out of the usual. Members of the environment cluster were able to incorporate their advocacies within the context of ABI and the national budget. Engagement in the ABI complemented CSO work on the field. (SWP 2010, 3–4)

SWP–ABI's engagement with budget-making processes was comprehensive. According to Janet Carandang, Head of the SWP Secretariat,

> We analyzed various programs and budgets of the national government and proposed policy, program and budgetary recommendations; we undertook lobbying and networking with a broad range of actors like policymakers, the media and grassroots groups, and we built our capacities along the way. (SWP 2010, 23)

5.4 Formidable Challenges to Citizens' Participation

SWP–ABI's engagement with the national budget process was never a steady or linear process. Support for their budget advocacies also depended on the political dynamics among policymakers and which side of the political aisle they were on. For example, in the formative years of SWP–ABI, its strongest allies were part of the political opposition during the Gloria Macapagal Arroyo (GMA) administration. These SWP–ABI legislative allies who were part of the political opposition during President Arroyo's regime, authored many of the budget bills supported by SWP–ABI and other citizens' groups. However, when President Benigno S. Aquino III assumed the presidency in the next administration, the SWP–ABI advocates found that many of their ally legislators during the GMA regime were now allies of the new President and were part of his majority group in Congress. It was in this context that SWP–ABI members noticed that many of these same legislators, who were proactive in championing many of their advocacies in the past, were now more cautious in pushing for the same budget reforms. Upon reflection and informal discussions with some of these legislators, one important reason that emerged for this more cautious approach was not so much a "change of heart" of their stand on these issues. Rather, it was because they now found themselves ensconced within the new ruling coalition in Congress, and there was increased political pressure on them to uphold the proposed budget of the President. For SWP–ABI, this signaled the need to diversify their allies in Congress across political persuasions and work with any and all legislators on an issue-to-issue basis depending on whoever was receptive to their specific advocacies.

As earlier mentioned, there was also formidable resistance from a good number of legislators to institutionalizing citizens' participation in the budget process. The legislative measure "Allowing the Active Participation of Bona Fide People's Organization in Public Hearings in Congress Annual Budget Deliberations," which was originally filed in 2006 and refiled in subsequent Congresses, continues to languish. The participatory approach of granting CSOs a formal role in the budget process was never enacted into law due to opposition by many legislators. In fact, some of the legislators were downright hostile, as they viewed CSOs' involvement as encroaching on their turf. At worst, it would mean making some of them vulnerable to the public exposure of less-than-aboveboard deals. According to Jessica Cantos, Co-Convenor of SWP and former Chief of Staff of Congressman Erin Tañada,

> The provision for citizens' participation in the budget process was included in a Budget Reform Bill under the Duterte Administration. There were reports, however, that the inclusion of this specific provision in the bill was met with resistance from some government officials and was eventually deleted. (Cantos 2021)

Furthermore, SWP–ABI advocates began to wonder: once the budget was enacted into law, how was it utilized? This was of particular concern for those budget items that they had successfully pushed for through their champion legislators and which were eventually incorporated into the GAA—referred to by them as "budget wins"—through effective lobby work and which constituted political gains in and of itself. Rene Raya, also one of the convenors of SWP, wondered about "the integrity of the money trail" when he asked,

> Were the increased budgets for national programs, say in health and education, properly and efficiently channeled to the end-users such as local schools, clinics, including barangay health stations and in a timely manner? (Raya 2021)

This line of questioning led SWP–ABI to the notion of "paper victories," that is, getting certain budgets increased in the GAA, but which may not have much substantive meaning if not properly allocated and released to its targeted end-users, such as local schools and barangay health stations. As such, the SWP–ABI piloted a "budget tracking" exercise on a limited basis in 2008. For example, the SWP–ABI Education Cluster was one of those who tried to follow the "money trail," so to speak, by checking whether the budgetary allocations from the national budget were properly allocated to local schools—in this case, those in Dumaguete, Negros Oriental. At that time, the Education Cluster researchers tried interviewing some local school officials but, unfortunately, were met with significant difficulties. According to Raya, they soon found out that the national budgetary allocations for elementary schools were allocated at the Division level of DepEd and that the GAA only reflected specific allocations for national high schools. This made it difficult to determine how much funding from the GAA was actually funneled to each of the public schools at the local level.

Furthermore, the researchers found out that their interviewees were only willing to answer the more qualitative questions, such as on the effects of increased school budgets on school performance. They were wary of answering questions that had to do with hard budgetary amounts (e.g., How much was your budgetary allocation from the national government? Did you receive the entire amount? When did you receive it?). One respondent even asked with hostility what authority did the researchers have to ask questions on public funds. Given the degree of resistance and in some cases, even downright hostility, on the part of the respondents, some researchers reported feeling unsafe and decided not to pursue this line of questioning. After several attempts of trying to get information on hard budget data of local schools allocated by the GAA, the SWP–ABI Education Cluster researchers gave up and concluded that there was a pervasive lack of trust and transparency to share information of this nature by the respondents. As such, their objective of ascertaining the integrity of the money, in this case, the education budget for some local schools in Dumaguete, remained unmet.

It must be noted, however, that this is one only case and therefore generalizations cannot be made. Moreover, the SWP–ABI network also found out that there have been instances when certain budget items already embedded in the GAA were not always released to the end-users for various reasons. This is another issue that needs to be further explored in future research.

The need to strengthen CSO capacities to effectively undertake budget advocacy on a sustained basis remains an ongoing challenge. On the one hand, the advocates across the various clusters have relatively "mastered" the intricacies of the specific agency budgets that they are focused on. Many of them now have the ability to effectively work with agency officials and legislators, a process that has somehow been "ritualized" given the annual engagement of SWP on the national budget. However, the role that Briones played—providing holistic analysis of the entire national budget, including engaging the macro-economic assumptions that underlie it, and problematizing how this squares with overall developmental and fiscal trends from the lens of poverty and inequality—is a capacity that further needs honing among the current crop of budget advocates. Performing these tasks entails going beyond sectoral budget engagements and developing the ability to contextualize budget-making within a broader political and economic framework that is rights-based, developmental, welfare-enhancing and inclusive. This remains a work in progress.

SWP can be credited for sustaining citizens' engagement with the national budget, even during periods without financial support for this endeavor. This speaks of their abiding commitment to push for increases in pro-poor, pro-people, and pro-environment public spending, efforts that would have been significantly amplified in the context of sustained financial support.

5.5 Conclusion

Outcomes that reduce poverty, inequality, and promote inclusive development are more likely to happen when subaltern groups, through various channels and strategies, can influence public policymaking at national and local levels. Historically speaking, while the Philippine policy climate has not been conducive for ordinary citizens to actively participate in the national budget-making process, the country's 1987 Constitution and many other laws that have been issued since then have recognized their value in public life. SWP, through its ABI initiative, has engaged the Philippine bureaucracy for over a decade and a half— in particular, select government agencies as well the legislature (both the House and the Senate)—by pushing for budgetary increases, as well as decreases, of certain government programs. This is done with a view to making the national budget more responsive to the needs and interests of poor and socially excluded groups, protecting the environment, as well as reducing wastage and corruption. During the budget legislation phase, this further requires advocating to legislators certain realignments in the budget allocation so that some programs are prioritized while reducing the budgets of others.

Throughout the years, SWP–ABI has advocated for budgetary realignments in order to prioritize certain programs and activities of various government agencies during the budget preparation and legislation phase. When certain agency officials are receptive to incorporating SWP–ABI proposals into the departmental budget plans or when some legislators decide to champion such proposals, it facilitates the chances of integrating these measures into the GAA. This constitutes a political victory from the perspective of SWP–ABI advocates. However, they also realize that these budgetary increases in the GAA do not automatically mean that these funds will be eventually channeled to the envisioned end-users, such as barangay health stations and local schools.

SWP–ABI advocates found out that some budgets embedded in the GAA, for various reasons, were never released and downloaded by the national government. Second, even if these were released, the integrity of the fund releases still need to be ascertained. As in SWP–ABI's limited budget-tracking research for education for some schools in Dumaguete show, questions about the correctness of the amounts of fund released and the timeliness of its release to end-users remained unanswered due to the opaqueness of the budget execution process and the lack of respondents willing to give information on actual amounts received. To address this concern, SWP can explore proactively engaging the Commission on Audit (COA) given the latter's initiative to officially involve citizens in auditing the use of public funds through their Citizen Participatory Audit program.

Indeed, the fight over public funds—who wins, who loses—is, at heart, an extremely politically contentious process. Tipping the balance of power so that more public funds are allocated to underserved and marginalized groups will depend on a number of factors, including how developmentally and equity-minded are policy-makers, the state of play and the balance of forces within the Executive and with the legislature, the political strength of civil society and social movements, the quality of partnerships forged between and among these groups, how socially engaged media practitioners, academe and other stakeholders are, among others. As shown by the SWP–ABI experience, certain legislative allies may be champions of their budget advocacies in one political season, but not necessarily the next. As such, the need to diversify political allies among legislators and those in the Executive on an issue-to-issue basis serve as a more pragmatic approach to networking and alliance-building.

As the SWP–ABI experience underscores, politicians have successfully resisted the enactment of bills that would institutionalize citizens' participation in the budget-making process. This has occurred across three administrations: Macapagal-Arroyo, Aquino, and Duterte. As of this writing, the second Marcos administration is continuing this trend. This shows that the prevailing ideational and policy climate for citizens' participation in budget-making in the legislature generally remains unconducive.

SWP–ABI advocates realize that, because the fight over public funds is essentially resolved in favor of who has political power, it is not enough to propose a well formulated developmental, social-equity-driven agenda and budget plan. Such an initiative must be backed up by a formidable political force. The latter can be done by strengthening and expanding SWP–ABI linkages with subaltern organizations with a robust mass base. Likewise, strengthening and diversifying alliances with key sectors—the media, academe, bureaucrats, and legislators—will strengthen their flank and multiply platforms for budget advocacy.

The need for active citizens' participation in the budget-making process to promote a people and environment-centered development agenda remains more urgent than ever given the intensifying social, economic, and environmental crises that pervade Filipino peoples' realities. The continuing relevance of SWP–ABI work is clear. The urgent task of strengthening their political and technical capacities, and working with other CSO groups, social movements, and developmentally minded policymakers, to generate the political momentum needed to make the public budget work better for people and planet are obvious next steps.

5.6 Government Documents Cited

Executive Order No. 292. "Instituting the Administrative Code of 1987." 25 July 1987. https://www.officialgazette.gov.ph/1987/07/25/executive-order-no-292-s-1987/

The Constitution of the Republic of the Philippines. 1987. https://www.officialgazette.gov.ph/constitutions/1987-constitution/

Republic of the Philippines. General Appropriations Act. 2018.

Republic of the Philippines. General Appropriations Act. 2019.

Republic of the Philippines. General Appropriations Act. 2020. Republic of the Philippines. General Appropriations Act. 2021.

G.R. No. 209287. "Araullo v. Aquino." 1 July 2014. https://lawphil.net/judjuris/juri2014/jul2014/gr_209287_2014.html

Presidential Decree No. 1177. "Revising the Budget Process in Order to Institutionalize the Budgetary Innovations of the New Society." 30 July 1977. https://lawphil.net/statutes/presdecs/pd1977/pd_1177_1977.html

Republic Act No. 7160. "An Act Providing for a Local Government Code of 1991." 10 October 1991. https://www.officialgazette.gov.ph/1991/10/10/republic-act-no-7160/

References

Briones, Leonor Magtolis. 2008. "May Pera Na Ba? mobilizing domestic financial resources for development." In *Finance or Penance for the Poor, Mobilizing Resources for the MDGs: The Five-Year Review of Financing for Development*, edited by Filomeno S. Sta. Ana III, 7–20. Quezon City: Social Watch Philippines.

Cantos, Jessica R. 2021. Interview by Maria Victoria Raquiza and Roja Salvador. September 2021.

DBM (Department of Budget and Management). 2013. *A Brief on the Special Purpose Funds in the National Budget.* https://www.dbm.gov.ph/wp-content/uploads/DAP/Note%20on%20the%20Special%20Purpose%20Funds%20_Released%20-%20Oct%202013_pdf

Diokno-Sicat, Charlotte Justine, Maria Alma Mariano, Angel Faye Castillo, and Ricxie Maddawin. 2020. "Assessment of the performance challenge fund and the seal of good local governance: perceptions from municipalities." *Philippine Institute for Development Studies Discussion Paper Series* No. 2020–05 (March 2020): 1–59. https://.pids.gov.ph/CDN/PUBLICATIONS/pidsdps2005.pdf

Guce, Czarina. 2020. "Citizen participation, transparency, and accountability in local governance: Reflections and lessons learned after 25 years." In Vol. 5 of *Local Matters: Experiences in Autonomy, Decentralization, and Beyond—A Book of Readings*, eds. Erwin A. Alampay and Gerard Martin C. Suarez, 71–94. Quezon City: Center for Local and Regional Governance.

Raquiza, Maria Victoria R. 2021. "An alternative engagement with the national and local budget process." In *Building Peoples' Movements in Southeast Asia: Towards an Alternative Model of Regionalism*, 19–23. Quezon City: UP CIDS. https://cids.up.edu.ph/wp-content/uploads/2022/02/2nd-sea-conference-alternatives-proceedings-online-ver.pdf

Raya, Rene. 2021. Interview by Maria Victoria Raquiza and Roja Salvador. September 2021.

SWP (Social Watch Philippines). 2010. 2017. *Enhancing Civil Society Participation in the Health Budget Towards Building Active Citizenship for Development, Transparency, and Accountability.* External Evaluation Report 2013–2015 for Save the Children.

SWP (Social Watch Philippines). 2010. 2018. *ABIyahe: Assessing the 10-Year Journey of the Alternative Budget Initiative Advocacy in the Philippines* (2006–2016). Evaluation Report for Christian Aid.

SWP (Social Watch Philippines). 2010. *Alternative Budget Initiative: Three- Year Experience.* Quezon City: Social Watch Philippines, United Nations Development Programme, and National Economic and Development Authority.

SWP (Social Watch Philippines) and Save the Children Philippines. 2020. *Investing in Children through Child Sensitive Alternative Budget January 2017–December 2019*. Project Evaluation Report for Save the Children.

United Nations. 2002. *Report on the International Conference on Financing for Development, Monterrey, Mexico, 18–22 March 2002*. New York: United Nations. https://www.un.org/development/desa/financing/document/report-international-conference-financing-development-monterrey-mexico-18-22-march-2002

UNRISD (United Nations Research Institute for Social Development). 2010. *Combating Poverty and Inequality Structural Change, Social Policy and Politics*. Geneva: United Nations Research Institute for Social Development.

Ma. Victoria R. Raquiza is an associate professor at the National College of Public Administration and Governance, University of the Philippines Diliman. She has been specializing in teaching courses on poverty, inequality, social policy, the developmental state and governance, and the Sustainable Development Goals (SDGs) at both graduate and undergraduate levels since 2009. She obtained her Doctor of Philosophy from the City University of Hong Kong in 2016. She has extensive experience in writing and publishing on issues related to poverty, social protection, social equity, good governance, the SDGs, and women's rights and gender equality-related matters at national, regional, and international levels. In addition, she served as a commissioner of the Philippine Population and Development Commission from 2021 to 2022 and is currently a member of the Advisory Council of the Internal Affairs Service, Philippine National Police (2019-present).

She has also been actively involved with several non-government organizations, serving as a member of the Boards or Executive Committees of organizations like the Social Weather Stations (2020–2023) and Social Watch Philippines (1999–present). Moreover, she has represented NGOs in numerous official country delegations to various UN conferences and served as a consultant for a Gender Audit for member NGOs of the Network of Coordinating Regional Support Organizations (NCRSO) in India.

Roja Estrelita Salvador has a Master of Arts degree in Health Policy Studies from the University of the Philippines-Manila and is a Fellow of Social Watch Philippines.

Part III
People-Centered Agrarian Systems

Chapter 6
Power from the Grain: Bringing Back the Lost Heritage of Sorghum in Nusa Tenggara Timur, Indonesia

Ananeza P. Aban

Abstract Not known to many, the province of Nusa Tenggara Timur (NTT), Indonesia, hosts a well-grounded and powerful community-led food movement that aims to bring into fore neglected traditional knowledge and sustainable practices. The key is sorghum. Sorghum, a traditional and principal food crop of this province in the ancient times, more nutritionally dense and available, and broadly adaptable in hot and dry agroecologies, has been reduced to animal fodder. In fact, eating other staple grains that is not rice has been equated to being economically vulnerable. This thinking is rooted in the national rice program introduced during the Suharto regime in the late 60s that both altered the cultivation and consumption of sorghum and other local crops such as millet, barley, and corn in NTT.

Keywords Community-led food movement · Nusa Tenggara Timur · Sorghum cultivation · Sustainable agriculture · Traditional knowledge

Sorghum (*Sorghum bicolor [L.] Moench*), the forgotten staple food crop in the southeastern part of the Indonesian archipelago found its rebirth through the efforts of a local peoples' food movement that is grounded in their determination to combat food insecurity and malnutrition in the province. Local food producers in the province of Nusa Tenggara Timur (NTT) also made this effort to reclaim their lost cultural identity, an agricultural heritage that was eroded for decades in order to accommodate a government program embellished as a national strategy to achieve food self-sufficiency.

Farmers belonging to the Perhimpunan Petani Sorgum untuk Kedaulatan Pangan are speedily reviving the cultivation of many local varieties of sorghum across NTT, particularly in the regencies of East Flores (including Larantuka and the islands of Solor and Adonara), Ende, Lambata, Manggarai, Rote, and Sikka. The physical geography of NTT provides an optimum environment for this food crop that is tolerant to hot and arid agroecologies. Throughout the province, rainfall is a major factor to

A. P. Aban (✉)
Ghent University, Ghent, Belgium
e-mail: ananeza.aban@ugent.be

J. M. C. Sy and A. P. Aban (eds.), *Reimagining Development in Southeast Asia*, Asia in Transition 21, https://doi.org/10.1007/978-981-97-9140-8_6

crop production (Pellokila, Field, and Momuat 1991), while soil moisture can limit yield (Fitzpatrick and Chapman 1988, quoted in Pellokila, Field, and Momuat 1991). NTT is characterized by low rainfall having only around three to four months of rainy season and prolonged drought, with 92.32% or about 1.5 million hectares of total land area identified as dry land (Mukkun et al. 2018). It is situated in the volcanic region of the Lesser Sunda Islands of Indonesia where land is fertile but water is scarce.

Climate change now becomes a huge threat to food security due to greater loss of soil moisture, reduced groundwater, and increasing reduction of rainfall. NTT has a majority population reliant on small-scale subsistence farming, thus minor shocks including climate change will significantly impact the population's livelihoods (World Food Programme 2013). The contributing factors include lack of access to safe drinking water, indigents' limited access to food, poor sanitation, and limited access to health care (World Food Programme 2013).

Sorghum, upland rice, and millet were the main cereals throughout NTT before maize took center stage as the principal staple of the province. These food crops go well in areas where rainfall is below 900 mm (mm). Later on, sorghum, tubers, legumes, and other crops have become supplemental crops, especially in periods when maize harvest fails (Pellokila, Field, and Momuat 1991).

Local sorghum varieties that are adaptable to the agro-climatic conditions of NTT are seen to have high diversity, coming from inherited varieties from parents and those new genotypes that are products of cross-breeding (Mukkun et al. 2018). Since their reintroduction of sorghum, local farmers in the province now possess seven local varieties of sorghum, commendable for human consumption. Meanwhile, Mukkun et al. (2018) discovered more than 30 species of sorghum, with almost all species having great potential as high-yielding varieties suitable for human food consumption and diversification.

6.1 Sorghum: The World's Important Grain, an Agricultural Heritage

To begin with, Ethiopia is the heartland of sorghum that is grown by entirely almost all subsistence farmers (Grain 2007; Teshome et al. 1999). However, some scholars mentioned that its early domestication may have occurred in northeastern Africa or the Egyptian-Sudanese border (Mann et al. 1983, quoted in Hariprassana and Patil 2015). Africa is also home to the largest diversity of sorghum. For over 5,000 years, it has been nurtured and spread out along trade routes across the continents (grain 2007), making it an important food crop of people in the semi-arid tropics of the African and Asian continents (Hariprasanna and Patil 2015). It is actually the fourth most important world cereal, which is only surpassed by wheat, rice, and maize in terms of production area (Teshome et al. 1999).

In India, this is the third most important grain, a cornerstone of poor peasants' strategies to assert their food sovereignty in the Medak District of Andhra Pradesh by controlling over their seeds and food after their painful saga with the Green Revolution of the 1960s, when government strongly pushed production of new high-yielding rice and wheat varieties, consequently devaluing traditional crops (Grain 2007).

In Flores[1] society, sorghum is likewise a key traditional food in local cultural identity, as it is broadly adaptable to its land and climate (Loretha 2018). As a staple food of the locals, it has become a vital ingredient to their socio-economic struggle. In decades past, it was an indicator of becoming resilient in the midst of limited resources, their answer to food insufficiency when rice was scarce or rather unaffordable.

The impulse to replant sorghum began from the spontaneous enthusiasm of one woman—Maria Loretha. It all started when she relocated to NTT with her family to avoid the detrimental effects of the Asian financial crisis affecting Indonesia under Suharto in the late 90 s and to live an alternative rural life. One day, she ate a plate of steamed sorghum garnished with coconut flakes offered by her neighbor. Immediately, she "fell in love" with it. Thereafter, she hunted for seeds in the far-flung villages and recovered them from the elders. Without any hesitation, she propagated the seeds and distributed them to farmers who were willing to grow them (Loretha 2018; McKinley 2012). The beginnings were a bit tough for her. Farmers, mostly dependent on maize and cashew for their livelihood, are not keen on reviving sorghum. The motivation sharply started only during a prolonged drought in NTT. The Catholic Church in Larantuka was also instrumental in calling all its laity to plant sorghum as a measure to ensure food in times of calamity (Loretha 2018).

This preliminary experience of Maria Loretha reflected that rural culture of seeds conservation and exchange still thrives among farmers of NTT, despite the odds they confronted in the face of agricultural modernization and development. UP CIDS Program on Alternative Development's (AltDev) field visits to the villages of Tanjung Bunga and Kawalelo, both in the capital town of Larantuka in East Flores Regency, revealed that the culture of farmer-to-farmer seed exchange still thrives as depicted in the diversity of sorghum varieties that has been conserved and protected by communities throughout the years. The current agricultural practice among the communities articulates the multi-functionality of seed exchange by surfacing the interdependency among the neighborhood of farmers and the value of social relationships where every farmer is associated with one another through this bond of culture, knowledge, and diversity (Grain 2007).

This whole experience of bringing back sorghum has been a journey of NTT farmers to achieve "biodiversity-based farming" which agricultural biodiversity expert Teshome (2006) described as the stage when farmers are doing their own seed selection, especially since seed is the linchpin of farming; they manage their

[1] NTT is a province in the eastern part of Indonesia and consists of more than 500 islands. Flores is one of the largest and dominant islands in NTT.

own cropping and are free to cultivate a mixed variety to encourage the continuous exchange of genes and maintain a dynamic system (Grain 2007).

6.2 Understanding Indonesian Agriculture and Food Policy Under Suharto's New Order

Given the vast dry zones and scarcity of rain, the natural landscape of the province of NTT is therefore not compatible for wet paddy cultivation that requires continuous water flooding. The International Rice Research Institute (IRRI), which was instrumental in the lowland rice propagation in Indonesia and elsewhere in Asia, explained that good water management for lowland rice means continuous flooding of the fields for seven up to ten days before harvest. This is to help ensure sufficient water and to control weeds from growing as these rice varieties are extremely sensitive to water shortage, owing to its semi-aquatic ancestry. On average, in order to produce 1 kg of lowland rice, 1432 L of water is needed. However, the amount of water needed depends on the soil type. Total seasonal water input to paddy fields varies from as little as 400 mm in heavy clay soil with shallow groundwater tables to more than 2000 mm in sandy or loamy soils with deep groundwater tables (IRRI n.d.). The data suggests that lowland rice varieties are water-intensive, compared to sorghum and the other traditional root crops of local farmers.

The loss and revival of sorghum in Flores underscores an era in the country that witnessed the abandonment of traditional crops and eventual neglect of farmers' welfare along with their agricultural knowledge and capacity. It is therefore imperative to have a thoughtful review of the historical dimension of agricultural development in Indonesia that shaped the fate of sorghum and its farmers. This devastating episode of Indonesian rural life eroded local seed varieties of traditional crops and resulted in the enduring marginalization of non-rice producing peasantry, whose majority remain impoverished in this eastern part of the country. Rural sociologist Sediono Tjondronegro (2013) argued that there is "hidden poverty" in NTT with 62.7% of the rural population categorized as poor, and with 26% of them classified as very poor. That same year, the World Food Programme reported that poverty level in NTT was 23.35%, but citing a 2011 data.

The New Order of the Suharto Regime, which started in the late 60 s until the Asian financial crisis in the late 90 s that triggered the call for his ouster and his eventual resignation, has intricate connections with the situation of Indonesian farmers in general. His regime constituted an important juncture in Indonesia's agriculture that indicated a pivot in the formulation of policies on rice production (Hansen 1972). A productivist approach on food policy was in place that stressed food security and food self-sufficiency as the foundation for economic development through the state's investment on agri-science, inputs, and farm infrastructure (Vel et al. 2016). Rice self-sufficiency and rice price stabilization were crucial drivers of Suharto's political stability. Collier, et al. (1982) posited that this national rice policy became

Fig. 6.1 Women farmers of Perhimpunan Petani Sorgum untuk Kedaulatan Pangan (*Source* Maria Loretha)

equivalent to national food strategy, and with that, there was but a scant attention to a general food strategy for all the islands (Fig. 6.1).

Rice farming concentration, together with the outpour of much-needed resources, was in the island of Java, geographically located in the western part and where wet paddy cultivation is most adaptable (Hansen 1972). But this meant devaluation of other traditional crops and therefore neglect to the outer islands including the Lesser Sundas where NTT belongs. Intensifying rice productivity was somehow perceived by the state power as a paramount solution to food insecurity. On the contrary, for dry land farmers in the eastern part who saw the disappearance of their other nutritious crops such as sorghum, dry paddy varieties, *jewawut* (millet), *jelai* (barley), and maize, it was a disaster. Instead of harnessing the productivity of these varieties, Suharto's government urged and supported farmers to only grow one kind: rice, which also dramatically intensified malnutrition issues in these dry zone communities (Ashoka 2013).

Known as the BIMAS (*Bimbingan Massal or* mass guidance) program, the strategy for rice intensification under Suharto was done by incentivizing rice farmers through provision of high-yielding lowland seed varieties, fertilizer, pesticides, credit programs for farmers, and domestic procurement through its state-owned enterprise, the Bureau of Logistics or BULOG (Hansen 1972; Sajogyo 1973; Timmer 1975; Vel, McCarthy, and Zen 2016; Tjondronegoro 2013). The role of BULOG then was to stabilize rice prices and, in the early part of its inception, included rice ration for those in the civil service and the military (Hansen 1972; Timmer 1975). Until the late 1970s, the public campaign for rice consumption was widespread (McKinley 2012).

Deliberately altering their way of life, including food intake habits, sorghum consumption was rather misconstrued as a symbol of poverty. Sorghum cultivation has not been as intensive as before and eventually reduced for the purpose of animal fodder (Mukkun et al. 2018; Susetyo 2017). Flores people who ate sorghum were consequently identified as those belonging to the poor and marginalized class who cannot afford to buy rice. Whereas on the other side of the spectrum, it was their traditional approach to food security and primary health care and a calculated response to rice insufficiency in the province. Maria Loretha explained that one of the challenges that the Flores rural society faced is a changing mindset (Susetyo 2017) cast in the mold of government food policies that favor rice and perpetuate the cultural shaming of eaters of the "inferior crop" as poor (ibid.). She added that over time, only the elders in the village have memory of sorghum. For the younger generation below 40, it is a foreign crop to them (McKinley 2012).

The condition of sorghum farmers also mirrors the lingering hardship of Indonesia's many rural populations. The saga of rural Java deserves attention here. In 1973, rural sociologist Sajogyo presented a seminal study about the modernization of rural Java but without development, which points out to government's very narrowly targeted wet-rice intensification, its failure to understand the rural agrarian structure and to enforce a comprehensive agrarian reform program.[2]

6.3 Pitfalls of the Green Revolution in Indonesia

Rice cultivation is millennia-old and a longstanding tradition way before the Dutch colonial times in Indonesia, when chemical fertilizers and pesticides were non-existent in agricultural production (Timmer 1975). While rice holds an important place in Indonesian cuisine and everyday diet, as it is increasingly preferred for consumption during this stage of agricultural modernization, it also has its share of drawbacks.

Agrarian scholar Ben White (2018) described that the abandonment of agrarian reform during Suharto's regime also coincided with the beginning of Indonesia's Green Revolution. The Indonesian rural community witnessed the intensive use of food crop fertilizer during this Green Revolution period which rose from nearly zero in 1959 to almost 200,000 tons in 1970 (Timmer 1975).

Sajogyo (1973, 9–11) discussed that the first wave of agricultural modernization in rural Java started during Sukarno's regime known as the Guided Democracy era. The "paddy center" program (1959–62) was initiated as a national policy for rice sufficiency. The state-owned enterprise PERTANI was established for an integrated

[2] Representing Bogor Agricultural University (now IPB University), Sajogyo wrote this paper as a contribution to the study on changes in agrarian structure organized by the United Nations Food and Agriculture Organization, 1972–1973.

program that delivers agricultural technologies to rice farmers. He described this period rather as a chemical fertilizer "revolution" imposed to Java's rice farmers which resulted to farmers' financial loss and inability to comply credit repayment, and therefore did not significantly contribute to the national rice stock. The second wave was the *BIMAS* rice intensification program from 1964–67, when the Bogor Agricultural University spearheaded the prototype field research project. This made way for the third wave, during the early years of Suharto's New Order regime (1968–70).

Significantly beginning this period, the rice production program integrated several elements of foreign relations influence, such as the introduction of new and imported high-yielding/IR[3]-rice varieties which are more responsive to the imported chemical inputs, and greater dependency to foreign capital from developed countries. This signaled the start of the government's foreign loans plan for Suharto's Five-Year Development Plan (Sajogyo 1973; Tjondronegoro 2013).

The BIMAS, however, was criticized as too much "planning from above" by economic planners and a "pushed down" strategy to the village "headmen" before it reaches to the farmers. Farmers felt to have "strait-jacketed" by BIMAS for having no chance to participate in the local planning (Sajogyo 1973, 11).

During this transition period, many farmers felt animosity over this government-imposed program and expressed this by refusing to plant the IRRI-developed high-yielding rice varieties and to utilize fertilizer and pesticides. This is mainly because several of these early high-yielding varieties were unsatisfactory to their farming productivity (Hansen 1972; Timmer 1975). Factors such as plant disease, insect infestation, consumer rejection, and the cost of the needed agrochemical inputs have discouraged farmers from using these rice varieties (Timmer 1975). However, this is not to discount administrative deficiency on the part of government and its top-down or coercive institutional intervention (Hansen 1972).

Hansen (1972) expounded that the bureaucratic distribution of pre-packed agrochemical input[4] reflected the government's lack of confidence on the farmers' capacity to perform their role as rational decision-makers. It reduced farmers' freedom of choice in the selection of fertilizer and pesticides compatible with their respective rice fields. This was even made worse by the introduction of aerial spraying of pesticides that were operated by foreign agrochemical corporations contracted by the government. The reliance on this external support seemed an answer to the administrative shortcomings. But in reality, it further deprived the peasantry of the opportunity to make guided decisions by transferring decision-making to the public bureaucracy and corporations that added an element of coercion. Farmers with contiguous plots were compelled to enroll in the spraying program.

---

[3] IR is the code for international rice.

[4] The author also termed this the 'packet formula' in which the government prescribed the kind and quantity of fertilizer and pesticides and distributed these to farmers in the form of a packet. It was intended that each farmer would receive a standard quantity of the agrochemicals.

Hansen (1972) further argued that while yield surged by late 1969, it was achieved through enormous political, economic, and environmental costs. There were subsequent complaints of the harmful effects of aerial spraying to humans and natural resources. Fiscal problems became a heavy burden as the rice repayment rate among farmers was abysmal, giving the government the difficulty to balance its ledgers.

The fourth wave: 1970–72, introduced the "improved" national *BIMAS* which saw the formation of BUUD (*Badan Unit Usaha Desa or* village-unit organizations) (Sajogyo 1973). But the year 1970 was the ultimate demise of this rice campaign. In the end, a weak rural bureaucracy was unable to compel farmers to follow government directives which are a rigid enforcement of an unpopular program (Hansen 1972).

Using a class differentiation approach (Sajogyo 1973; White 2018), only farmers of the upper class have benefited from the BIMAS program, who at that time, composed one-third of the rural population in Java. Indicative of which was their better credit repayment and closer ties to the village government bureaucracy. The majority, who were two-thirds of the rural farming population, needed to struggle for their subsistence. Sajogyo explained: "if farms are too small (<0.5 has) it is the group of lessees, renting-in or sharecropping in land, who are the weakest group, in danger of even not having enough rice to feed one's household" (1973, 25).

This terrain of poverty among rural Javanese depicts chronic land dispossession and indebtedness. After large farmers profited from high market prices of the high-yielding varieties, they bought lands from poor farmers who then were reduced to landless farm laborers (Tjondronegoro 2013). White (2018) explained that by 1981, citing a re-study[5] in nine Javanese villages he conducted with fellow agrarian scholar Gunawan Wiradi, the rise of a few agrarian elites juxtaposed the increasing landlessness of the many. This resulted in "about half of all households without ownership rights to rice fields, and about 40% without cultivation rights" (ibid.1119).

6.4 Continuing Dilemma of the Food Program

Rice price volatility that started to build up in the 70 s did not substantially provide positive gains to the rice farmers for the long term who remain lurching on compounding economic hardship. There was a failure of providing a true incentive for farmers to actively participate in the rice intensification program. This was because government top-down food policies stagnated rice productivity as reflected in its reliance on imports and global trade policies, depicting an era in Indonesia lacking rural-oriented policy (Timmer 1975). In particular, the 1973 procurement policy which was an attempt to improve rice procurement, instructed the BUUD/village units to purchase rice harvest from the farmers, process in BUUD's own facilities, and sell them to BULOG. But purchase was at a lower price resulting in widespread reluctance of farmers to sell here. With a scenario of soldiers deployed to

[5] See White, Ben, and Gunawan Wiradi. 1989. "Agrarian and non-Agrarian Bases of Inequality in Nine Javanese Villages.".

enforce the order, such procurement was rather done at gunpoint, quickly replacing the supposed farmers' incentives (ibid.).

By the early 1990s, national rice production could not anymore cope with the increasing consumption of the country's growing population (Mc Carthy et al. 2012). This was because the agriculture sector was no longer the springboard toward national development. Government priorities were the oil and gas sector, mining, forestry, and the manufacturing industry that attracted foreign investments (Tjondronegoro 2003).

This became prominent after an agreement with the International Monetary Fund to liberalize rice policy that led to the imposition of seasonal import restrictions that surged rice prices by almost 80% (McCulloch 2008, quoted in Vel et al. 2016), to the disadvantage of rice farmers who receive lower farm gate price and the consumers who buy rice at higher rates.

There are three main reasons that unceasingly beset rice production: low productivity, conversion of paddy fields to urbanization or more rewarding land use, and harvest failure as a result of climate change and bad irrigation infrastructure (Vel et al. 2016). Within this landscape of erratic rice production, the idea of reviving traditional crop cultivation was still not under the radar of Suharto's government. Instead, it developed a grand plan to establish a one million-hectare rice project in Central Kalimantan dubbed as the "mega-rice project." The project was seen as a failure and unsustainable, and thus it was later terminated (McCarthy et al. 2012).

6.5 Post-Reformasi and Beyond

Despite significant political changes after Suharto stepped down from power, the Post-*Reformasi* political paradigm of the ruling political parties was rather aimed at debt repayment, stimulation of investment, and economic growth. Agrarian reform was not central to improving the country's economy. To that effect, importation was the government's solution, making Indonesia the world's largest rice importer (Tjondronegoro 2003). The proceeding five (5) state leaders after Suharto had no concrete plans about agricultural lands for national food sufficiency (Tjondronegoro 2013). Instead, food policies passed by state leaders who succeeded Suharto were seen to have competing and mutually incompatible frameworks resulting in contradictory legislations. Indonesians witnessed the proliferation of large-scale agribusiness plantations[6] such as those of oil palm and sugar that rather serve the demand of the market and do not reflect an analysis of structural political-economic causes of food insecurity in the country (Vel et al. 2016).

Maintaining its ranking in world rice importation, the United States Department of Agriculture in 2012, dubbed Indonesia as the third largest rice producer, but also the world's seventh rice importer (ibid.).

[6] Authors cited the problems with the conversion of ricelands into oil palm plantations in North Sumatra and the replacement of subsistence agriculture to sugar cane plantations in Sumba, NTT.

Fig. 6.2 Sorghum fields in NTT, Indonesia (*Source* Maria Loretha)

6.6 Cross-Cutting Benefits of Sorghum Toward Strengthening the Local Food Movement

Giving up expectations from the government to detour and reconsider the propagation of traditional food crops, sorghum farmers decided to take the lead. The following are the positive outcomes of their initiative.

6.6.1 *Ecological Gains*

Taking pride in their contribution to biodiversity conservation, farmers explained that planting sorghum has cross-cutting effects. It is their approach to restoring the ecological system by transforming dry and barren lands into productive fields which benefit both themselves and the ecosystem. Across NTT, Perhempunan members are producing organically grown crops because, by default, their sorghum does not require chemical fertilizer. As a reciprocity of nature, sorghum productivity is by far profitable for them (Fig. 6.2).

6.6.2 *Health Benefits*

Stunted growth among children under five in most eastern provinces is above the World Health Organization's critical threshold of 40% (Ministry of Health Survey

2010, quoted in World Food Programme 2013). Alarmingly, the highest rate in Indonesia is in NTT that reached 58.4%. Moreover, the percentage of children under five who are underweight among children under five is 29.4%. These figures indicate that high micronutrient deficiencies are prevalent in NTT (World Food Programme 2013).

For farmer Bonifacius Kolah, the president of the local farmers group in Likotuden, sorghum has addressed the malnutrition of their children, as a result of daily consumption within the household. Sorghum, especially those of darker colors (red and black), has higher nutritional value. The phytochemicals found in sorghum show high antioxidants beneficial for human health. It also has anti-carcinogenic properties when consumed regularly in diet (Mukkun et al. 2018).

6.6.3 Creation of an Empowering Space for Peasant Women

Women farmers mostly composed the local organization Perhempunan. In a visit of UP CIDS AltDev to Tanjung Bunga and Kawalelo, both in East Flores Regency, sorghum farmers are predominantly women. Sorghum cultivation has given them opportunities to provide enough food for the family and additional income. It has also provided them space to meet other women for mutual support, share knowledge and resources such as seeds, talk about their lives, and more importantly, plan and decide about the sustainability of their farms, harvest, their future. It is a chance for them to break the barriers to social participation that is usually the domain of men in traditional peasant societies. With sorghum, these women hold the key to their self-determination and economic independence allowing themselves to end the dependence on agrochemicals and from the devastating impact of drought and a cycle of poverty (Global Greengrants Fund 2018). Kolah said that while he is at the top of the leadership, majority of the leaders are the mothers. The reason is that they are the most assertive in the community and are more committed to run their local organization. Decision-making has been collective. There are many times when the women's voices prevail since they are the majority. The influence of Maria Loretha, the pioneer in the sorghum movement, a woman, is also very strong.

6.6.4 Economic Benefits

Describing their hard work, farmers expressed that it takes almost forever, basking under a blazing sun during harvest season. But for them, a good harvest always compensates for the labor-intensive character of sorghum farming. They harvest sorghum three (3) times a year (Ashoka 2013) compared to wet paddy rice that can only be grown and reaped once during the rainy season. So far, there was no case of crop failure among the communities. With the support of the Catholic Diocese of

Larantuka,[7] these farmers manage an average of one-hectare sorghum farm. Each hectare produces an average of three (3) tons of grain. Their yield is primarily for household consumption while the excess is sold to the local market (Baran 2016).

Monocropping is not a practice of these small-scale farmers. Aside from sorghum, they also intercrop their plots with traditional crops earlier mentioned in order to diversify their produce, complement their diet, and ensure buffer stocks for domestic consumption and income.

In the rock-strewn, semi-arid hilly landscape of Kawalelo in Larantuka that reaches out to the sea and the twin volcanoes of Lewotobi in a short distance, farmers have expanded their farm project with a target of transforming their village into a sorghum pilot area. Post-harvest facilities for sorghum have gradually been set up.

With a significant sorghum surplus from the community's daily consumption, Maria Loretha said that they have started selling in small scale and at a fair price to the neighboring islands such as Bali. But for the long term, Maria Loretha said they aspire to contribute to ensuring food security in the region by selling sorghum to Southeast Asian countries as a statement to the world that among the people in NTT, it is their basic food.

Along with Maria Loretha and Kolah is the collective leadership of farmers Yosep Mukin of Adobera in Tanjung Bunga in East Flores, Vibronia Peni of Wuakerong in Lembata, Lamanepa Elo of Sikka, Benediktus Panbur of West Manggarai, Alfin of Rote, and Ahmad Boro of Pledo in Adonara Island. With their commitment to take the lead in sorghum propagation, the mission to ensure safe and healthy food for their local communities through sorghum has been achieved.

References

Baran, Melkhior Koli. 2016. "Indonesian diocese points way to beat the drought." *UCA News*, June 21. http://www.ucanews.com/news/indonesian-diocese-points-way-to-beat-the-drought/76363.

Collier, William L., Gunawan Wiradi Soentoro, Effendi Pasandaran, Kabul Santoso, and Joseph F. Stepanek. 1982. Acceleration of rural development of java. *Bulletin of Indonesian Economic Studies* 18 (3): 84–101. https://doi.org/10.1080/00074918212331334230.

Global Greengrants Fund. 2018. "Seeds of Change: Sorghum in Indonesia." *Global Greengrants Fund*, November 9. https://www.greengrants.org/2018/11/09/SORGHUM/.

GRAIN. 2007. "Sorghum: A Crop to Feed the World or to Profit the Industry?" Grain, April 2. https://www.grain.org/en/article/558-sorghum-a-crop-to-feed-the-world-or-to-profit-the-industry.

Hansen, Gary E. 1972. "Indonesia's Green Revolution: The Abandonment of a Non-Market Strategy toward Change." *Asian Survey* 12(11): 932–46. https://doi.org/10.2307/2643114.

Hariprasanna, K., and J.V. Patil. 2015. Sorghum: Origin, Classification, Biology and Improvement. In *Sorghum Molecular Breeding*, ed. R. Madhusudhana, P. Rajendrakumar, and J. Patil, 3–20. New Delhi: Springer.

IRRI (International Rice Research Institute). n.d. "How to Manage Water." *Rice Knowledge Bank, International Rice Research Institute*. http://www.knowledgebank.irri.org/step-by-step-production/growth/water-management.

[7] The diocese's support was through the Yaspensel Foundation, its social development arm.

Loretha, Maria. 2018. Interview by Ananeza P. Aban and Angeli Fleur Nuque. *Documentation of Alternative Practices and Field Visits in Nusa Tenggara Timur, Indonesia* (June).

McCarthy, John F., Jacqueline A.C.. Vel, and Suraya Afiff. 2012. Trajectories of land acquisition and enclosure: Development Schemes, virtual land grabs, and green acquisitions in Indonesia's outer Islands. *The Journal of Peasant Studies* 39 (2): 521–549. https://doi.org/10.1080/03066150.2012.671768.

McKinley, Lara. 2012. "Maria Loretha and the Sorghum solution: Part 2." Oxfam Australia, October 1. https://www.oxfam.org.au/2012/10/maria-loretha-and-the-sorghum-solution-part-2/.

Mukkun, L., H.J.D. Lalel, N. Richana, M.B. Pabendon, and S.R. Kleden. 2018. The diversity of local Sorghum (*Sorghum bicolor* L. Moench) in Nusa Tenggara Timur Province. *IOP Conference Series: Earth and Environmental Science* 144: 1–7. https://doi.org/10.1088/1755-1315/144/1/012065.

Pellokila, Ch., Simon Field, and E. O. Momuat. 1991. "Food crops development in NTT." In *Nusa Tenggara Timur: The Challenges of Development*, eds. Colin Barlow, Alex Bellis, and Kate Andrews, 121–144. Canberra: Department of Political and Social Change, Research School of Pacific Studies, Australian National University.

Ashoka Philippines. 2013. "Maria Loretha." *Ashoka Philippines*. http://www.ashoka.org/en/fellow/maria-loretha.

Sajogyo. 1973. "Modernization without development in rural Java." Paper contributed to the Food and Agriculture Organization. Bogor Agricultural University. Indonesia.

Susetyo, Grace. 2017. "Maria Loretha: Championing Sorghum, Flores' Crop of Hope." *Indonesia Expat,* April 26. https://indonesiaexpat.biz/travel/maria-loretha-championing-sorghum-floress-crop-hope/.

Teshome, Awegechew, J. Kenneth Torrance, Bernard Baum, Lenore Fahrig, D.H. Lambert, and J. Thor Arnason. 1999. "Traditional farmers' knowledge of Sorghum (*Sorghum bicolor* [Poaceae]) Landrace Storability in Ethiopia." *Economic Botany* 53(1): 69–78.https://doi.org/10.1007/BF02860794

Timmer, C. Peter. 1975. "The political economy of rice in Asia: Indonesia." *Food Research Institute Studies* 14(3): 197–231. https://doi.org/10.22004/ag.econ.135509.

Tjondronegoro, Sediono M.P.. 2013. "An agricultural development legacy unrealised by five presidents, 1960–2014. *Jurnal Masyarakat Indonesia* 39 (2): 379–395.

Tjondronegoro, Sediono M.P. 2003. "Land policies in Indonesia." Working Paper. World Bank and Bogor Agricultural Institute. https://documents.worldbank.org/curated/en/777651468260378738/pdf/%20374350IND0Land0policies1PUBLIC1.pdf.

Vel, Jacqueline A. C., John F. McCarthy, and Zahari Zen. 2016. The conflicted nature of food security policy: Balancing rice, sugar and palm oil in Indonesia. *Anthropological Forum* 26 (3): 233–247. https://doi.org/10.1080/00664677.2016.1190919.

White, Ben. 2018. Marx and Chayanov at the margins: Understanding Agrarian change in Java. *The Journal of Peasant Studies* 45 (5–6): 1108–1126. https://doi.org/10.1080/03066150.2017.1419191.

World Food Programme. 2013. "WFP Indonesia: East Nusa Tenggara (NTT) Profile." *World Food Programme Food Security and Vulnerability Atlas Indonesia*. January 2013. http://wfp.org/countries/Indonesia.

Ananeza P. Aban is currently a Ph.D. researcher of the Department of Conflict and Development Studies and a recipient of the Special Research Fund doctoral scholarship of Ghent University in Belgium. She also serves as a senior research analyst of UP Center for Integrative and Development Studies Program on Alternative Development. Among her major tasks are coordinating the field research across Southeast Asia, co-editing the Alternative Practices monograph series (2020, 2023), and ensuring the international linkages of the program. She finished her Master of Community Development degree from the University of the Philippines in Diliman.

Chapter 7
Alternative Land Practice in Thailand: A Study on the Southern Peasants' Federation of Thailand

Supatsak Pobsuk

Abstract This study examines the land struggles of the Southern Peasants' Federation of Thailand (SPFT) in Surat Thani province, Thailand, highlighting the counterpoint with the country's centralized land management system, which has led to unequal land distribution and subsequent conflicts. The author introduces the concept of community rights as a viable alternative to strengthen the land rights movement in Thailand, particularly for the SPFT. The narrative includes the history of the SPFT's activism and also details the SPFT's alternative strategies to challenge the centralized system and resist capitalist pressures. These strategies encompass innovative land governance and management practices, community economic initiatives, grassroots democracy, networking efforts, the establishment of community constitutions, the promotion of local cultures, and women empowerment. By recognizing land as a fundamental right for peasants, the SPFT successfully utilizes it as a means of production for sustainable livelihoods, thereby reinforcing it as a safety net and social capital for local communities.

Keywords Community rights · Land governance · Land struggles · Southern Peasants' Federation of Thailand · Sustainable livelihoods

7.1 Introduction

Fight for land.
Fight for right
Fight for liberty
Fight for equality
Fight for justice
Revolution for the Democracy
Motto printed on the Southern Peasants' Federation of Thailand's t-shirt

S. Pobsuk (✉)
Focus on the Global South, Bangkok, Thailand
e-mail: supatsakp@gmail.com

J. M. C. Sy and A. P. Aban (eds.), *Reimagining Development in Southeast Asia*, Asia in Transition 21, https://doi.org/10.1007/978-981-97-9140-8_7

Land conflicts between the Thai state and the people, especially peasants, in Thailand have lasted over time since the period of modernization. Land contestation in Thailand is rooted in a land management system unilaterally centralized by the state and the dominant development discourse which aims at economic growth underpinned by neo-liberal capitalism (Hall et al. 2011; Thai Land Reform Network and Local Act 2010). In this context, land tenure has been classified into two forms, consisting of state-owned and private lands (Thai Land Reform Network and Local Act 2010; Aphron 2014) through different types of land titles (Lamb et al. 2015; Lubanski 2012). In this sense, state-owned land is allocated for national resources, e.g., conservation forest, national parks, wildlife sanctuary, infrastructure development projects, economic land concession, etc. Moreover, private land titles allow land to be commodified in a market economy.

This study aims to explore alternative economic, political, social, and cultural practices in relation to land management led by local communities. The study starts by contextualizing land governance in Thailand which results in inequitable land distribution, i.e., land concentration and eventually land conflicts. It discusses further the concept of community rights as an alternative development underpinning land right movements in Thailand, which includes the SPFT. All in all, this study illustrates how local communities employ alternative practices as strategic tools to countervail a centralized land management in Thailand. Those practices show how the local people have become empowered to manage their resources and to initiate participatory development and democracy.

7.2 Land Governance in Thailand

Historically, the Thai state had centralized land management and ownership since it was reigned by an absolute monarchy. During that period, monarchs, lords of principalities, and nobles appropriated taxes, in-kind and labors, from land slaves and subjects (Vandergeest and Peluso 1995). Even though the Thai state has attempted land reform numerous times, lands have not been distributed to people equally. The first attempt of land reform in Thailand was by the People's Party in 1933 (B.E. 2576) after the Siam Revolution through unifying lands and redistributing them equally to people; however, such reform was resisted by the elites who amassed the vast majority of land plots (Ariyaviriyanant 2009). In 1954 (B.E. 2497), the Thai state issued the Land Code, which established the land legislation in Thailand, including the issuance of land title deeds and utilization certificates. Particularly, it determined that one limited holding a maximum 50 *rai* (approximately 8 hectares) of land to prevent land accumulation and speculation. In addition, the code stirred people to use their land productively; otherwise it would be turned to state land if it were left barren for some consecutive years. However, the limitation of land holding was revoked by the announcement of the National Executive Council No. 49 in 1959 (B.E. 2502), as it stated that the limitation hindered the prosperity of the national economy (Ariyaviriyanant 2009; Lubanski 2012; Lamb et al. 2015). Arguably, the limitation of land accumulation has been against the principle of market economy.

In terms of forestland management, the Forest Act 1941 (B.E. 2484) defines forest as "land which is not acquired by any persons under the Land Code" (Section 4). In other words, the Thai state owns all land which has not been identified as belonging with formal documents to any individuals. However, the National Reserved Forests Act 1964 (B.E. 2507) allows business sectors to use degraded forest areas on economic concession, for a maximum of 30 years (Section 16). In 1966 (B.E. 2509), the Thai state also granted private sectors legal logging concessions for economic uses which led to the mass destruction of forest areas in Thailand. There has been a decrease of 52 percent of forest areas since 1961 (B.E. 2504) and thus, the concessions were eventually banned in 1989 (B.E. 2532) (Seub Nakhasathien Foundation 2017). In addition, to respond to the global demand of industrial crops, the Thai state facilitates private sectors to grow cash crops such as cassava, rubber, oil palm, and eucalyptus on state- owned lands, especially degraded forest areas for export-orientated agricultural products. As a result, land prices have increased and lands have been grabbed and transformed to monoculture plantations (Hall 2011).

After the people uprising in 1973 (B.E. 2516), the farmers' mobilization between 1973 and 1976 (B.E. 2516–2519) successfully propelled the Government to issue the Agricultural Land Reform Act in 1975 (B.E. 2518) and established the Agricultural Land Reform Office (ALRO). The act and the roles of the ALRO are to redistribute unused land through expropriating and purchasing from private landowners or allocating state-own land to landless and insufficient land households for agricultural purposes; the allocated land cannot be sold or leased to others. The act also set out that one should not have land over 50 *rai* or eight hectares (Lubanski 2012). However, it is arguable that the act and the roles of the ALRO have been ineffective to expropriate land from private landowners, but the office has bought land from those who have been willing to sell. Most of the land has been alloted from state-owned land, especially degraded forest (Thai Land Reform Network and Local Act 2010). Even though land from the ALRO, also known as Sor Por Kor (SPK), is issued as an individual land title, many studies discovered that land titles have been sold from entitled recipients to land speculators and investors for land accumulation through the corrupted system, despite the fact that it is prohibited by laws (Bello et al. 1998; Thai Land Reform Network and Local Act 2010).

The Thai state formalizes land tenure through land titling processes (see Table 7.1 for land titles and uses in Thailand). The prominent land titling program in Thailand was launched by the World Bank with loans amounting to USD 183.1 million from 1984 to 2004 (B.E. 2527–2547) to support Thailand's Department of Land in issuing land titles to landholders by using cadastral surveying and mapping technology in the issuance of land titles. Moreover, it intended to create land tenure security as a means of reducing poverty, increasing the capacity of individuals, especially agrarians, to access finances by using land as collateral, and increasing land productivity through an efficient drive of a free market (Leonard and Narintarakul Na Ayutthaya 2003; Bowman 2004; Lubanski 2012). By the time of its completion, the project had issued approximately 13 million titles to landowners in Thailand (Bowman 2004). Arguably,

Table 7.1 Types of land titles and uses in Thailand

Land title	Description	Legislation/Agency	Note
Sor Kor 1 (SK1)/Sor Kor 2 (SK2)	Land claim certificate	1954 Land Code/ Department of Land (DOL)	Right to farm in the forest, need proof of pre-1954 land use
Nor Sor 2 (NS2)	Pre-emptive rights certificate	1954 Land Code/ Department of Land (DOL)	Granted on condition of use, reservation license and only transferable by inheritance
Nor Sor 3 (NS3)	Certificate of Utilization	1954 Land Code/ Department of Land (DOL)	Granted on condition of use, can be used as collateral and saleable after 30 days of a public notice
Nor Sor 4 (NS4) (Chanote)	Fully land ownership certificate	1954 Land Code/ Department of Land (DOL)	Full title, private ownership and transferable; saleable
Sor Por Kor (SPK)	Agricultural usage certificate	1975 Agricultural Land Reform Act/ Agricultural Land Reform Office (ALRO)	Distributed to landless/ poor agrarians for agricultural use and non-transferable
Sor Tor Kor (STK)	5-year agricultural usage certificate	1985 National Reserved Forest Act/ Royal Forest Department (RFD)	Issued for agricultural use in degraded forest land, limited to 15 Rai and non-transferable but by inheritance
Community land title	Land use document for communal land	2010 Regulation of the Prime Minister's Office/Community Land Title Office (CLTO) under the Prime Minister's Office	Issued to a community entity for using and managing state land; non-saleable but can be returned to community entity for further management
Community forestry	Land use document in forest land	2019 Community Forest Act/Royal Forest Department (RFD)	Allows communities to manage communal forest land and use natural resources, particularly degraded forest and non-saleable
Community Land Title	Land use documents for communal land	2010 Regulation of the Prime Minister's Office on/Community Land Title Office (CLTO) under the Prime Minister's Office	Issue to a community entity for using and managing state lands, non-saleable but return to community entity for further management

there have been three critical landmarks in terms of land governance in Thailand. First, the Thai state has simplified equivocal and unorganized parcels into formal categories of land which the state can properly control and utilize through taxation according to imposed rules and regulations. In other words, land and landowners have been governed under invented statecraft binding to the nation-state system (Scott 1998; Leonard and Narintarakul Na Ayutthaya 2003). Second, the titling program can be seen as the market-led land governance (Chiengkul 2015) in which an individual land title fully enables a landowner to sell a land as well as use land as a collateral, as needed, in a market economy (Leonard and Narintarakul Na Ayutthaya 2003; Thai Land Reform Network and Local Act 2010; Hall et al. 2011). Third, the land titling program has disregarded customary land tenure at the community level such as communal land (Leonard and Narintarakul Na Ayutthaya 2003).

Nonetheless, many studies pointed out that there was a corrupted system in issuing land title deeds, Sor Por Kor 4-01 as well as leasing and using state-own lands. In addition, the issuance of land title deeds has resulted in the concentration of land, because a small number of affluent Thais who are able to gain information and capitals as well as have connections to authorities have more purchasing power in easily accumulating and concentrating land for speculation. In addition, the laws have not restricted landowners to hold many land title deeds, which have engendered the accumulation of land in Thailand. As a result, the price of land has escalated over time in a land market which hinders landless and small landholder peasants from accessing land as a means of production. According to a study on farmer registration and census data in 2003, 2013, and 2018 (Attavanich et al. 2018), more than 50% of 5.76 million households have agricultural land less than 10 *rai* (1.6 hectares) per household. The study also showed that the scale of agricultural landholders has decreased over time.

7.3 Concepts

The SPFT has employed the concept of community rights as an alternative development whereby local communities are the main actors in creating their own development pathway. The SPFT has proposed an alternative land management model to demand the decentralization of land management from the state to the local people. It is an attempt of local people to participate in the development processes and share power from the central state. Ultimately, it is a counter-discourse which emphasizes political, social, cultural, environmental, and ethical issues in relation to the well-being of people, rather than focusing on economic growth (Ospina and Jimenez 2017). All in all, alternative development has primarily put emphasis on development from below, self-determination, and people-centered development which ultimately demonstrates how people participate in governance through participatory democracy.

Land reform movements in Thailand have used the idea of community rights, which have promoted collective land and natural resources management. The idea of self-determination is also the key idea where local people demand the right to livelihood and community rights to common properties, including land, forest, and water from the government and propose the community legitimation against mainstream development paradigm, focusing on economic growth. Local communities have employed many strategies and tactics to negotiate with the government for them to be involved in decision-making processes of national development. Those strategies and tactics include public demonstration, civil disobedience, petition, and negotiation.

Arguably, community rights have two dimensions, including community cultures and the rights of communities to participate in governance (Rakyutidham 2011). The former demonstrates the ideal rural societies where people live in a harmonious society and help each other to develop their community. The latter has called for the enhancement of people's power to negotiate the power of state and capital. In this sense, local people need to share their grievances and ideas to create a sense of collectivism and solidarity. Collective ideologies and actions have been initiated and developed through community constitution and regulations to instill such ideas into local people. In the case of the SPFT, this process is needed to forge new settlers into a newly established community.

In terms of land management, the idea of complexity of rights proposes that there are overlapping forms of rights established in one area, between community and individual rights (Ganjanapan 2001, cited in Chiengkul 2015). The Community Land Title (CLT) is introduced to people thereby community members own land collectively while they have the right to farm on their plot of land. In addition, community members are able to access and utilize communal land through sharing labor to make use of the land in the most productive way.

7.4 Methodology

This study was conducted as a qualitative research in four SPFT's communities, including Klong Sai Pattana, Nam Deang Pattana, Perm Sub, and Kao Mai Communities located in Chai Buri District, Surat Thani Province on October 1–5, 2018. The study applies various methods, including documentary research, focus group discussion, in depth interviews with key informants, and non-participation observation. The data collection Tables 7.2 and 7.3 below show the methodologies and sample sizes.

Table 7.2 Data collection table for focus group discussions

Date	Community	Male	Female
03–04 October 2018	Klong Sai Pattana	2	3
04 October 2018	Nam DaengPattana	4	6
05 October 2018	Kao Mai	2	3
05 October 2018	Perm Sub	5	3
Total		13	15

Table 7.3 Data collection table for key informant interviews

Date	Position
01–02 October 2018	The SPFT coordinator
04 October 2018	A representative of women group in Santi Pattana community
03–04 October 2018	SPFT secretariats

7.5 The Case of the Southern Peasants' Federation of Thailand

In contemporary history, the struggle for land in the south of Thailand started between 2001 and 2002 (B.E. 2544–2545), when a people's organization network in Surat Thani province became involved in working with the Government to analyze the encroachment of people in forest lands. They discovered that landless peasants and workers, including those who lost their lands during the period of crop booms and those who had limited land for farming, encroached and cultivated in forestland.

Moreover, this opportunity brought the people's organization network to access information related to the land concessions by private companies in the South, based on their right as stated in the Official Information Act 1997 B.E. 2540). Situated in the South of Thailand, Surat Thani Province is 529 km (328 miles) from Bangkok, the capital city of Thailand. In the province, Thai and foreign companies had concessions to use 200,000 *rai* (approximately 32,000 hectares) of forest gardens and reserved forest lands. Of those areas, the concessions of 60,043 *rai* (approximately 9,606.88 hectares) were already expired. In the case of palm oil companies, they expanded cultivated land over land concessions they got and also continued to cultivate and harvest palm oil, despite expired concessions (The Southern Poor People Network 2006). It is arguable that the Government did not enforce stringent law, rather than favoring a few Thai and foreign palm oil companies to expand oil palm plantations by predominately accumulating state-owned land for export-oriented agriculture production during the period of crop booms (Hall 2011). According to the Southern Poor People Network (2006), those palm oil companies owned cultivated lands between 1000 and 40,000 *rai* (approximately 160 to 6400 hectares). The process has excluded landless peasants and workers as well as small-scale landholders from accessing land as a means of production. As a manifestation of Marx's primitive accumulation (2003), peasants have been deprived of means of production by commodifying and privatizing land. Moreover, the process of land concession neglects collective and customary rights on natural resources and environmental commons (Harvey 2003).

Consequently, a land rights movement named the Southern Poor People Network (SPPN) emerged to demand governmental agencies to investigate expired land concessions of palm oil companies in 2003.

Later, the network worked with governmental agencies, especially the Ministry of Natural Resources and Environment to respond to land conflicts in the South. The SPPN investigated and found that more than 70,000 *rai* (approximately 11,200 hectares) of land concession for palm oil plantations already expired. During that time, government officials and local authorities responded passively to the issues. The network decided to occupy land to demand the Government to expropriate land from those companies and redistribute it to landless peasants and workers. Even though the network tried to negotiate with the Government through the National Human Rights Commission of Thailand, government armed officials cracked down on groups of people under the SPPN in October and December 2003 (The Southern Poor People Network 2006).

The Southern Peasants' Federation of Thailand (SPFT) was established in 2008 (B.E. 2551) after the first ripple of land occupation movement was ended in 2003 by the aggressive violence committed by the Government. The SPFT reviewed the lessons learned from the past experiences so as to visualize the problems and challenges and eventually re-established a land rights movement in Surat Thani province. Based on lessons learned, the SPFT discovered that the previous movement primarily prioritized temporary benefits, i.e., land occupation and agriculture products on land, rather than achieving a long-term goal, i.e., the right to land for landless peasants and workers. The SPFT analyzed that the movement did not have a collective ideology and goal and lacked consciousness of the land rights struggle. More importantly, the movement lacked mutual disciplines and regulations (The SPFT Coordinator, pers. comm., October 1, 2018).

Since 2008 (B.E. 2551), the SPFT has gathered landless peasants and workers in Surat Thani province who have seen land as a means of production and as the foundation of life. This group has been excluded by the centralized land governance and the socio-economic development in Thailand. Most of them are agriculture workers, especially in rubber and oil palm plantation while some of them formerly owned and farmed land but sold it to others. Many members worked in urban areas and migrated to seek economic opportunities in Surat Thani Province (Aphorn 2014). The SPFT community members realized that land is essential for their subsistence and livelihoods. For them, the land is their social safety net. Thus, they have joined the effort of the SPFT in struggling for land rights.

In accordance with the lessons learned, the SPFT constructed solid rationales, with strong ideologies, goals and an organizational structure. Namely, its rationales consist of three critical rights of landless peasants and workers: (1) the right to land reform for a just society; (2) the community rights to land and natural resources management; and (3) the right of landless peasants and workers to new community settlements. The SPFT coordinator expressed that all rationales have been inspired by land and peasant movements in Thailand and the international communities, especially the Farmers' Federation of Thailand (FFT) during 1974 and 1976 (B.E. 2517 and 2519) and the Landless Workers Movement in Brazil, also known as the *Movimento dos Trabalhadores Rurais Sem Terra* (MST). In addition, the concept of community rights that has risen among social movements in Thailand during the 1980s, legitimized by the 1997 Thailand constitution, supports the SPFT's struggles on land rights (The SPFT Coordinator, pers. comm., October 1, 2018).

Moreover, the SPFT has clear goals in compliance with its rationales, including land rights, peasant rights, and democracy. For its short-term goal, the SPFT aims

Fig. 7.1 Official logo of the Southern Peasants' Federation of Thailand (SPFT)

to achieve equitable land distribution through land reform. This is to ensure landless peasants' and workers' access to adequate housing and food security. For its mid-term goal, the SPFT aims at building food sovereignty through communities' rights to land and essential natural resources management, agricultural reform and agroecology system… For its long-term goal, the SPFT aims at establishing a democratic society and creating new peasant cultures through solidarity and equality (SPFT 2016).

For the strategies and approaches, the SPFT has promoted the community land title (CLT) campaign by advocating the Governments of granting peasants the communal land management. In order to achieve land reform, the SPFT has further advocated three relevant measures, including: (1) the legislation of a progressive land tax; (2) the establishment of a national land bank; and (3) the protection of agricultural lands for peasants. These approaches will be discussed in the next section. In addition, the SPFT is a member organization in the People's Movement for a Just Society (P-Move),[1] which has advocated community land and natural resources management for marginalized people in Thailand at the national level.

[1] The People's Movement for a Just Society (P-Move) is a local network organized by local organizations, such as Northern Peasants Foundation (NPF), Isaan Land Reform Network (ILRN), Four Regions Slum Network (FRSN), Bantad Mountain Land Reform Network (BMLRN) and Southern Peasants' Federation of Thailand (SPFT), to raise grievances of marginalized people across the

In terms of the organizational structure, the SPFT has established internal structure and responsible persons elected by members of networking communities, including Political, Academic, Women, Agriculture and Cooperative, Campaign, Cultural, Media and Information sectors, and Secretariats (The SPFT Coordinator, pers. comm., October 1, 2018). Moreover, the SPFT has established organizational regulations through its general assembly which are guidelines ways for SPFT community members to pursue. Overall, such internal regulations are based on the concepts of democracy, equality, justice, solidarity, accountability, transparency, and communal tenure and benefits (SPFT 2016). These are the community constitution that community members conform to. The organizational committees and community committees are elected through votes by means of democracy to prevent the emergence of a dominant leader inclined to corruption.

7.6 Community Profiles

7.6.1 Land Reform, Liberty, Human Rights, Democracy, and Justice—The SPFT Motto

As of October 2018, there are five communities in Surat Thani Province networking under the SPFT umbrella, including the Klong Sai Pattana, Nam Deang Pattana, Perm Sub, and Kao Mai communities located in Chai Buri District and the Santi Pattana community located in Phra Saeng District. These communities are newly settled communities where landless peasants and workers have gathered to seek for land as a means of production and aimed to pursue land and natural resource management in compliance to communal tenure underpinned by the concept of CLT. These communities occupied land owned by state agencies, such as ALRO and RFD as well as expired land concessions of private sectors, such as palm oil companies.

According to the interviews with community members, they started assembling and occupying land between 2006 and 2013 after the first attempt of land occupation in the early 2000s. They assisted government agencies to investigate expired and illegal concessions on oil palm plantations invested by both international and national investors. Even though the SPFT and its community members successfully assisted the government agencies in issuing lawsuits toward palm oil companies, the community members have not been permitted to legally settle in those lands, instead of temporarily staying. It has shown the insincerity of government agencies in tackling land conflicts and inequitable land distribution.

It is noticeable that they have employed land occupation to claim the community rights in lands for peasants. A community member in Kao Mai explained, "We

country and advocate the Government for more equitable distribution and management of natural resources (e.g., land, water, forest) (Lubanski 2012).

occupied land to put pressure on the government since the government lacked motivation in land reform and equitable land distribution. We had used formal ways by coordinating with the government to investigate expired land concessions of palm oil companies but there was no progress in expropriation" (A Kao Mai community member, pers. comm., October 4, 2018). However, community members who have occupied land are seen by the government, palm oil companies, and neighboring communities as squatters violating the law.

During the land occupation and settlement, community members have encountered different types of violence, including assassination, forced eviction, arbitrary arrest and detention, destruction of properties and crops, intimidation, and judicial harassment. The international and national news have covered the incidents; all interviewed communities expressed that they have been threatened by unidentified armed groups which the community members believe that those are connected to palm oil companies and local influential groups to some extent. Apart from incidents of intimidation such as random gunfire into the communities at night, destruction of community houses and crops, and trespasses of local influential groups with soldiers, four assassinations occurred with members of the Klong Sai Pattana community. In the four incidents, four community members were killed.[2]

These incidents were reported nationally and internationally by media, and the cases have been taken into account as the assassination of land rights defenders.[3] Unfortunately, only two cases, those of Chai Bunthonglek and Supot Kalasong, were given justice. However, it has appeared that none of the perpetrators have been sentenced because all legal cases were acquitted, as the court claimed, because of insufficient evidence (Fortify Rights 2018). To prevent loss of community members, security measures have been employed by setting up checkpoints around communities. For example, there are four security posts located at Klong Sai Pattana where

---

[2] Incidents related to community members of the Klong Sai community from 2010 to 2016 include the following:

1. January 11, 2010: Mr. Somporn Pattaphum was shot dead during dinner at his home.

2. November 19, 2012: Ms. Montha Chukaew and Ms. Pranee Boonrat were shot and killed while they were on their way to a local market. Their bodies were mutilated by gunmen to intimidate other community members.

3. February 11, 2015: Mr. Chai Bunthonglek was shot dead by two gunmen when he was visiting his relative's house.

4. April 8, 2016: A gunman attempted to shoot Mr. Supot Kalasong when he was driving to the community. In his case, he was rescued by other community members and took him to a hospital immediately.

[3] For news related to the violent incidents, see Alisa Tang, "After Violent Struggle for Land, Thai Campaigners Face Challenge to Community Farming," *Reuters*, July 26, 2016, https://www.reuters.com/article/us-thailand-landrights/after-violent-struggle-for-land-thai-campaigners-face-challenge-to-community-farming-idUSKCN10603B; Alisa Tang, "Thai Land Rights Campaigners' Most Powerful Weapon? Access to Information," *Reuters*, July 14, 2016, https://www.reuters.com/article/us-thailand-landrights-idUSKCN0ZU00U; Alisa Tang, "Four Murders and a Bloody Battle for Land Rights in Thailand," *Reuters*, July 26, 2016, https://www.reuters.com/article/us-thailand-landrights-idUSKCN0Z9006; and *Al Jazeera*, "Harassed by Palm Oil Company, Thai Village Defends Land," August 9, 2017, https:// www. aljazeera.com/indepth/features/2017/06/harassed-palm-oil-company

all community members are shifted to be community security guards every 12 h to monitor security matters 24 h a day, thus ensuring that community members are safe. The community limits one community entrance and exit, and the gate is opened at 6:00 a.m. and closed at 6:00 p.m., except for an emergency situation. To commemorate the struggle of land rights defenders, the SPFT built a memorial monument situated in Klong Sai Pattana community. This monument includes a white foundation striking with a coil spring represented as a continual evolution and three stars on top: a red one which stands for struggle; green for prosperity of peasants, and; yellow for morality (Fig. 7.2).

In addition to these violent incidents, judicial harassment has been employed by palm oil companies to prosecute community members who have been charged with three criminal offenses, including the offenses of trespass, mischief, and criminal

Fig. 7.2 The SPFT built a memorial monument situated in Klong Sai Pattana community to commemorate the plight of land rights defenders (Photo courtesy of Supatsak Pobsuk)

association by palm oil companies. For example, 15 members in Nam Deang Pattana community have been charged with three criminal offenses (Protection International 2017). Even though all communities under the umbrella of P-Move are mediating under land conflicts with the government agencies, criminal cases have been ongoing against the community members. While the SPFT requested a justice fund from the Ministry of Justice to assist community members in legal fees, community members expressed that the Justice Fund Office has not responded to the needs of local people timely. On the contrary, community members explained, "We have never accessed justice since we have settled here. When our community was threatened by outsiders affiliated to palm oil companies, even the local police have never taken our cases" (A Nam Deang community member, pers. comm., October 4, 2018).

Amidst the intense conflicts in SPFT communities, the SPFT approached governmental agencies, such as the National Human Rights Commission of Thailand (NHRCT) to file complaints on human rights-related issues and the Community Organisations Development Institute (CODI) under the Ministry of Social Development and Human Security (MSDHS) to request assistance for community development. When the communities encountered violence and intimidation, the SPFT, as a representative, approached different human rights organizations at the national and the international level to petition and voice their grievances. Not only have news related to their land struggle been published in Thailand, but the issues have also spread at the international level through international media such as Reuters, Aljazeera, etc. The SPFT usually presents itself in the public sphere to raise awareness regarding the land rights of landless peasants and workers. For example, the SPFT, in collaboration with a university in Surat Thani province, organized a public forum in relation to land struggle in the south of Thailand which gained attention from students and people. The SPFT believed that a public event is a good opportunity to share their struggles, ideas, and concepts regarding communal land and natural resources management (Fig. 7.3).

7.7 The Federation's Alternative Practices

> More than land and solidarity of peasants' community is food sovereignty and human dignity
> — The SPFT motto

The case study on the SPFT shows an alternative development approach by applying community rights to land and natural resources as a countervailing approach and practice to the dominant development discourse. This section discusses the SPFT's alternative approach in relation to economic, political, social, and cultural practices.

Fig. 7.3 The Southern Peasant's Federation of Thailand collaborated with a university in Surat Thani province to organize a public event disseminating information about people's struggles for land (Photo courtesy of Supatsak Pobsuk)

7.7.1 Alternative Land Practices

The SPFT intertwines their struggle for land and agroecology with a critical change in the top- down model of centralized development in Thailand. The SPFT communities have seen land as a fundamental right of peasants. As such, land serves community members as a means of production, i.e., a foundation of community security and a source of livelihoods and food. In addition, the right to land and agriculture can be seen as a safety net and social capital so community members ensure that land and natural resources will be maintained for their descendants. The SPFT has managed their land in compliance with the concept of community land title (CLT). The communal tenure is the idea underpinning the concept. It is where all community members own land collectively, which is beyond the existing forms of land tenure in Thailand, namely state-own land and private tenure. It is arguable that the CLT is a concrete product of the concept of community rights which aims to empower people in a democratic society. Moreover, the CLT is a tool to decentralize power to local people in making decisions on issues related to their everyday life, especially livelihoods. Democratically, community members participated in establishing regulations of land uses within their community. The collective land ownership guarantees that land is used and owned by local people in a productive way, particularly for agriculture. This way, land cannot be sold to outsiders but is restored to a community where all members decide how to utilize available land. Through an agreement, community members manage land collectively while they have the right to utilize their plot of land. Ultimately, the CLT is a bottom-up approach to resist centralized land management and the commodification of land.

Along with the CLT, two proposals, including a progressive land tax and a national land bank, have been promoted and advocated for addressing land disparity. A progressive land tax will mitigate the concentration of land and land speculation, as the measure requires those who own unproductive land to pay revenue in accordance with the amount and scale of land they owned. A national land bank, funded by a progressive land tax, will purchase or lease land and distribute it to landless communities. Community recipients will reimburse the land bank through its cooperative (Lubanski 2012; Hayward 2017). Since 2011, the Thai Government established the Land Bank Administration Institute (Public Organisation)–LABAI to be an organization that collects information regarding unproductive land owned by the Thai state, local government administration, and private sectors, where the institute will then purchase or lease those lands for landless peasants or small landholder peasants. [4] Another mission of the LABAI is to establish a national land bank as a land fund to assist peasants who need agricultural land for their subsistence.

In terms of community land management, the SPFT communities agreed to utilize land into six categories as follows: (1) individual agricultural plot; (2) collective organic farm for food crops; (3) individual residential land; (4) reserved forest land; (5) land for livestock; and (6) public used land. For example, Klong Sai community members agreed upon the land utilization plan where each household member has one *rai* (0.16 hectare) for housing and 10 *rai* (1.6 hectares) for an individual agricultural plot, including market crops such as rubber, palm oil, bamboo, banana, etc., as well as food crops. This way, community members are able to both secure their subsistence livelihoods and generate incomes. There is a 20-*rai* communal farming for both market and food crops where community members practice non-chemical farming method contributing to peoples' livelihoods and maintaining environmental sustainability. The SPFT has promoted food crops, especially agroecological rice vegetables crops in its communities for food security and food sovereignty. In a collective farm, community members work together on the plantation of food crops and economic crops and share profit from selling their products. For example, the Kao Mai community plants four *rai* of bamboos (0.64 hectares) where all community members invest, plant, harvest, and sell products together. The members earn profits by 60% while another 40% will be set aside for the next plantation, community administration cost, and other urgent community matters agreed upon by community members. The community spares 10 *rai* of land for livestock in which, the community members have also learned how to produce organic manure for their agroecological land. Furthermore, there is approximately 35 *rai* (5.6 hectares) for public space such as a community pond, a community hall, a community kitchen, an herb garden, and a commemorated monument, etc. (Fig. 7.4).

The SPFT established two mechanisms, the Agricultural Cooperative of the Southern Peasants' Federation of Thailand (ACSPFT) and the land fund, to support

---

[4] Read more about the Land Bank Administration Institute (Public Organisation) at http://www.labai.or.th/.

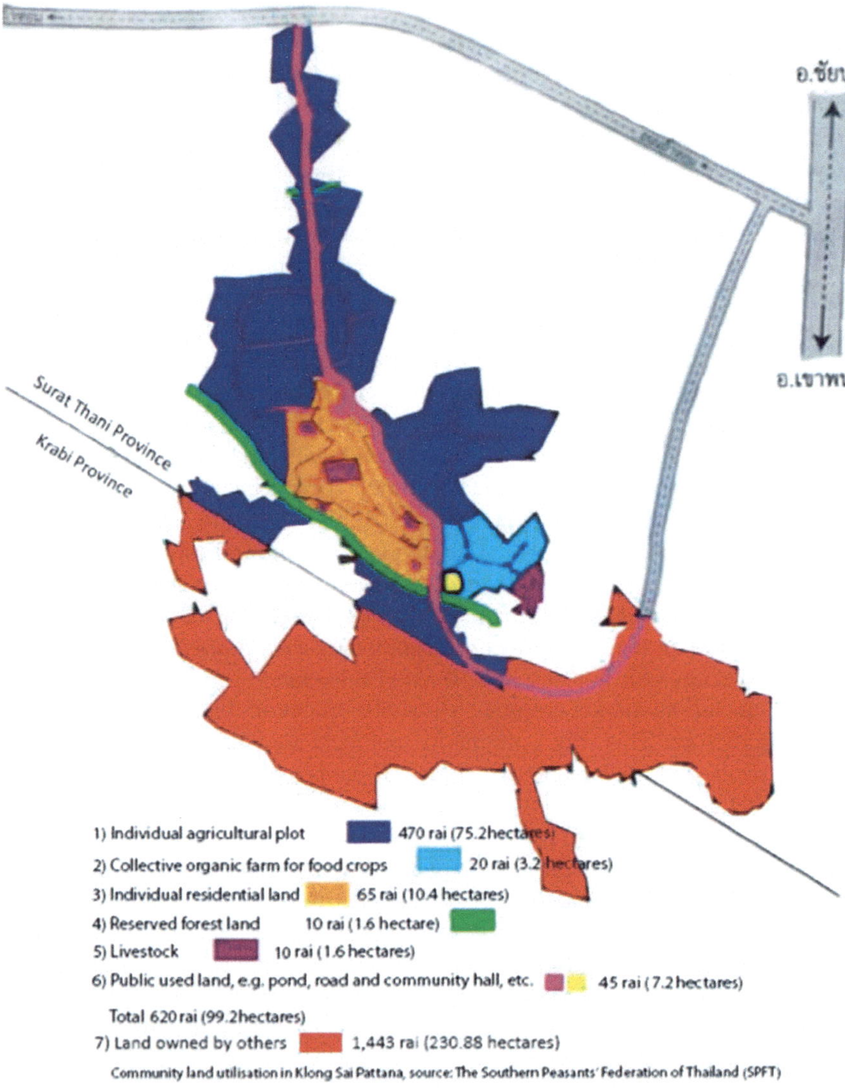

Fig. 7.4 Community land utilization in Klong Sai Pattana illustrates how community members use land in a productive way based on the interests of the community (*Source* Southern Peasants' Federation of Thailand)

its communal land management in compliance with the CLT. The ACSPFT serves as a community entity for administering community land in future. In addition, it is represented as a cooperative business center which collects agricultural products from community members and distributes locally on their behalf. For example, the SPFT has set up an oil palm collection center on behalf of the ACSPFT in the Perm Sub community to bargain a price of oil palm, thus ensuring that community members are not unfairly treated by the market. This way, the SPFT creates an alternative production–distribution practice. The cooperative is a place for community members to save their money and access financial services such as loans. The community land fund is a means for communities to save money as a seed fund for securing a community land. Both mechanisms have ensured socio-economic and land securities for local communities.

Moreover, the SPFT communities have promoted the participation of women in community development. In terms of community economic practices, women take a leading role in food processing and selling products through the ACSPFT. Currently, pickled bamboo shoots and banana chips are famous products made by SPFT community members. Particularly, pickled bamboo shoots are processed without chemicals but by using water and salt to pickle bamboo shoots. The community products, including food processing products and agricultural products, have been promoted by the district office as organic products (Fig. 7.5).

As the SPFT Coordinator said, "I guarantee that our community products are produced locally by organic methods unlike other communities in this district. Our products are good enough for Thai authorities to invite us to showcase at the district office promoting clean and organic food. While the district officials gain a reputation, they ignore our struggles for land" (The SPFT Coordinator, pers. comm., October 1,

Fig. 7.5 The processed food products and agricultural products made by the SPFT communities showcase how SPFT community members diversify food options and use them to gain recognition from outsiders (Photo courtesy of Supatsak Pobsuk)

2018). Nonetheless, the SPFT has seen such an opportunity as a way to promote its communities and to gain recognition from outsiders.

Alternative land management practiced by the SPFT communities counters the land development model introduced by the ALRO. The ALRO has proposed that a small-scale farmer should have five *rai* (0.8 hectares) for an agricultural plot and one *rai* (0.16 hectare) for housing, the so-called *5 + 1 model*. The SPFT argued that this top-down model is deficient in local participation because the state agencies have not consulted with community members in deciding on their community development in accordance with particular contexts. For example, the ALRO aims to apply the *5 + 1 model* to all SPFT communities without a consultation of community members. Especially in the Kao Mai community, a community member expressed that "the authorities haven't talked about land and community development with us. They told us that five *rai* for farming is enough but they have no idea about our context and people. Also, they proposed to dig a 200-*rai* pond [32 hectares] for a community water source. No one in our community wants it as such the scale of a pond is too big beyond our need. The authorities haven't concern about our livelihoods since they will construct new roads and other public utilities, crosscutting our farm plots and destroying our cultivated products (A Kao Mai community member, pers. comm., October 4, 2018). The SPFT communities also questioned the criteria of landless people who need an agricultural plot, as they realized that those registered in the government system are people who do not need land for farming, but they are gaining benefits through corrupted networks in provinces. Those who comply with the governmental model need to follow regulations set by the ALRO. The ALRO requires people who receive allocated land to plant cash crops promoted by the government, such as cassava, rubber, sugar cane, and maize. Given the lack of clear communication and consultation with local communities, it was reported that 10% of beneficiaries nationwide returned their entitled land to the ALRO (Khaosod 2017).

7.7.2 Political Alternatives

The SPFT has politically been established as a new social movement that aims to create a political space for negotiating controlled powers, in this case state and capital (Johnston 2014). The SPFT has raised their grievances and requested changes in land policies and increased the bargaining power of local people regarding land and natural resources management. As mentioned, the SPFT is a group of landless peasants and workers who lost land as their means of production through continued primitive accumulation and accumulation by dispossession (Harvey 2003) which marginalized them at the edge of development. The case of SPFT has explicitly illustrated grassroots democracy and participatory development. It shows how local people organize themselves, build formal and informal networks and uses different types of approaches and strategies to resist and challenge injustice. This highlights how they are active agents who know what they want and how to manage and use their resources in a way of benefiting themselves and their community (Kerkvliet 2009, cited in Sangkhamanee 2013).

As newly settled communities, the SPFT needs solidarity among its communities. The community regulations act as a community constitution which harmonizes people originating from different locations in new settlements. The community regulations are in line with the SPFT constitutions which focus on solidarity, justice, morals, collective land management, and security measures. It is arguable that SPFT community members live together through common orders and disciplines. However, the SPFT has promoted a democratic manner within their communities to prevent a predominant community leader who tends to be corrupt and authoritarian. The SPFT has set up a political culture which encourages all community members to participate in decision-making processes regarding community issues, from minor issues to community strategies. It is stated in its constitution that "the SPFT aims to create a democratic environment in communities and to encourage all members to participate in decision-making and community development processes."

In addition, the SPFT has established internal mechanisms to monitor the function of community regulations. A checks and balances mechanism has been formed. Those mechanisms include a morning check-in and monthly meeting. For example, members in Klong Sai Pattana Community hold a daily gathering at 7:00 a.m. to report security situations, disseminate news and information, exercise, sing a community anthem, and solve community issues. A community member shared: "When we have a community issue, we brainstorm, discuss, and seek for a solution together. We make sure that we listen to every voice as it is significant. We will reach a resolution collectively with a community consensus" (A community member in Klong Sai Pattana, pers. comm., October 3, 2018). The SPFT communities hold community meetings twice a month. For example, in Santi Pattana Community, the monthly meeting is held every 8th and 20th of the month to discuss problems and lessons learned, share ideas and information, and seek out solutions for community development.

In the SPFT structure, a political sector has been set up to monitor and analyze political and socio-economic situations in Thailand in order to formulate its strategic plans, tactics, and policy. Moreover, the SPFT has realized that its community members should understand and crystallize the context of their struggle for land so that they can develop collectively their political class identity which builds community unity and eventually provides a positive impact toward the SPFT mobilization. In this case, a political school has been forged to share ideas and concepts among community members. The SPFT's political sector has conducted political education programs in communities regularly to build political knowledge and capacities of community members, especially the young generation. The SPFT has regularly engaged academic scholars at universities to conduct critical research in their communities and organize a public event to raise awareness of the struggle for land and inequality in Thailand.

Like other social movements, the SPFT has shown their resistance through protest, which is a form of civil disobedience and recognized as direct participatory democracy. They expressed their grievances on the ineffectiveness of the Thai bureaucratic system in formulating and implementing land and natural resource management.

The SPFT has seen the politics of protest as a resistance tool to bargain power with governmental agencies and to demand development from below. Allied with P-Move, the SPFT has joined public demonstrations on the streets and at governmental offices to demand policy changes and concrete actions toward their problems. The SPFT has also filed their petitions to and requested formal meetings with responsible governmental agencies when they encountered threats. It is worth noting that SPFT is a well-organized social movement which indicates that local people are not passive and submissive to injustice.

7.7.3 Alternative Social Practices

In SPFT communities, community committees have been set up to look after their community members as well as to operate community services and businesses. There are nine committees established in each community, including committees of coordination, registration, finance, agriculture, development, healthcare, women and youth, security, and culture. These committees are elected by community members through a vote to take a lead in particular community issues and coordinate with the central SPFT.

Particularly, the SPFT has promoted the participation of women in communities and empowered them to be involved in community development, as stated in its constitution that the SPFT members should respect human and women rights and promote gender equity in their communities. Women in SPFT communities have been empowered to participate in the struggle for land. A woman from an SPFT community involved in reporting the implementation of the Convention on the Elimination of All Forms of Discrimination against Women (CEDAW) in Geneva, Switzerland in 2017 expressed that "not many people realize how women related to land issues. Women are restricted from access to land which is a source of food for their family. Women are a food distributor and a primary caretaker of their household. If women cannot access land, I believe that the other problems such as domestic violence will happen in their family" (A woman in Santi Pattana Community, pers. comm., October 4, 2018).

Women's access to land secures community food sovereignty. The SPFT's women groups have been formed to work on the community food production in collective lands for food crops such as rice, vegetables, banana, bamboo, coconut, etc. and raising chicken and duck for eggs. Moreover, gender division of labor has been employed by community members by dividing production roles on collective farms. For instance, men work on preparing the land and mowing while women work on seeding and planting. Community members sell all harvested products for communities and use them during community activities. In addition, women groups take a leading role in processing food products such as banana chips, pickled bamboo shoots, etc. The women in each community hold a monthly meeting whereby they discuss how they can participate in community development and initiate activities to empower women in communities (Fig. 7.6).

Fig. 7.6 Community members in Klong Sai Pattana community harvesting rice on the communal paddy which was transformed from a large-scale oil palm plantation; sharing labor among community members has been used to promote peasant culture and build collective peasant identity (Photo courtesy of the Southern Peasants' Federation of Thailand)

As newly settled communities, they have initiated a community welfare system as a safety net. With the ACSPFT, community members have welfare savings, one baht per day which will be allocated to community welfare, including healthcare and education. For example, community members will receive transportation costs based on the distance for traveling from their community to a hospital. As small-scale peasants, it is quite difficult for them to access formal financial services. The SPFT has thus created another community-based savings to provide microcredit to its members. Community members can save up at least 100 baht per month, and the members will earn their dividend by 7% per year, whereas 3% will be administration cost of the ACSPFT. Furthermore, those who want a loan service can access it with a low-interest rate. The SPFT continues to develop its welfare schemes to support its communities' members. A woman from Santi Pattana Community shared, "in the future, we are thinking to initiate community healthcare services, community care centers for children and elders to ensure our community members have social protection as necessary" (ibid.).

7.7.4 Building Community Cultures

It is significant to build a spirit of solidarity among new settlers, especially through cultures. However, the SPFT represents itself as a group of landless peasants and workers and reinvents a peasant culture. The SPFT believes that such culture forges harmony among community members, in which the SPFT attempts to show the importance and value of land to peasants' life. From the struggle for land as life, the movement has attempted to develop a sense of political class-based identity among their community members through collective activities, especially sharing labor on collective farms. As mentioned, collective farms are the place where community members have rotated to work on preparing soil, seeding, planting, harvesting, selling, and sharing.

Moreover, the SPFT has encouraged community members to participate in community activities such as the morning assembly for sharing information, exercising and singing SPFT songs, monthly meetings for discussion and consultations, and other community activities such as religious rituals and sports activities. In such a way, the SPFT believes that a harmonious and stable society can be established in the communities and that this can strengthen them in terms of negotiating with the state and outsiders. The SPFT has also organized activities, particularly community anniversary activities, and have usually invited neighboring communities to join. They see participations as a tool to gain acknowledgment from other communities and to eventually make their struggle visible and understandable in the eyes of the public. For example, the Santi Pattana community organized their 11th community anniversary on October 28, 2018 where they invited nearby communities to participate in sports activities and performances. As a community member expressed, "we used community activities such as sports and performances to build our solidarity and harmony. Beyond that, we normally invite our friends from other communities to join and to learn how we live. In this way, they understand who we are and why we needed to occupy land" (A community member in the Santi Pattana community, pers. comm., October 4, 2018).

The SPFT expressed that after they engage other communities in SPFT activities, the attitudes of outsiders have positively changed. A community member expressed that, "in the past, others called us a mob. We felt uncomfortable when we went outside our community as others were not welcome. After they have known us more through our activities, they call us neighboring "communities" (A community member in Klong Sai Pattana, pers. comm., October 3, 2018).

To some extent, the SPFT communities are not closed communities, as they welcome conditionally new community members. The SPFT welcomes landless peasants who really want to work on agriculture and live in peasant cultures. A new community member or family will be on three-month probation, where they receive 1 *rai* (0.16 hectare) for settling and planting their food crops. During the probation period, a new member or family will work voluntarily on community activities such as security guarding, community meeting, and collective plantation, etc., whereby community committees review how they conform to the community regulations and adjust to agrarian cultures.

7.8 Conclusion

Through this study on the Southern Peasants' Federation of Thailand (SPFT), the following points can be recapitulated to see how peasants in Thailand struggle for land.

First, landless and small-scale peasants in Thailand have struggled for land over time because of the domination of land management by the centralized power of state and economic growth paradigm. It is arguable that land reform in Thailand has not successfully been formulated and executed. As a result, equitable land distribution has never been reached, and landless peasants find it difficult to access land as a means of production.

Second, the concept of community rights has been employed to legitimize people's struggle for land. Such a concept has promoted the participation of community people in decision-making. It also proposes development from below and encourages the idea of self-determination. Eventually, local people aim to create participatory development and democracy. Community land title (CLT), which is a concrete manifestation of community rights, has been introduced to counter the predominant land tenure, including state-owned and private land, by proposing that land can be owned and managed collectively by community members.

Third, due to ineffective land management in Thailand, the SPFT used land occupation as a strategy to claim their right to land; however, community members encountered violence and judicial harassment. The SPFT hoists flags of land rights for landless peasants, human rights, democracy, and food sovereignty to counter the dominant discourse on land management in Thailand. They have developed and practiced numerous alternative practices, including land management, grassroots democracy, community constitution and community cultures as strategic tools to countervail mainstream development paradigm, normally focusing on economic growth.

Lastly, the emergence of the SPFT has called for equitable land distribution in Thailand. It has proved that land management should be contextualized, as there are alternative practices on land management. Local people are empowered to manage their resources based on their interests and knowledge. Moreover, they demand to be involved in decision-making regarding development and bargain power with the Thai state.

References

Aphron, Vithaya. 2014. "Community land titles and livelihood security for southern farmers." [In Thai.] *Journal of Sociology and Anthropology* 33(1): 105–39.

Ariyaviriyanant, Veerawat. 2009. *Development of Land Reform in Thailand: A Case Study on the Passing of Land Reform Law (B.E. 2475–B.E. 2518).* [In Thai.] [Doctoral dissertation, Chulalongkorn University].

Attavanich, Witsanu, Sommara Chantarat, and Boontids Sa-ngimnet. 2018. "Microscopic view of Thailand's agriculture through the lens of farmer registration and census data." [In Thai.] PIER Discussion Paper No. 09 (May 30). Bangkok: Puey Ungphakorn Institute for Economic Research. https://www.pier.or.th/wp-content/uploads/2018/05/aBRIDGEd_2018_009.pdf.

Bello, Walden F., Shea Cunningham, and Kheng Poh Li. 1998. *A Siamese tragedy: Development and disintegration in modern Thailand.* London/New York: Zed Books.

Bowman, Chakriya. 2004. "Thailand land titling project." Case study presented at *Scaling Up Poverty Reduction: A Global Learning Process and Conference*, Shanghai, May 25–27.

Chiengkul, Prapimphan. 2015. *Hegemony and counter-hegemony in the agri-food system in Thailand (1990–2014).* Ph.D. diss.: University of Warwick.

Fortify Rights. 2018. *Thailand: Follow-up Submission to the U.N. Human Rights Committee on Thailand's Compliance with the International Covenant on Civil and Political Rights (ICCPR).* Belfast, ME: Fortify Rights. https://www.fortifyrights.org/downloads/Follow-up_Submission_to_the_UN_Human_Rights_Committee_on_Thailands_Compliance_with_the_ICCPR_July_2018.pdf.

Hall, Derek. 2011. Land grabs, land control, and Southeast Asian crop booms. *The Journal of Peasant Studies* 38 (4): 837–857. https://doi.org/10.1080/03066150.2011.607706.

Hall, Derek, Philip Hirsch, and Tania Li. 2011. *Powers of Exclusion: Land Dilemmas in Southeast Asia.* Singapore: National University of Singapore Press.

Harvey, David. 2003. The 'New' imperialism: Accumulation by dispossession. In *Socialist register 2004: The New Imperial Challenge*, ed. Leo Panitch and Colin Leys, 63–87. London: The Merlin Press.

Hayward, Daniel. 2017. "Community land titling in Thailand: The legal evolution and piloting of titling policy." Thematic Study Series No. 7 (October). Chiang Mai and Vientiane: The Regional Center for Social Science and Sustainable Development, Mekong Land Research Forum, and Mekong Region Land Governance.

Al Jazeera. 2017. "Harassed by palm oil company, Thai Village Defends Land." *Al Jazeera*, August 9. https:// www. aljazeera.com/indepth/features/2017/06/harassed-palm-oil-company-thai-village-defends-land-170605105935706.html.

Johnston, Hank. 2014. *What is a social movement?* Cambridge/Malden, MA: Polity Press.

Khaosod. 2017. "Farmers returned 10% of agricultural land as required to put more investment and forced doing agriculture." [In Thai.] *Khaosod,* October 16. https://www.khaosod.co.th/economics/news_572994.

Lamb, Vanessa, Carl Middleton, Rebecca Sofia Leonard, and Nga Dao. 2015. *Access to Productive Agricultural Land by the Landless, Land Poor and Smallholder Farmers in the Four Lower Mekong River Basin Countries.* Phnom Penh: Oxfam in Cambodia.

Land Reform Network and Local Act. 2010. *A Report on the Study "Land Reform Process and Sustainable Land Management by People."* [In Thai.] Nonthaburi: Land Reform Network and Local Act.

Leonard, Rebeca, and Kingkorn Narintarakul Na Ayutthaya. 2003. "Thailand's land titling programme: Securing land for the poor?" Paper presented at *Politics of the Commons International Conference, Regional Centre for Sustainable Development,* Chiang Mai University, Thailand, July 11–14. https://dlc.dlib.indiana.edu/dlc/bitstream/handle/10535/1951/Rebeca_Leonard.pdf.

Lubanski, Jason. 2012. "Land is life: A policy advocacy case study of the Northern Thailand land reform movement." Master's capstone paper, SIT Graduate Institute. http://digitalcollections.sit.edu/cgi/viewcontent.cgi?article=3588&context=capstones.

Miyake, Yuki. 2003. Contesting rights over land: a case study of landless farmers' social movement in Lamphun Province. *Articulating Development and Strengthening Local Practice* 7: 11–14.

Ospina, Olga Lucia Castillo, and Juan Masullo-Jiménez. 2017. "Alternative development is no longer an alternative—Post-development could be." *Filosofía de la Economía* 6(2): 99–119.
Protection International. 2020. "The Southern Peasants' federation of Thailand under attack." *Protection International,* May 2. https://www.protectioninternational.org/en/node/408.
Rakyutidham, Atchara. 2011. "Mainstreaming alternative development." [In Thai.] *Journal of Sociology and Anthropology* 30(2): 13–43.
Sangkhamanee, Jakkrit. 2013. Democracy of the desired: Everyday politics and political aspiration in the contemporary thai countryside. *Asian Democracy Review* 2: 5–37.
Scott, James C. 1998. *Seeing like a state: How certain schemes to improve the human condition have failed.* New Haven, CT/London: Yale University Press.
Seub Nakhasathien Foundation. 2017. "Lessons Learned from the Logging Concessions in Thailand." [In Thai.] November 18. https://bit.ly/2PJ8Wsh.
Southern Peasants' Federation of Thailand. 2016. *The Southern Peasants' Federation of Thailand Booklet*. [In Thai.]. Unpublished document.
Tang, Alisa. 2016a. "*Thai Land Rights Campaigners' Most Powerful Weapon? Access to Information.*" Reuters, July 14, 2016. https://www.reuters.com/article/us-thailand-landrights-idUSKCN0ZU00U.
Tang, Alisa. 2016c. "*After Violent Struggle for Land, Thai Campaigners Face Challenge to Community Farming.*" Reuters, July 26. https://www.reuters.com/article/us-thailand-landrights/after-violent-struggle-for-land-thai-campaigners-face-challenge-to-community-farming-idUSKCN10603B.
Tang, Alisa. 2016b. "*Four Murders and a Bloody Battle for Land Rights in Thailand.*" Reuters, July 23, 2016. https://www.reuters.com/article/us-thailand-landrights-idUSKCN0Z9006.
The Southern Poor People Network. 2006. *Unveiling the Truth of Forest Garden Concession Investigation.* [In Thai.] Surat Thani: The Southern Poor People Network.
"Thai Forest Act 1941 (B.E. 2484)." [In Thai]. http://www.forest.go.th.
"Thai National Reserved Forests Act 1964 (B.E. 2507)." [In Thai.]. http://web.krisdika.go.th/data/law/law2/%BB27/%BB27-20-9999-update.pdf.
Vandergeest, Peter, and Nancy Lee Peluso. 1995. "Territorialization and state power in Thailand." *Theory and Society* 24(3): 385–426.

Supatsak Pobsuk Supatsak Pobsuk has been working in the field of development for more than 15 years. He works with international, national and local actors to support vulnerable populations including refugees, migrants and grassroots communities in Southeast Asia. His research interests include land, forest, natural resources, community rights, social movement, migration, trade and investment, democracy and political economy. He is the author of several publications on democracy, refugees, migration and land rights. Supatsak studied international development studies and political science. He worked with Focus on the Global South in 2018–2020 where he wrote this article. Currently, he is an independent researcher.

Chapter 8
Rediscovering the Sacred Among Agrarian Societies: The *Tara Bandu* Case in Timor Leste

Ananeza P. Aban

Abstract The cultural practices of Timorese societies that are intricately embedded within their natural surroundings have been periodically disrupted during the Portuguese and subsequent Indonesian occupation. During these periods of unrest, people were denied, or to some extent, discouraged to express their ritual life, initiate their own traditional resource management and conservation efforts, as well as maintain social cohesion. However, they survived these stages of volatile political landscape, punctuated by multiple episodes of mobility, forcible relocation to new territories, and series of resistance underlying this dark past. Furthermore, the brutality of their history has not prevented them from reviving traditions with such fortitude after the Independence. Post-conflict Timorese agrarian societies regenerate an ancient customary practice known as the tara bandu to forward their land claims pending a national agrarian policy that recognizes communal land rights. Beyond the performance of an ancient ceremony lies the potential of this cultural practice and the strength of the symbol for effective community decision-making, collective action, and enforcement system. In the words of a peasant leader, tradition may not always be good but it may also hold possible solution to complex community conundrums.

Keywords Communal land rights · Customary practices · Indigenous agriculture · Tara Bandu · Timorese agrarian societies

8.1 Introduction

Southeast Asia is a wellspring of non-mainstream or alternative practices of communities that have thrived for many years traversing the economic, political, social, and cultural features of human life. Many peasant communities in the region waged their struggle for agrarian reform, not only to produce food, but also for the pursuit

A. P. Aban (✉)
Ghent University, Ghent, Belgium
e-mail: ananeza.aban@ugent.be

J. M. C. Sy and A. P. Aban (eds.), *Reimagining Development in Southeast Asia*, Asia in Transition 21, https://doi.org/10.1007/978-981-97-9140-8_8

of sustainable living. Their diverse and time-honored alternative practices are products of their resistance to the state's development agenda that has long neglected and marginalized them. The land claims they have courageously endeavored are intricately linked to their history and identity as cultural bearers of agricultural sustainability which is an important heritage of the region.

Timor Leste is one country within this geographical location that showcases a peasant union which in its constant drudgery to realize a peoples' version of agrarian reform, still upholds the potential of their cultural practices through the strength of the symbols for effective community decision-making, collective action, enforcement system, and sustainable management of their agrobiodiversity. The study is focused on the peasantry in Ermera, the country's largest coffee farming community, and home to one of the world's finest variety of *Arabica* coffee beans. In this rural village, an oral tradition locally known as *tara bandu* is vibrantly practiced on a district (or municipal) level, amidst a backdrop of widespread poverty and contentious landholdings that remain concentrated to the *mestizo* families in Timorese society. These are the families who have dubiously acquired lands since Portuguese colonial times.

Tara bandu is a system of customary law that the local communities apply to natural resources management. In Tetun language, *tara* means hanging, while *bandu* means prohibition. Through a community-organized ritual and hanging of a symbolic representation of a prohibition at an established altar, a prohibited activity is announced publicly. The typical prohibitions are: burning of forests, the cutting of trees, collection of forest products, agricultural harvests, and hunting and fishing in a forbidden zone (Miyazawa 2013, 519).

This case study tries to provide a deeper understanding on why putting primacy on the symbolic in trying to respond to community concerns elicits a higher level of compliance among the peasant class of Timor Leste as juxtaposed to the imposition of government laws or regulation.

This case also tries to rediscover the relevance of the commons principle that clarifies the fundamental problems of privatized ownership accompanying land distribution which extensively rejects the foundations of the collective life of humanity (Houtart 2011). As opposed to the prevailing and mainstream market-oriented development model that features cutthroat competition, corporate take-over of human and natural resources, and the destruction of ecosystems, Houtart (2011, 20–33) presented the four (4) foundations of humanity's collective life: (1) balanced relationship with nature; (2) economy or production of the material basis of life; (3) social and political collective organization; and (4) the importance of culture in interpreting reality. Putting "the commons" in the context of Timor Leste, the mutual reinforcement of the social, ecological, and the sacred has been becoming invisible in land law formation due to the pressures of market-oriented production.

This study highlights the collective wisdom of the community that carefully entwines the role of the peasant union, the village leadership under the *xefe de suco* (village chief), and the *lia nain* (cultural elder) in the pursuit of a just agrarian society.

Various literatures explicitly described *tara bandu* for local regulations on social relations and natural resource management. But this study decides to rearticulate *tara bandu's* contribution to the peasant movement to solve the quandary of landlessness, to correct the historical injustice inflicted on the Timorese, and to bravely challenge the threat of dispossession pending land policies that protect their interest as food producers. This stance follows the perspective of Ermera's peasant union that categorizes *tara bandu* as a politico-cultural practice that plays an important and foundational role in their struggle for agrarian reform.

8.2 Brief Historical Background of Timor Leste

The dynamic political past of the Timorese that was characterized by forced migration, displacement, and return to their ancestral places has significantly contributed to the shaping of their cultural values and practices. McWilliam and Traube (2011, 2) tried to interrogate the complexities among Timorese rural population's way of "making and remaking their ties to the land." The diversity of people in Timor is a product of multiple histories of mobility, the venture to new territories, and the series of resistance underlying these periods. In 1701, the first Portuguese governor of then East Timor[1] was appointed. By 1901, the Portuguese Parliament passed the colonial law that "vested all land not used for permanent residence or cultivation in the Portuguese state," powerfully marking their colonial territorialization in East Timor (Fitzpatrick 2002a, 146–7, quoted in Fitzpatrick and Barnes 2010, 209).

At the break of the Second World War, Japanese wartime occupation has also contributed to periods of upheaval, widespread displacement, and coercive resettlement which were also accompanied by numerous accounts of Timorese resistance (McWilliam and Traube 2011; Fitzpatrick and Barnes 2010) where as many as 100,000 Timorese became casualties from the period 1942–1945 (Gunn 1999, 236, quoted in Fitzpatrick and Barnes 2010).

Right after Portugal ended its colonial control over East Timor in 1975, Indonesian military forces invaded this island and subjugated the population for more than two decades (Fitzpatrick and Barnes 2010). The dark legacy of the Indonesian occupation (1975–1999) is gauged based on the intensity to which the local people were displaced and resettled as a strategy for control and surveillance. This adds to the gruesome statistics of human rights abuses, starvation, and extensive damage to property, aside from killings by the Indonesian militia (Fitzpatrick and Barnes 2010; McWilliam and Traube 2011; Nevins 2005, quoted in Batterbury et al. 2015). These indiscriminate atrocities, however, were confronted through the guerrilla movement formation of the men and women resistance fighters under Fretilin[2] and its armed wing Falintil,

---

[1] Former name of East Timor Leste.

[2] FRETILIN—Frente Revolucionária de Timor-Leste Independente (Revolutionary Front for an Independent East Timor) was founded as a resistance movement that fought for the independence of East Timor from Portugal and eventually from Indonesia. Its military wing is FALINTIL—Forças

who found their refuge in the jungles of Timor and got popular support from the masses (Capizzi, Hill and Macey 1976; Niner 2013).

In 1999, the Indonesian troops left Timor Leste, which marked the end of their rule after the Timorese voted for their independence in a referendum (Fitzpatrick and Barnes 2010). Interestingly, because of the locals' deep sense of culture that is closely attached to their land, resources, and a strong belief of the sacred powers or guardians *(lulik)* behind it, they have even attributed in part the victory of their resistance movement to this "alliance between the human and non-human realms" (McWilliam and Traube 2011, 11). On May 20, 2002, the independent state of Timor Leste was established (Fitzpatrick and Barnes 2010). However, after more than a decade, Timor Leste still faces challenges in its nation-state building, particularly with the absence of a comprehensive policy on land distribution for its largely peasant population.

8.3 Land Claims and Land Grabs in Timor Leste Under Foreign Rule

Land, cultural values, and community structures are intimately tied to one another in Timorese rural society. This reality actually supports Shinichi's (2014) argument that in both urban and rural settings in developing countries, land is of cardinal importance to the survival of the populace. This is because peoples' livelihoods are heavily centered to land which is also their critical resource in politics and a foundation for their identity formation. However, the changing political regimes of colonization to military occupation have contributed significantly to the compounding problems of land ownership in post-independence Timor Leste. First, it was an era of providing the right to use *(aforamento)* vast tracks of land to a few who were allies or had business agenda with the Portuguese colonial rulers. The second wave is the massive displacement and dispossession of the people under Indonesian power. Land policies under the Indonesian rule have accepted titles issued during the Portuguese era (da Silva and Furusawa 2014). Dovetailing Shinichi's (2014) position, this particular exodus from their places of origin has disrupted to a great extent their natural life connections, the sources of identity and survival. These sprawling ancestral territories bring forth a deep connection to the Timorese rural livelihood strategies that are characterized by mobility through the hunting and gathering approach which represents the diversity and flexibility of subsistence farming practice. But Indonesian law enforcers dramatically prevented these occupied population from their natural movement within their accustomed spatial territories. As a result, this rather abnormal condition has triggered resistance to Indonesian occupation (Pannell in McWilliam and Traube 2011).

Affirming Timor Leste's historical background, there are two sources of land claims that are contradictory to each other: (1) titles issued by the two previous

---

Armadas da Libertação Nacional de Timor-Leste (The Armed Forces for the National Liberation of East Timor).

regimes and (2) customary rights. Critically, these titles issued from the past political regimes are not always acceptable for many Timorese (da Silva and Furusawa 2014) who have been the actual victims and survivors of these historical injustices underlying these previous land ownerships.

8.4 Land Rights and Agrarian Reform in Post-independence Period

This multiple inter-connectivity of life, land, and the spirit is vital and mutually reinforcing that can be defiled in its entirety by any government policy passed (Batterbury et al. 2015; Harson 2015). Until today, the concept of social order that withstood foreign invasion for the longest time is based on customary systems embedded in Timor Leste's social landscape, derived from narratives of origin and first possession (Fitzpatrick and Barnes 2010). These overarching principles of ancestral first possession, albeit socially constructed arrangements, remain resilient because of their viability to maintain social order and authority (ibid). Communities always derive from their ancestral origin in terms of identifying the original settlers or landowners. People assert their relationship to the origin groups, or clans, or their membership to the sacred house, or possession of a sacred object in reference to their rights and access to land (Fox 1996:132 in Fitzpatrick and Barnes 2010) or for communal land tenure arrangements. As a reinforcement of this custom, "this property relations are constituted around ritualized norms of original authority" (Fitzpatrick and Barnes 2010, 206). In the words of Ermera peasant union members interviewed, their collective participation in community rituals made them understand better what belongs to them. People in the village know who collectively owns which.

Given this, land and social connection deserve a critical lens to carefully observe, investigate, and understand contemporary Timorese society within the backdrop of a changing, politically transformed cultural landscape (McWilliam and Traube 2011).

The point however is that this distinctive character of cultural solidarity as applied to land contradicts the neoliberal spirit behind the concept of privatized ownership that is being pushed in Timor Leste. Formal land titling has become a contested practice which in the process has been associated with putting market value to land, a concept that is foreign and "unthinkable" for the local farming communities (Peñaflor 2011, quoted in Batterbury et al. 2015). In the course of drafting the land law, there was opposition from the ruling class and foreign advisors to fully recognize the communal and customary land management narrative (Batterbury et al. 2015).

Meanwhile, in the absence of a clear-cut framework and a strong people-oriented land reform policy, land disputes and land grabs have been uncovered. da Silva and Furusawa (2014) cited multiple land dispute and land grab cases across the country; some involved big businesses. Another chronic land dispute case was identified in a coastal plain within the Viqueque District because of entangling claims among individuals who based their hereditary claims on "histories of actual possessory acts" that are associated with the state and regime change rather than customary authority (Fitzpatrick and Barnes 2010). And worth mentioning again is Ermera District, where Timorese descendants of elites granted *aforamento* still claim ownership of large coffee farms in the area. Notwithstanding, this district is a witness to a "popular land reform" movement under the leadership of the peasantry in the post-Independence era (da Silva and Furusawa 2014).

This persistent social disorder caused by perennial land disputes and land grabs reflects a deficit of a clear focal point for cooperative land allocations founded on customary principles of origins (Fitzpatrick and Barnes 2010). This actually demonstrates a dilemma within contemporary agriculture policies where self-reliance principles in agriculture (as reflected in subsistence farming) are made less visible in order to prioritize market-oriented production (Shephered 2013 in Batterbury et al. 2015). Other scholars also argued that the global phenomenon of tenure formalization did not even show positive results as it rather facilitated the potential for land grabs by foreign and corporate interest (Anderson 2008; Sikor and Lund 2009, quoted in Batterbury et al. 2015). Emphasizing historical contextualization, McWilliam and Traube (2011, 14) observed that "the impact of Indonesian resettlement policies and political prescriptions based on idealized agrarian citizens continues to frame much government policy and planning." This, Pannell (cited in McWilliam and Traube 2011, 14) added, "threatens to erase the multiple social interests and ancestral connections to forests and other non-cultivated spaces."

Therefore, as da Silva and Furusawa (2014) stressed in developing land policies in Timor Leste, utmost recognition of the cultural should be done because land claims are related to families, clans, and the *uma lisan* (origin groups). Without in-depth anthropological studies, the notion of land registration will only rouse further conflicts, more complicated as discussed earlier. Moreover, Batterbury et al. (2015) argued that the formulation of an agrarian policy should not bypass viable and grounded solutions and initiatives of Timorese communities that portray their capacities for effective governance.

In 2012, the parliament passed the land law package: the Law on Special Regime for the Determination of Ownership of Immovable Property, the Law of Expropriation, and the Law on Real Estate Financial Fund (da Silva and Furusawa 2014).

However, former President Jose Ramos-Horta[3] vetoed these laws after observing that the rights of the people, particularly their communal property and traditional tenure, were not sufficiently protected and recognized (da Silva and Furusawa 2014). These proposed land laws reflect the leaning toward a Western-influenced land management principle at the expense of the customary land (Batterbury and Longbottom 2007, quoted in Batterbury et al. 2015; da Silva and Furusawa 2014).

In June 5, 2017, the Special Regime Law[4] came into force. But for the Uniaun Agrikultores Munisipiu Ermera (UNAER) and their institutional partner K'dadalak

[3] Jose Ramos-Horta received the 1996 Nobel Peace Prize along with Catholic Bishop Carlos Filipe Ximenes Belo. Both worked for Timor Leste independence from Indonesia.

[4] The title of the law is Rejime Espesiál Ba Definisaun Titularidade Bein Imovel.

Sulimutuk Institutu (KSI), the majority agrarian population in the country still awaits a robust political solution to the land problem. Unsatisfied by the passage of this law, they demanded the Timor Leste government to instead prepare a comprehensive national agrarian reform policy and program that expresses the state's accountability in upholding the farmers' collective and customary rights over the land. Land for the Timorese, according to KSI director Jenito Jewata Santana, mirrors the political position of the government. This implies that whatever the government agenda on land is, reflects its fundamental bias for whom development should be (Santana 2018).

In their observation, the supposed land distribution for farmers is unclear in this new land law. Its emphasis is on land use, instead of providing a definitive discussion about land ownership that can serve as a benchmark in establishing who stands to benefit from it and who has the obligation to protect it. What sparks intrigue is the recognition of landownership during the colonial times that is historically understood as unjust and runs contrary to the Timorese Constitution that acknowledges collective ownership, local knowledge, laws, and practices, all implicit in their tradition and cultural values (Santana 2018).

They strongly argued that contrary to individualized or privatized land titling, collective land ownership that exudes the power and effectiveness of collective decision-making among Timorese societies remains the core solution to the country's problem over land dispute.

With this new land law, they also felt apprehensive as it likewise opens the floodgate for corruption because it welcomes market access to land as a profitable property, referring to what Houtart (2011) explained about the capitalist logic of commodities. They warned that land certification under this law should be seen with caution as it can be simply acquired when there is no dispute.

This synergy of the peasants' struggle for the recognition of their customary rights over land and the Timorese' self-assertion of their identity as managers of their resources and society has found consolidation in the ***tara bandu***, an ancient communal practice that survived this historically volatile political landscape of Timor Leste.

8.5 Importance of *Tara Bandu* in the Struggle for Agrarian Reform

Segueing to the cultural aspect, ethnographic studies (Fox 1980, quoted in McWilliam and Traube 2011, 4) revealed that from the period when Timor Leste was under Portuguese colonization, Timor rural society "to a significant degree, remained organized around predominantly ancestral principles and protocols, where the influence of the colonial state had a significant but limited impact on practices of social reproduction."

Ritual life may have been disrupted but not completely destroyed during the Indonesian occupation (Belun and the Asia Foundation 2013; McWilliam and Traube 2011; Palmer 2017). This reality illustrates the resilience of the occupied population to recreate their various forms of resistance. Despite the challenge of forced migration, their "attribution to origin places encouraged symbolic strategies for adapting to displacement" (McWilliam and Traube 2011, 12).

Tara bandu, as mentioned earlier, is a traditional Timorese mechanism to enforce peace and reconciliation, regulate social exchange practices, and facilitate environmental management, through the power of public agreement. It has undergone stronger resurgence after Timor Leste's independence suggesting the reinforced link between the community and their natural, cultural, and economic resources (Belun and The Asia Foundation 2013, da Silva and Furusawa 2014; Miyazawa 2013; Palmer 2017).

Part of the *tara bandu* process is the holding of the community-wide ritual which is a multi-day event that is followed by the hanging of culturally significant items (e.g., head of a sacrificed animal, leaves of a tree species) from a wooden shaft in a specified area indicating prohibition or ban of a certain practice (Belun and The Asia Foundation 2013; da Silva and Furusawa 2014; Palmer 2017). The ceremony is usually coordinated by the *xefe de suco* (village chief). During these days, this village leader coordinates with the district head, concerned government officials, NGOs, and the Catholic Church. The ceremonial activities are: ritual speech, dancing, drumming, singing, betel nut exchange, and animal sacrifice (Palmer 2017). The policies within it have been categorized as: (1) regulating people-to-people relations; (2) regulating people-to-animal relations; and (3) regulating people-to-environment relations (Belun and The Asia Foundation 2013).

Aside from the banning of activities to protect natural resources (Miyazawa 2013), *tara bandu* is also enforced to stop practices with negative social impacts such as stealing from another's farm and lavish parties (Belun and The Asia Foundation 2013; da Silva and Furusawa 2014; Palmer 2017).

Because *tara bandu* has become well entrenched in the socio-cultural structures of the locals, there is a higher level of compliance with the agreed prohibitions than may be expected in the imposition of government laws and regulation. The sanctity of the ritual and belief of the presence of the *lulik* (sacred) are also vital factors for this community-wide compliance (Palmer 2017). One of Ermera's village chief Luis dos Santos from Gleno explained that, departing from maintaining *tara bandu* as an oral tradition and an organic process of community transformation, the practice has been elevated into a *Tara Bandu* Code that signifies formalization by the Ministry of Law and underwent a long consultation process with stakeholders. These written regulations are the basis for enforcement and evaluation as well as the establishment of a formal monitoring committee (Dos Santos 2018). Being a village chief, Dos Santos coordinates all of these.

8.6 Transforming Ermera District Under the Peasant Union and the *Tara Bandu*

Following the end of the Indonesian occupation, farmers in Ermera District occupied and distributed lands, also known as the "popular land reform" (da Silva and Furusawa 2014), as a unilateral expression of their capacity to implement agrarian reform on their own terms.

During those years of fighting for land acquisition, UNAER, together with KSI, organized farmers in the district for mobilizations and dialogues with government officials to defend their rights over the land as *mestizo* families continued their claims over huge coffee farms in the area. Batterbury et al. (2015) wrote that these *mestizo* families in the country have acquired these lands from the Portuguese rulers on condition of their non-involvement in anti-colonial activities.

UNAER has led the distribution of the 28 huge coffee plantations in Ermera to the farmers since 2000. UNAER currently has a stronghold of around 24,000 members only in Ermera, making them a big and expanding peasant movement in the country.

For UNAER, this is their version of agrarian reform animated by equal land distribution that aims to secure farmers' ownership of the lands from land grabbing under the past regimes and to implement agroecology. This radical effort is to reclaim these lands which for them are preconditions to their continuing political struggle that traces back to their fight against colonization.

UNAER's stimulus for agroecology is the farmers' overdependence on the coffee economy, a paradox behind their continuing underdevelopment. Coffee cultivation finds its roots under the reign of Portuguese Governor Affonso de Castro in the late nineteenth century, that marked a transition from the "tributary" economy toward the development of an export plantation economy, an era depicted to "solidify, extend, and rationalize colonial control" (McWilliam and Traube 2011, 7).

Arguably, in post-conflict Timor Leste, coffee as a significant export commodity still failed to alleviate marginalized coffee producers from poverty. Khamis (2015) claims that the current coffee fair-trade initiative as a practice of ethical consumption is problematically woven into the fabric of global neoliberal systems making Timorese coffee growers among the most vulnerable in the global coffee economy. Since the dismantling of the International Coffee Agreement in 1989, there are no assurances and regulations that are meant to protect the coffee-producing nations from fluctuating market prices while corporations in coffee-consuming nations, engaged in retail and roasting, enjoy the bigger profit. Thus, the supposed morals of "social equity and environmental responsibility" within this fair-trade mechanism could not transcend the "sensual pleasures of coffee consumption" bounded by the cultural parameters of liberalized trade (Khamis 2015, 494).

Internal challenges within the country are also imminent: the presence of middle buyers that purchase coffee from small farmers at a very low price due to poor coffee harvest quality and transport costs; the farmers' lack of security over the land and the constant threat of land grabbing have become an impediment to improving existing coffee farms; the state's sluggish pace in prioritizing agriculture development vis-à-vis agrarian reform (Harson 2015). Similar to other rural areas in the country, large land holdings in Ermera claimed by the *mestizo* families are still extensive. Some of them are actually absentee landowners because they are either in the capital, Dili, or based in Portugal, Australia, or elsewhere overseas.

While a strong peasant union strongly asserts its claims within the political space in the district, KSI confirmed that Ermera remains a land dispute area thus, posing a challenge for UNAER to advance their land struggle. Only 25% of the people own land while 75% of the land is still controlled by these *mestizos*. This reality, however, provided more reason for the peasant union to assert their presence in the area and get the best out of what they have so far achieved, while waiting for the passage of a land law that will uphold their interest and political agenda.

To ease the rural population's economic dependence on coffee and to realize their full potential in land management, UNAER President Alberto Martins Guterres explained that, with the help of KSI, they began to introduce horticulture as a sustainable farming approach: tree nursery and planting of endemic tree species to intercrop with coffee trees and maintain forest diversity; poultry raising; and planting of staple food such as maize and rootcrops (cassava and sweet potato) to ensure food supply and supplement the seasonal income from coffee production (Guterres 2018). Vanilla cultivation was also introduced to integrate with coffee trees. Vanilla, a tropical vine commonly used as a flavoring agent, is considered a high-value crop in Timor Leste that is sold to the local market and for export. Non-conventional farming approaches such as the soil trap technique and land terracing are also practiced (Santana 2018).

In Ermera, land distribution, sustainable land management, and *tara bandu* complement one another to reach the goal of a more just and sustainable community (da Silva and Furusawa 2014). Guterres said that UNAER's mission is "to obtain a just agricultural society and sovereignty, live in the spirit of collectivism, social solidarity, and sustainability from generation to generation." As a strategy to realize this mission, the peasant union's practice of *tara bandu* in Ermera widely covers the entire district of 52 villages. To signify the importance of this tradition, its initial ritual ceremony was attended by no other than former Timor Leste President and 1996 Nobel Laureate Jose Ramos-Horta.

Guterres discussed that there are two (2) things that make *tara bandu* important for them: (1) to secure the political agenda of the union which is agrarian reform; (2) to promote sustainable agriculture. *Tara bandu* also demonstrates traditional/cultural evidence to powerfully secure the community's right to land. The *lia nain* (community cultural expert) holds both indigenous and historical wisdom to identify which part is owned by a certain clan, *uma lisan,* or family. Given that, the *lia nain* will fortify the union's fight for the land that belongs to the community and not to someone else (Santana 2018). *Tara bandu,* for the peasant union, is about sustainable agriculture that highlights the relations of people to ecology or natural resources.

Aside from the land problem, Ermera locals used to be stereotyped for their extravagant festivities that last for days or weeks. Alarmingly, this practice drains their seasonal income from coffee that should have been allocated for family essentials such as children's education. Ermera also used to have a reputation of being a notorious district where conflicts breed among the young people as an effect of being school drop-outs.

On how *tara bandu* started in Ermera, Santana explained that KSI conducted consultation with the peasant union in light of the recurring and complex socio-economic problems in the community. The locals identified *tara bandu* as an alternative strategy in addressing these problems of mismanagement of household income, disintegration of families as a result of financial distress, and low rate of education. They later lobbied to the district and village political leaders and the Catholic Church for the possibility of initiating the *tara bandu* process, despite cynical reactions from some segments.

Tara bandu in Ermera is seen as a "cultural revolution" that somehow changed the mentality of people to think about a better future (Santana 2018). This paradigm shift eventually yielded substantial gains. Given their poor economic condition and other vulnerabilities, the *tara bandu* becomes a tool to educate the people of the fact that they hold solutions to their own problems.

Social benefits have been identified as the following: reduction of conflicts within the community and strengthening mutual respect and solidarity among the people, especially those who practice horticulture. Political stability in the community has been experienced that has manifested in revitalized grassroots leadership and democratic governance. Financial savings from not overspending on traditional ceremonies stand out as a major economic benefit. Communities began to re-direct their spending to important family expenditures. This has been collectively appreciated as a strengthening of local culture for a positive communal life that preserves natural resources and continuously protects the *lulik*.

To describe the transformation in their village, Guterres said that many young people are now in school until the university level. Some studied at the universities in Dili and abroad. A proof of its success in ensuring the youth's education is UNAER's Secretary Leonardo Fatima Soares who has finished his course in the university and is currently managing the pre-school program of the peasant union.

Soares in an interview explained that the youth, who spend more time in school and who are conscious of the hefty penalty for violating *tara bandu* on overspending, have been observed to have diverted their spending on liquor and parties to personal needs. Youth conflicts have also subsided as they become pre-occupied with more productive activities such as horticulture and studying (Soares 2018). The increased presence of motorbikes and the retrofitting of houses using permanent building materials have been identified as indicators of the improved financial management of the locals and their compliance to the *tara bandu* (Guterres 2018).

Now, Ermera, according to Guterres, is a model. First, because UNAER is now becoming a leading peasant organization at the district level. Second, because of the implementation of *tara bandu*. The peasant union is in the process of expanding to neighboring districts of Aileu, Manufahi, and Suai toward building a movement for popular agrarian reform, especially in light of a pending agrarian reform policy. The target is to establish the peasant union in all the districts of Timor Leste. UNAER aspires to expand to a national federation, while currently maintaining its linkage with the international peasant movement La Via Campesina.

Ermera's peasantry also reflects how the locals have understood and internalized the principle of inclusiveness toward the achievement of communal reconciliation and sustainable peace. da Silva and Furusawa (2014) explained that farmers are prepared to embrace the rights of others who were displaced and relocated by the foreign rulers, as they are likewise victims of conflict and injustice.

Ermera's *tara bandu* has been implemented for eight (8) years. On the 8th year, the community will hold a district-wide evaluation process before they will ever decide whether to renew the *Tara Bandu* Code or not.

In order to maintain the relevance of the *tara bandu*, Guterres said that the challenge is to continue educating the community about the benefits of this practice. The *xefe de suco* has a critical role. Guterres explained that in order to ensure that the interest of the farmers are protected, political governance at the village level should therefore be coming from the peasant union who fully comprehends the value of this customary practice in relation to the land struggle.

8.7 Conclusion

Ermera's *tara bandu* and the leadership of the peasantry highlight the limits or even the failures of the Western or market-oriented concepts of land titling that cannot just serve as a blueprint for land policy formation, especially in the Global South where collectivism is strong and vibrant. This study insists that the state has the social responsibility to urgently prioritize agrarian reform (as this is central in any development agenda) that incorporates strong participation of grassroots communities in the policy formation and implementation, along with high recognition on the viability of collective land ownership that reflects customary values, social justice, and sustainable peace. The *tara bandu*, together with the maintenance of other cultural practices that benefit communities, should occupy a vital space in the formal governance system of the country.

References

Batterbury, Simon, Alex Cullen, Demetrio do Amaral de Carvalho, Balthasar Kehi, Lisa Palmer, and Thomas Reuter. 2015. Land access and livelihoods in post-conflict Timor Leste: No magic bullets. *International Journal of the Commons* 9 (2): 619–647. https://doi.org/10.18352/ijc.514.

Belun, and The Asia Foundation. 2013. *Tara Bandu: Its Role and Use in Community Conflict Prevention in Timor-Leste*. Dili: Belun and The Asia Foundation.

Capizzi, Elaine, Helen Hill, and Dave Macey. 1976. FRETILIN and the struggle for independence in East Timor. *Race and Class* 17 (4): 381–395.

da Silva, Antero Benedito, and Kiyoko Furusawa. 2014. Land, state and community reconstruction: Timor-Leste in search of a sustainable peace. In *Confronting Land and Property Problems for Peace*, ed. by Takeuchi Shinichi, 212–241. Oxon/New York: Routledge.

Dos Santos, Luis. 2018. Interview by Ananeza P. Aban and Angeli Fleur Nuque. July 2018.

Fitzpatrick, Daniel, and Susana Barnes. 2010. The relative resilience of property: First possession and order without law in East Timor. *Law & Society Review* 44 (2): 205–238. https://doi.org/10.1111/j.1540-5893.2010.00402.x.

Guterres, Alberto Martins. 2018. Interview by Ananeza P. Aban and Angeli Fleur Nuque. July 2018.

Harson, Siktus. 2015. Toiling for Beans. *UCA News*, July 2. https://www.ucanews.com/news/toiling-for-beans-in-timor-leste-the-struggles-of-coffee-farmers/73865.

Houtart, François. 2011. *'Common Goods' to the 'Common Good of Humanity'*, ed. Renato Sabbadini, trans. Victoria Bawtree. Brussels: Rosa Luxemburg Foundation.

Khamis, Susie. 2015. Timor-Leste coffee: Marketing the "golden prince" in post-crisis conditions. *Food, Culture & Society* 18 (3): 481–500. https://doi.org/10.1080/15528014.2015.1043109.

McWilliam, Andrew, and Elizabeth Traube. 2011. Land and life in Timor-Leste: Introduction. In *Land and Life in Timor-Leste*, ed. Andrew McWilliam and Elizabeth Traube, 1–22. Canberra: ANU Press.

Miyazawa, Naori. 2013. Customary law and community-based natural resource management in post-conflict Timor-Leste. In *Land and Post-Conflict Peace Buildiing*, ed. Jon Unruh and Rhodri C. Williams, 511–532. London: Earthscan.

Niner, Sara. 2013. Bisoi: A Veteran of Timor-Leste's independence movement. In *Women in Southeast Asian Nationalist Movements*, eds. by Susan Blackburn and Helen Ting, 226–249. NUS Press.

Palmer, Lisa. 2017. Timor Leste: Embracing resource governance through ritual in a post-conflict society. In *Routledge Handbook of the Environment in Southeast Asia*, ed. Philip Hirsch, 483–495. London/New York: Routledge.

Santana, Jenito Jewata. 2018. Interview by Ananeza P. Aban and Angeli Fleur Nuque, July.

Shinichi, Takeuchi. 2014. Introduction: Land and property problems in peacebuilding. In *Confronting Land and Property Problems for Peace*, ed. Takeuchi Shinichi, 1–28. London/New York: Routledge.

Soares, Leonardo Fatima. 2018. Interview by Ananeza P. Aban and Angeli Fleur Nuque. July 2018.

Ananeza P. Aban is currently a Ph.D. researcher of the Department of Conflict and Development Studies and a recipient of the Special Research Fund doctoral scholarship of Ghent University in Belgium. She also serves as a senior research analyst of UP Center for Integrative and Development Studies Program on Alternative Development. Among her major tasks are coordinating the field research across Southeast Asia, co-editing the Alternative Practices monograph series (2020, 2023), and ensuring the international linkages of the program. She finished her Master of Community Development degree from the University of the Philippines in Diliman.

Chapter 9
Agroecology Trends and Initiatives on Seed Saving/Banking in the South East Asia Region: The La via Campesina Experience

Raphael Baladad

Abstract The corporatization of agriculture across Asia has intensified rural marginalization, displacing small food providers and undermining traditional food systems. As land, water, and forests are privatized for profit, rural communities face escalating threats, from land grabbing and forced displacement to hunger and loss of cultural food practices. Despite mounting global food demands, governments continue to prioritize corporate agriculture, free trade agreements, and industrial food production over sustainable, community-led alternatives. Against this backdrop, food sovereignty has emerged as a powerful counter-narrative, advocating for the rights of small-scale food providers to determine their own agricultural policies, protect traditional knowledge, and resist neoliberal exploitation. Rooted in redistributive agrarian reform, agroecology, and seed sovereignty, this movement mobilizes grassroots actions to reclaim food systems from corporate control. Case studies from Indonesia, the Philippines, and Thailand highlight how social movements, including La Via Campesina-affiliated organizations, strategically engage in policy advocacy, direct action, and community-led initiatives to advance food sovereignty. While significant challenges remain, including policy co-optation and internal movement pressures, the struggle for food sovereignty represents a critical site of resistance and transformation. By reclaiming the right to land, seeds, and local food systems, rural movements in Southeast Asia are reshaping agricultural futures based on social justice, ecological sustainability, and self-determination.

Keywords Agrarian reform · Agroecology · Food sovereignty · La Via Campesina · Land rights · Rural movements in Southeast Asia

---

R. Baladad (✉)
Focus on the Global South, Manila, Philippines
e-mail: raphael.baladad@focusweb.org

J. M. C. Sy and A. P. Aban (eds.), *Reimagining Development in Southeast Asia*, Asia in Transition 21, https://doi.org/10.1007/978-981-97-9140-8_9

9.1 Introduction

The dominance of corporations across Asia has aggravated the economic and political marginalization of peasants, fishers, indigenous peoples, pastoralists and other people working in rural areas against the backdrop of neoliberal globalization. As land, water and forests are transformed into profitable assets, the countryside has become an arena where elite interests clash with those of communities over territories and natural resources. In the name of development, the struggle for rights, recognition and social justice by small food providers are sidelined by governments to create better business environments and ease the flow of capital.

In a context where global food demands have increased dramatically along with mounting social pressures to attain food security, protecting local food systems has become critical when the majority of the world's food supply still comes from small-scale food providers.[1] Governments, however, have continuously shifted policies and programs away from addressing obstacles to rural development toward broadening investments for corporate/industrial agriculture, expansion of free trade regimes under the World Trade Organization (WTO) or mega free trade agreements (FTAs) and integration to regional/subregional value chains to meet national food supply needs. Rural communities remain mired in poverty and vulnerable to a myriad of threats; from displacement due to land grabbing, armed conflict and climate change, to indebtedness, loss of livelihoods and hunger.

The World Food Programme estimated in 2018 that about 821 million of the world's population are hungry, with approximately half situated in the Asia–Pacific region.[2] Around 75% of the world's hungry live in rural areas and depend almost completely on agriculture for food, employment and income. The United Nations Food and Agriculture Organization (FAO) has also estimated that half of the world's hungry are from small-holder farming communities; another 20% are from landless farming communities; about 10% depend on fishing, herding and forest resources, and 20% live in slums in and around cities, many of whom are migrants from rural areas. Though telling, these numbers do not account for small-scale food providers affected by mega infrastructure projects, extractive industries and other forms of development aggression. Beyond hunger and poverty, small-scale food providers also face the loss of traditional and culture-based methods for food production, as well as traditional seeds' varieties and plant genetics, as they fall into the traps set by corporations who promise better yields and incomes.

Addressing systemic problems requires equally systemic and radical reforms. Where threats against small food providers remain unabated, the need to challenge economic, political and social structures that disempower, erode or capture local food systems become an essential for human survival. Equally important is the need to surface viable alternatives that reclaim the development discourse away from profiteering and commodification of food and natural resources.

---

[1] The Food and Agriculture Organization (FAO) estimates that smallholder farms produce over 80% of the world's food supply.

[2] The WFP estimates the figure at 486 million in 2019.

9.2 Food Sovereignty as an Alternative

Various social movements in the region have been pushing for Food Sovereignty as an alternative to market-oriented and neoliberal models of food production. With growing protests against corporate and trade-driven agriculture that threaten local food systems, Food Sovereignty is gaining traction as a paradigm that outlines the people's right to define their agriculture and food production methods, systems and policies. Since its introduction at the World Food Summit in 1996 by La Via Campesina (LVC),[3] Food Sovereignty has evolved into a holistic campaign that includes redistributive agrarian reform; the defense of land and natural resources against privatization and financialization; the promotion of agroecology, saving and exchange of native seeds, cooperative marketing; and the protection of the rights of peasants and small-scale food producers. In 2007, more than 500 representatives from key land-based social movements around the world adopted the "Nyeleni Declaration" to promote Food Sovereignty as the "right of peoples to healthy and culturally appropriate food produced through ecologically sound and sustainable methods."[4]

While the food security framework adopted by many governments focuses on the provision of food for all by whatever means necessary, Food Sovereignty offers an alternative framework and presents an opportunity for small food providers to resist the corporate capture of food and agriculture, bring autonomy and dignity back to small-scale sustainable food producers and build food systems that are resilient to the challenges of climate change (Patel 2009). Aside from opposing free trade regimes, initiatives under Food Sovereignty aim to promote small-holder farming, to transform conventional farming systems back to its traditional roots and to counter the impacts of Green Revolution technologies.[5] Hence, social movements across the world have been promoting agroecology and seed saving/banking as expressions of Food Sovereignty (La Via Campesina 2003).

---

[3] La Via Campesina (LVC) is an International Movement which coordinates peasant organizations of small- and middle-scale producers, agricultural workers, rural women, and indigenous communities from Asia, Africa, America, and Europe. Currently, LVC has a membership of 182 local and national organizations across the globe.

[4] The Declaration was adopted at the LVC Forum for Food Sovereignty held in Febraury 2007 at Selingue, Mali.

[5] The Green Revolution was a set of research technology and transfer initiatives on food production in the 1950s to the 1970s to increase agricultural output. It was also associated with the adoption of chemical fertilizers, pesticides and other production inputs as well as new technologies for hybridized seeds.

9.3 Peasant Agroecology and Community Seed Banking/Saving as Resistance[6]

While Food Sovereignty is seen as a critical alternative to the concept of food security, advocates prefer not to present agroecology as an alternative at all. In their view, despite the corporate capture of various food systems–as evidenced by the cooptation of organic agriculture and fair trade systems—as well as the prevalence of conventional farming, agroecology remains a dominant model of food production. The glaring failure of agricultural industrialization and Green Revolution technologies to eradicate global hunger and malnutrition, however, have led rural social movements to reclaim losses in traditional modes of agriculture. Moreover, various movements see agroecology as a mode of resistance, not only against prevailing neoliberal models of food production, but against violations of political, social, economic and cultural rights perpetuated by elites, corporations and governments. Beyond training and knowledge-rebuilding on traditional farming methods, the practice of agroecology is also pushed by movements as an approach to reclaim community rights on land, water and forest resources; influence government policies and programs on agriculture; establish autonomy and self-determination in food production; and to broaden solidarities among rural communities.

In a context where access to traditional seeds and plant genetic varieties are threatened by intellectual property regimes imposed by corporations that seek to control seed diversity and supply, as well as to attract investments on biotechnology, seed sovereignty and community seed saving or banking have become centers of struggles for food sovereignty. The Intellectual Property Rights (IPR) chapter in the Regional Comprehensive Economic Partnership (RCEP)—the most prominent mega free trade agreement being negotiated by the ten ASEAN member states as well as Australia, China, Japan, New Zealand and South Korea—will force countries to comply with UPOV 91, an international convention that has been highly criticized by farmers and peasant movements for restricting usage rights and encouraging privatization of traditional seed varieties. The strong push towards corporate agriculture and agribusiness, and a concerted effort to undermine or criminalize the rights of farmers/peasants to seeds have resulted in various campaigns for community-based seed saving/banking, exchanges and dispersal as a form of resistance.

9.4 The La via Campesina South East Asia Experience

Through campaigns and initiatives by social movements, Food Sovereignty is slowly building momentum at the level of practice. While its underlying principles have yet to win the societal support needed to overturn the food policy discourse, rural

[6] Peasant agroecology is the type of agroecology practiced by peasants and small food providers, where agroecology, similar to organic agriculture, is now being co-opted by corporations and large agro-industries.

social movements have worked towards fleshing out policy frameworks and social-economic structures needed for mainstreaming. On the other hand, contextual realities faced by stakeholders to Food Sovereignty create difficulties on scaling gaps in knowledge, practice and movement building needed to generate pressure for legislative reforms or to effectuate systemic changes.

In promoting agroecology and seed banking/saving in the South East Asia Region, this section looks into campaigns and initiatives by LVC member organizations from Indonesia, Philippines and Thailand as lenses to highlight key strategies, emblematic cases as well as challenges.

9.5 Indonesia—Serikat Petani Indonesia (SPI)[7]

Formed in 1998 from the convergence of several leaders under the Federation of Indonesian Peasant Unions, Serikat Petani Indonesia (SPI) is one of the largest peasant organizations in Indonesia that also served as the International Operating Secretariat of La Via Campesina from 2008 to 2013. SPI has led and supported various campaigns and struggles for land rights and has worked towards strengthening the solidarities between Indonesian farmers' organizations and other rural stakeholders towards the realization of genuine agrarian reform and food sovereignty.

Indonesia saw a rapid decline in small-holder agriculture and engagement in family farming, from 31 million in 2003 to 22 million in 2013. The country also witnessed a steady decline in the number of farmlands as the implementation of agrarian reform has lagged in rural areas, where 56% of farming families own only half a hectare of land. With no clear government policies that protect peasant's land rights, corporations have moved towards capturing a majority of Indonesia's countrysides, with more than two decades of aggressive land grabbing, land-use conversions as well as displacement of communities by mega infrastructure projects and the establishment of industrial plantations. In 2014, as an outcome of various policy engagements with the government, Indonesian President Jokowi Widodo pronounced that Food Sovereignty will be adopted to redefine agricultural policy towards self-sufficiency. Conversely in 2018, Widodo also started operationalizing the National Agrarian Reform Program (NARP) which created policy spaces for peasant movements to assert greater claims over farmlands and their utilization. However, these programs remain unable to scale policy gaps in addressing specific land-use issues, realigning national priorities for agricultural production, and in strengthening social and tenurial rights protection for farming communities have resulted in a disjointed and slow implementation.

[7] This section contains excerpts from Focus on the Global South's documentation of La Via Campesina's Asia Continental Agroecology and Climate Justice Encounter in June 2017 and interviews from Agus Ruli Ardiansyah, General Secretary of Serikat Petani Indonesia in February 2020.

Hence, rural movements such as the SPI have struggled to overturn policy debates on agrarian reform and food sovereignty that has been captured by corporate industrial narratives on rural development. Specifically, SPI worked towards influencing local and national levels policies, particularly on (1) Law of Protection and Empowerment of Peasants, otherwise known as Law No. 19 2013; (2) policies that protect the sustainable use of agriculture land; (3) various food laws that govern or regulate agricultural production systems; and (4) the Presidential Instruction on Agrarian Reform that aims for an effective implementation of the NARP. SPI has also contributed Judicial Reviews for the Constitutional Court against policies that threaten the rights of peasants and other small food providers, particularly on state investment laws, free trade agreements and other policies that aim to deregulate corporate interests in agriculture.

In the practice of promoting food sovereignty, SPI has worked towards harnessing the energies of communities towards the practice of agroecology, with a vision to overcome threats to land tenure and livelihoods by making lands productive. As expressions of resistance, SPI campaigned for Agroecology towards: (a) strengthening ownership claims in the struggle of agrarian reform through land occupations; (b) challenging prevailing models of industrial monocropping by enhancing yields through traditional and natural methods of farming; (c) decreasing farmer/peasant vulnerabilities to climate change and demanding greater programmatic support from the government for adaptation efforts; and (d) encouraging farmer/peasants to resist conventional modes of production that has tied rural communities to debt and poverty.

Food sovereignty and the practice of agroecology for SPI are also rooted in the movement's ideologies and its struggles towards dismantling the neoliberal economy that has captured the markets and production discourse on food, by increasing the productivity of family farms, promoting national food self-sufficiency and addressing rural poverty while ensuring environmental sustainability. In this regard, the organization's key activities include (a) cooperative building and linking communities across provinces and regions; (b) The establishment of community agroecology schools and demo farms; and, (c) the strengthening of youth participation through political education and encouraging their broader involvement in local and national campaigns.

For seed saving initiatives, SPI has established seed centers at the national and community levels which are built based on the capacities and needs of its member organizations–with each one having its own system of preserving and cultivating indigenous seed varieties. Situated in a rice-producing region, seed centers in East Java house and disperse varieties of rice. West Java and Yogyakarta seed centers on the other hand houses vegetable varieties, while hard crops are kept in Sumatra. Such seed centers according to SPI are crucial for the success of agroecology as it addresses challenges on seed availability faced by family farms at the onset of cropping seasons However, with the constant pressure by the WTO to junk protectionist policies on agriculture and given the threats of trade liberalization to local food systems, the challenge for SPI's seed centers is to provide evidence that incomes can be raised while production costs can be cut in half by using indigenous seeds. SPI's key strategies in this regard include: (a) establishment of community seed banks for indigenous varieties; (b) promoting traditional methods of cross-breeding of seeds; and (c) promoting community seed development. In addition, SPI has developed new rice varieties through indigenous methods of hybridization (i.e., the SPI-120, now being tested in several locations with a potential to yield 7–9 tons per hectare) and is also in the process of breeding new seed varieties for high value crops such as spinach, papaya and *kangkong* (water spinach) to help communities resist biotech seeds.

Apart from raising productivity through agroecology, SPI's seed centers also support the struggle against criminalization of peasants who are accused of infringing the proprietary rights of corporations in the reproduction and distribution of indigenous seed varieties. Policies such as Indonesia's Plant Variety Protection Act allow farmers to be imprisoned or fined up to USD 70,000 for reproducing indigenous seeds already captured by intellectual property regimes Though no farmer/peasant has been prosecuted by the said law, it constrains the free movement of seeds between various communities/districts.

In sum, SPI's agroecology and seed saving initiatives complement their campaigns for genuine agrarian reform. To them, ensuring productivity is an equally important struggle after land rights have already been claimed. Towards this end, SPI has formed *Kampung Repormang Agraria* (Agrarian Reform Villages) where the practice of agroecology, seed saving has led to success stories, management of water resources, crops, agricultural land and housing.

9.6 Philippines—Paragos Pilipinas[8]

Paragos is a national farmer's organization in the Philippines advocating for social justice in the countryside through agrarian reform and rural development and has been active in supporting land struggles and pushing for the effective implementation of the Comprehensive Agrarian Reform Program. Established in 1992, Paragos became a member of LVC in the same year and has campaigned for sustainable agriculture and agroecology.

Similar to other members of LVC, Paragos' campaign for agroecology is situated at the heart of land struggles and the pursuit of social justice through land redistribution. With the Philippines' agrarian reform program hounded by inefficiencies—in particular, on support services extended to peasant beneficiaries—agroecology becomes crucial to sustain agrarian victories and prevent reversals caused by the lack of farm productivity and indebtedness In a setting where incidences of corporate land grabbing and agricultural land-use conversions have been a common sight in rural areas, the practice of agroecology in the experience of Paragos strengthened community

[8] This section contains excerpts from Focus on the Global South's documentation of La Via Campesina's Asia Continental Agroecology and Climate Justice Encounter in June 2017 and interviews with Jaime S. L. Tadeo, Chairman of Paragos Pilipinas in February 2020.

solidarities and morale in the struggle for land rights as well as reinforced peasant ownership claims in disputed lands.

In 2010, Paragos was part of a coalition of movements that supported the passage of the Organic Agriculture Act, which sought to promote a more sustainable, ecological and cost-effective means in land cultivation amid the rising costs of conventional farming due to oil price hikes in the global market While this law created the needed spaces to promote agroecology, lobbies by large-scale organic producers however facilitated the inclusion of stiff regulations and protocols for certification. In terms of marketing, the law banked on expanding niche middle to upper class markets on organic food. Small food providers, however, faced challenges in competing with established and certified growers. Paragos was then part of a government initiative to conduct capacity building on organic production with local governments, before it decided to disengage and focus on promoting agroecology as a way of life in smaller communities.

For Paragos, agroecology is about safeguarding the health not only of consumers but also of producers who are exposed to the negative effects of chemical fertilizers and pesticides. Paragos also sees agroecology as vital in restoring natural soil nutrients lost due to decades of conventional agriculture, which largely contributed to the lowering of farm productivity and incoes that led to the decline of the Philippines' agriculture sector in the last two decades Further, Paragos believes that agroecology is an effective approach in encouraging the broader participation of rural youth in agriculture, by stressing the importance of securing healthy food for families and communities first beyond potentials for better income.

Paragos sees the importance of agroecology in sustaining long-drawn struggles for land. The experience of the village of Sumalo in the province of Bataan highlights how the practice of agroecology bound the community together despite threats of harassment, criminalization and forced evictions from a corporate entity that seeks to capture farmlands. In their struggle for land rights that has lasted more than two decades, community members of Sumalo practiced agroecology to mobilize people towards occupying lands enclosed by the corporation, and in utilizing other free spaces to grow crops in support of families who lost livelihoods due to the enclosures. The production of natural fertilizers from vermicast sold to neighboring farming communities has supported, at some level, the financial needs to sustain legal struggles and casework. Agroecology has also highlighted the role of rural women in Sumalo who are at the forefront of defending the community against human rights violations, as well as in surfacing alternative livelihoods and other income opportunities for affected families. Lastly, the agroecology initiative in Sumalo has gained support from local and provincial governments that have extended production subsidies and crafted local policies to strengthen agricultural activities in the area, which in turn reinforced community claims over the land.

While agroecology, in terms of adoption, has seen victories in communities embroiled in land struggles, Paragos experienced some challenges in promotion within communities post-land struggles and in areas absent of any external pressure that could bind people together. In the village of Macabud in the province of Rizal, despite community training exercises and the establishment of learning centers and demo farms, some farmers still preferred to retain conventional practices due to concerns brought by significantly lower yields in the initial years upon shifting to agroecology. In addition, because agroecology requires more farm work particularly in the production of natural inputs, growers would opt to sell at higher prices which created difficulties in marketing. In this light, Paragos gears towards establishing community markets (i.e., mobile vendors) and other informal hubs where producers can be linked directly to consumers. Through its training programs and other knowledge building activities, Paragos rebuilds traditional customs in farming that stress community cohesion and food self-sufficiency, thereby mainstreaming practices on agroecology and effectively countering perspectives on agriculture as solely a profit-generating enterprise.

While Paragos has yet to undertake a broader campaign for seed saving/banking, seed exchange activities and training for indigenous seed preservation have already been integrated into various initiatives to promote Food Sovereignty and agroecology in communities. The passage of the Rice Trade Liberalization Law (RTL) in 2019, however, has compelled its members to undertake measures to protect local producers, such as buying *palay* (unhusked rice) from affected communities and extending support for milling and storage. As cheap imported rice flood local markets through corporate distribution networks, local rice varieties are now under a greater threat of being eased out and lost In this context, along with its participation in broader campaigns to repeal the RTL at the national level, Paragos looks into establishing community rice seed centers and conducting training programs for household level storage as well as exploring options for seed dispersal in partnership with local governments, civil society and other rural social movements.

9.7 Thailand's Assembly of the Poor (AOP)[9]

The Assembly of the Poor (AOP) is a social movement in Thailand that represents the political voice of communities and marginalized members of society. Present in 36 provinces, AOP works towards empowering small food providers while supporting struggles for land, water and forest resources. Established in 1995, AOP became a member of LVC in the same year and has since worked towards dismantling the influence of neoliberal economic systems in agriculture as well as defending communities from abuses and other rights violations perpetuated by transnational corporations. AOP has also campaigned against the government's unjust land management policies that have discriminated against and displaced thousands of small food providers.

AOP's campaign for Food Sovereignty is connected to sustaining victories in the struggle for land rights and preventing reversals caused by the overdependence

[9] This section contains excerpts Focus on the Global South's documentation of La Via Campesina's Asia Continental Agroecology and Climate Justice Encounter in June 2017 and interviews with Baramee Chaiyarat, Chairman of Alliance of the Poor–Thailand in February 2020.

to conventional farming. Prevalent in almost all food producing provinces in Thailand, conventional farming led small food providers into debt. With limited options to sustain livelihoods, increasing costs of production, higher vulnerabilities to crop failure due to extreme weather events, lack of production support from the government and lower incomes in rural areas, communities are forced to resort to contract farming and other exploitative means of production (Quinn-Thibodeau 2016).

Food Sovereignty has also guided AOP's campaigns against free trade agreements, including the Trans-Pacific Partnership, and other policies that endanger small food providers, such as the establishment of special economic zones or large plantation estates. AOP has also struggled to defend local food systems against protectionist policies such as forest, fisher forest and wildlife sanctuary laws which corporations and the government itself exploit to modify tenurial arrangements and push communities out of territories.

In terms of promoting agroecology, AOP has two main thrusts: campaigns for program and policy reforms at the national level in the short term and campaigns for community empowerment and strengthening solidarities in the long term. With regard to the former, AOP is continuously in the process of holding dialogues to pressure the Thai government and its ministries into supporting local initiatives for agroecology–essentially through reforming land policies, allocating budgets and institutionalizing programs to expand such initiatives. At the same time, with respect to the latter thrust, AOP is facilitating the formation of cooperative estates where communities are encouraged to integrate and broaden agroecological modes of food production.

AOP's key strategies in promoting agroecology include: (a) the integration of trainings and workshops in communities entrenched in land struggles, where the practice of agroecology is used to build stronger land rights claims; (b) strengthening indigenous and traditional modes of production within forest dwelling communities under threat of displacement; and, (c) support for disaster-stricken communities, particularly those hit by droughts, where agroecology can provide options in rebuilding livelihoods. AOP is also actively promoting integrated farming systems such as the cultivation of alternative crops or fruits, sustainable fishing and organic fertilizer production in sites of land struggles. Within forest communities, AOP promotes (a) multiplication of indigenous forest seedlings; (b) forest food calendars; and, (c) training for backyard gardening and livestock raising. In disaster-stricken communities, AOP encourages the breeding and planting of traditional disaster resilient crops varieties and training for alternative livelihoods.

AOP has also worked towards surfacing community perspectives on food sovereignty and agroecology. In the small forest dwelling village of Chong Tuko, the community strongly followed indigenous and traditional means of food production and has rejected practices that harm the forest's biodiversity. Despite this, the entire village was displaced by the military under the guise of upholding forest protection policies in the area. After years of struggle, the provincial government was pressured by AOP into allocating around 100 hectares of land within the forest reserve. Now renamed the village of Kokedoi, the community thrives in its practice of agroecology and is able to market products to visitors, tourists, as well as nearby urban communities. In its campaigns, AOP has highlighted community perspectives in agroecology to counter prevailing government narratives that forest dwellers destroy forest areas and should be driven out. In this case, we see how agroecology was vital in reclaiming rights and spaces for small food providers in the discourse of environmental protection and sustainable development (Lassa and Priamarizki 2015).

AOP has also pushed for agroecology in four villages in the Ta Sae District in the Chumphon Province, and three villages in the Ban Tad Mountain range in Southern Thailand. Communities have successfully adopted and sustained agroecology initiatives in these areas, similar to Kokedoi, to respond to threats of displacement from forest protection policies. However, in other provinces where communities face few threats to their lands, living conditions and livelihoods, AOP faced difficulties in encouraging or convincing people to adopt agroecology or veer away from conventional modes of food production.

Although AOP is yet to gear towards the promotion of seed saving/banking, regular seed exchanges integrated into agroecology training and workshops have widened the appreciation of communities on seed sovereignty and the need to protect indigenous seeds. At the national level, AOP supports campaigns to resist policies that impose greater intellectual property regimes on seeds and to lessen the restrictions to small food providers in utilizing seeds already subjected under IPs. AOP, along with several farmers' organizations in Thailand have staged massive protests in 2006 and 2013 to derail FTA negotiations with the United States and the European Union.

9.8 Conclusion

The success of Food Sovereignty in challenging neoliberal food regimes in the foregoing cases has depended largely on the resistance of social movements to conventional and industrial agriculture, the push for policy reforms that protect and promote the rights of small-scale food providers and the practice of food sovereignty expressions such as agroecology and seed banking/saving. While agroecology and seed banking/saving helped communities to respond to threats on local food systems (such as cheap agriculture imports, land grabbing, displacement and indebtedness), they have also become vital in challenging dominant narratives on agriculture and food, amplifying community voices and demands and in building solidarities to confront systemic problems faced by small-scale food providers.

The experiences of SPI, Paragos and AOP illuminate the relevance of an age-old strategy in advancing campaigns by generating pressure from below to effect changes and reforms from above. Deepening the practice of agroecology and seed saving in the contexts of struggles has helped in redefining community perspectives on development and emerged new campaign directions for rural social movements in reclaiming political and economic rights as well as in pursuing social justice in the access to land, water and forest resources. Through negotiations, protests and other social actions that engage governments complemented by community actions that reinforce solidarity movements are able to influence policies and programs that protect local food systems from corporate interests, defend community rights and uplift living conditions in the countryside.

Another pressure, however, comes internally from the need to validate the effectiveness of campaigns, strategies and initiatives when the need to surface alternatives becomes critical to the survival of communities. While the need to scale gaps in building awareness and broadening societal support have hounded campaigns on Food Sovereignty in the region, great strides have already been taken by movements in reclaiming the right to have rights over food systems and in grasping the reins towards developing of vibrant and peasant-based rural economies.

References

Coulibaly, Ibrahima. Why We Need Agroecology. In *Agroecology: Putting Food Sovereignty into Action*. Why Hunger, n.d.

Lassa, Jonatan A., and Adhi Priamarizki. 2015. Jokowi's food sovereignty narrative: Military in the rice land? S. Rajaratnam School of International Studies, Nanyang Technological University, February 27, 2015.

La Via Campesina. 2003. Food sovereignty—Via Campesina English, January 15, 2003. https://viacampesina.org/en/food-sovereignty/.

Patel, Raj. 2009. Food sovereignty. *The Journal of Peasant Studies* 36 (3): 663–706. https://doi.org/10.1080/03066150903143079.

Quinn-Thibodeau, Tristan. Landless villagers in Thailand fight for their families and lead the fight for agroecology. *Why Hunger* (blog), September 26, 2016. https://whyhunger.org/category/blog/landless-villagers-in-thailand-fight-for-their-families-and-lead-the-fight-for-agroecology/.

Raphael Baladad is a graduate of Political Science from the University of Santo Tomas and is currently a Program Officer at Focus on the Global South, where he focuses on rural development issues, particularly Food Sovereignty, Agrarian Reform, and Peasant Rights. In addition to his professional work, he is an active member of Paragos Pilipinas and an agroecology practitioner in his personal time.

Part IV
Education and Knowledge Transfer for Local Development

Chapter 10
The Organizing Work of Ayta Mag-Indi Women in Porac, Pampanga, in Advancing Women's Rights and Community Governance

Venarica Papa, Justin Jette Roldan, Micah Magaro, Carl Eugene Balbas, Rita Capuno, Josie Abuque, Marivic Capuno, and Violeta Valenton

Abstract Indigenous Peoples have a long history of struggle in asserting their right to ancestral domain and self-determination. These challenges deepened their commitment to uphold and protect their culture and rights to ancestral lands. Community organizing has played a crucial role in advocating for and securing these rights. This enabled them to utilize democratic spaces in sharing native knowledge, enhance their skills and build strategic alliances. The Ayta Mag-Indi of Porac, Pampanga, highlighted the role of women organizing in strengthening their community to rise above challenges. The women leaders of Ayta Mag-Indi shared their journey in advancing women's rights and how this has contributed to the broader struggle of the entire Ayta community; from addressing the cases of violence against women in their community to revitalizing the role of women in community governance. Their strength rooted on Indigenous knowledge, system and practice, in their way of doing, reflected in their communal way of life, grounded on collective values and social solidarity.

Keywords Ancestral domain · Community governance · Community organizing · Indigenous knowledge systems and practices · Women's rights

The Philippines has many different indigenous communities. They are distinguished from the majority of the population by their opposition to colonial powers. According to history, the entry of many foreign groups led them to live in the mountains and find a way to survive there (Ty 2010). The Aytas are recognized as one of the earliest indigenous groups in the Philippines. Eventually, the mountains became the homeland of generations of Aytas and served as their land of ancestry (Shimizu 2001).

V. Papa (✉) · J. J. Roldan · M. Magaro · C. E. Balbas
University of the Philippines, Diliman, Quezon City, Philippines
e-mail: vbpapa@up.edu.ph

R. Capuno · J. Abuque · M. Capuno · V. Valenton
Ayta Mag-indi Community, Porac, Pampanga, Philippines

J. M. C. Sy and A. P. Aban (eds.), *Reimagining Development in Southeast Asia*, Asia in Transition 21, https://doi.org/10.1007/978-981-97-9140-8_10

Aytas treat the mountains as part of themselves. According to Fondevilla (in Shimizu 2001, 7), their home has its roots in the world, which is shaped according to the clouds, the rainbow, the sun, and the stars, and as its name suggests, Mount Pinatubo serves as the source of life for the Aytas. The mountains that serve as the ancestral lands of many Ayta groups are found in the Central Luzon region. One of these groups is the Ayta Mag-indi, who dwell in the municipality of Porac, Pampanga. Their ancestral lands span more than 18,700 ha. Encompassing this land are Sitio Pasbul, a sitio (site) in Barangay Camias, and the Barangays of Sapang Uwak, Camias, Diaz, Villa Maria, and Inararo, and different Ayta communities settle in these areas (James and Paton 2016; Capuno 2018).

10.1 The Eruption of Mount Pinatubo

Despite their opposition to foreign powers, the Aytas have been continuously threatened by the attempts to claim and steal their ancestral domain and even by Apo Namalyari's (Almighty) display of wrath.

One of the most haunting events that significantly affected the lives of Aytas was the eruption of Mount Pinatubo in 1991 (Seitz 2004). Before Pinatubo erupted, danger zones around the area had been identified one by one since April 1991. According to local officials and disaster coordinating councils, the declaration of an area as a danger zone signaled mass evacuations (Tayag et al. 1999). In June, earthquakes became more frequent especially near the volcano, leading authorities to stress the need to evacuate especially those living in the volcano's 10-km radius. At the onset, around 2000 evacuated. Because of the impending eruption of Mount Pinatubo, many more people evacuated from their communities. The first formal evacuation of communities happened in April 1991 (de la Cruz 2015).

Ayta communities living around the volcano were significantly affected by its eruption. Leaving their homes became difficult and heartbreaking because it not only serves as a source of livelihood but also is the home of Apo Namalyari (Fondevilla 1991).

According to the stories of Aytas, there might be another reason for the eruption of Pinatubo, which has been dormant for centuries, aside from the belief that it was caused by a force majeure. Before the volcano's eruption, the Philippine National Oil Company (PNOC) had been conducting explorations because it believed that the volcano could be considered a source of geothermal energy. According to Bienvenido Capuno's testimony, an Ayta who worked for PNOC, its operations began in 1984. He said that it began with surveying hot areas of the volcano (Lillo in Shimizu 2001). After the first part of the exploration, the PNOC pored holes on the land at a large scale.

The PNOC laid pipes on each of the three boreholes on the mountain, but they all melted. At that point, the PNOC realized that the volcano cannot serve as a source of energy. It attempted to cover these large boreholes by pouring cement. There was also an attempt to cover up PNOC's significant fault through a news blackout.

Amid PNOC's operations, Bienvenido Capuno said that the Aytas had been aware that continued drilling on the volcano would not be beneficial at all, but it was more difficult to halt the project because the government agreed with it.

One study points out that the PNOC indeed conducted preliminary surveys on Mount Pinatubo from 1982 to 1986 and deep exploratory drilling and well testing from 1988 to 1990. The PNOC was forced to stop the project after failing to prove their hypotheses 13 months before the first eruptive activities of Mount Pinatubo happened (Delfin et al. 1999).

10.2 Evacuating and Seeking a New Place to Dwell

The hardships that the Aytas felt became more significant due to the need to flee their communities at a large scale and due to an uncertain future for them. The need to walk and find shelter amid an erupting volcano greatly affected the elderly, the youth, and even women. There were narratives of mothers and fathers carrying their children while crossing the lahar. Some pregnant women walked to the next city and crossed rivers and mountains to reach a safe place. Some died due to extreme thirst and fatigue caused by walking long distances. And some families separated. Many of those aforementioned stories have proven the difficulty of their journey to find a safe, decent, and temporary shelter.

According to the testimony of Bappa Condring, an Ayta elder, during the time of the calamity, some Aytas went to their relatives in nearby sitios and barangays while some rode trucks and other modes of transportation to evacuate. Yet most Aytas in Porac walked to various designated evacuation centers in neighboring areas such as Sindalan in Angeles and San Fernando. Around 25,000 "unat" and "kulot" people were cramped in these evacuation centers. Afterward, half of the Ayta evacuees were brought to resettlement sites that were constructed for them (Seitz 2004). The Aytas refer to outsiders as "unat" because this characteristic clearly distinguishes them from non-community members.

Other Aytas reached other cities and provinces such as Nueva Ecija, Zambales, and other towns bordering Manila. Don Bosco School in San Fernando, Pampanga became the first evacuation center for Ayta communities in the provinces of Pampanga, Zambales, and Tarlac. After temporarily dwelling in Don Bosco, they transferred and lived in Magalang, an area around Arayat, for a few years before transferring again to a designated resettlement area called "Tent City" in Barangay Planas. The area was called "Tent City" because when these indigenous communities came, there were no houses constructed. They were expected to construct huts or tents on the land as temporary shelters. Because of the length of stay in "Tent City," the area was eventually called "Katutubo Village" ("Indigenous People's Village").

The community decided to stay in Katutubo Village because it was closer to public services such as schools and hospitals and to the municipality of Porac. It is situated a few kilometers from Barangay Camias, which form part of their ancestral domain. After authorities declared the barangay safe, they attempted to return to Camias, and they saw how it was buried under lahar. However, it never became a reason for them to give up on their dream to return and protect their ancestral lands. Together they strove hard in digging lahar, and together they dug the land until it became suitable for agriculture.

The majority of the Aytas decided to not leave their houses in Katutubo Village. As such, going up to Camias became a part of their way of life to continue planting and hunting, which have been their primary source of living in their ancestral lands.

They also discovered that as they returned to their ancestral lands, there were new challenges in the form of intruders that took with them development projects. These were logging companies, mining, surveyors, and other parties that have since attempted to trespass into their ancestral domain.

Some other individuals, who are mostly "straight-haired," constructed houses and property inside their ancestral lands. This became a sign for the Aytas to see the need to further protect and care for their ancestral lands against those who attempt to enter, claim, and exploit them.

10.3 Lessons Learned

The hardships that the Aytas experienced since the eruption of Mount Pinatubo are part of the things that gave them the determination to rise up and strengthen their unity. Their great love and appreciation for their ancestral lands that are extensions of their lives inspired Ayta Mag-indis to continue and rise up. To lose these lands means to breathe their last breath. Therefore, when some attempted to intrude on their ancestral lands, they would act collectively to fight for them. The community's unity in standing for their ancestral domain prevailed. Both men and women in the community joined arms to organize and collectively fight for their rights. This led to the formation of organizations in their community such as the Pagkakaisa ng mga Ayta sa Camias (PAC, *Unity of Aytas in Camias*) and women's organization UPKC or Ugat sa Pusod ng Kababaihan ng Camias (*Rooted from the womb of Camias women*).

10.3.1 Indigenous Organizing Work

Manalili (2012) contributed to the growth of literature on community organizing. He says that development with relevance is not progress that is solely for personal gain. For Manalili (2012, 131–32), the goal of community development is the betterment of the whole community, especially the poor. He also adds that genuine development comes when people serve as the principal actors in the process of development (134). Therefore, we can fully realize the relevance of the general aims of community organizing and development if we continue to deeply understand the significance of the hardships of the majority (142).

For the native Aytas, organizing work is nothing new because of their communal way of life. They have been accustomed to working together in all aspects of their life. Their unity in terms of protecting their culture, beliefs, and rights to their ancestral lands has become more prominent. In the section "Saligang Paniniwala sa Pag-oorganisa at Sama-samang Pagkilos" ("Principles of Organizing and Collective Action") in Manalili's (2012) book, social investigation is one of the basic organizational tasks. In the Aytas' context, they have realized the importance of organizing and developing their community, especially the struggle for the recognition of their right to self-determination and their right to their ancestral domain.

Organizing became the key to establishing relations with other organizations. While the most significant and primary topic of these linkages is defending their right to their ancestral domain, they have also included participation in many kinds of activities such as trainings and short courses that raise their knowledge and skills, orientations in coordinating with government agencies and institutions and with other sectors, training in establishing dialogues and utilizing democratic spaces for advancing their rights, among others that altogether help improve their skills and broaden their connections with other institutions. All of these have helped strengthen their organizations. It opened opportunities to create alliances and discuss experiences and struggles with other organizations and communities of Aytas in Central Luzon. These efforts resulted in expanding knowledge and skills in organizing, and they helped in empowering their communities. These helped enhance their capacity to lead and govern themselves.

The communal character of the Ayta community should not be treated with superficiality. Communal living entails trusting each other, instilling compassion and empathy, helping voluntarily, giving importance to the self as part of the community, and treating the community as something that binds together individual interests and welfare. The struggle of an individual Ayta is a struggle for all Ayta communities. These highlight practices that value the formation of alliances and networks with various groups especially in campaigning for human rights, struggling for their ancestral domain, pursuing peace, and others. Some of the best examples are Paruogan, which is comprised of an alliance of Aytas in Central Luzon and church organizations, and Mideng Kitamu Ha Lutang Ninuno (Stand for Our Ancestral Lands), which are organizations that were formed by their struggle against foreign corporations' and individuals' intrusion and land-grabbing in their ancestral domain.

10.4 Rising from Challenges

The Aytas are remarkably resilient and resourceful. They have many ways to emerge from the challenges that they face. They take advantage of opportunities to learn prevailing ways of developing their indigenous practices. They utilize democratic spaces, sometimes creating these spaces themselves, in asserting their native knowledge and their ways of cultivating them. For example, they saw the importance of education in protecting their ancestral domain and stopping discrimination that they experience. As such, the Aytas, especially those from the younger generations, have been pursuing formal education, pursuing courses they know their community needs most such as education and social work. Along with this, they strengthen their coordination with the Department of Education to ensure the inclusion of their indigenous knowledge and practices into the curriculum.

In relation to education, the Aytas also studied and continue to study the rights of indigenous peoples and the laws that protect them. Through this knowledge, they further strengthened their advancement and defense of their rights.

The Aytas also consulted and coordinated with those that they considered allies. Included in the strategy of continuous learning is the translation of documents into the Filipino languages to allow people in the community to become more aware and involved in the issues that affect them, their culture, and their ancestral domain. They also form alliances with different institutions such as schools, churches, and nongovernment organizations (NGOs). They also have their own initiatives in spreading awareness of their situation and experience. There was a time when a company conducted aggressive mining activities in their area, as a response, the Ayta themselves coordinated with universities to ensure that their experiences were documented and exposed to the public. They also spoke in open fora in one occasion, they have invited experts in anthropology as part of their campaign to stop mining in Barangay Camias. Their community has likewise actively participated in meetings where issues that affect indigenous Filipinos like them are discussed.

10.5 The Influence of the "Straight-Haired" on the Ayta Community

As the clutches of urbanization reach the Ayta community, and some "straight-haired" people's expropriation of some parts of their ancestral land, the Ayta community has since become more susceptible to external influences—and changes in their lifestyle became inevitable. According to them, the arrival of the "straight-haired" people in their community marked the beginning of their hardships.

While the ancestral lands that primarily sustain their needs still continue to exist, the Aytas started to have other needs that entail the need for money. For example, because of the desire to end discrimination against them, they have started to accept and value formal education. As such, they need money to send their children to school, signaling more expenses for tuition, allowances, fares, and more.

Leaders have also needed to spend money whenever they attend meetings and conferences concerning the rights of indigenous Filipinos. In addition, they also continue to coordinate with various organizations and allies, entailing the need for them to go on trips from their communities.

They sought many different ways to earn money. Some continue to go up to their ancestral lands to plant corn, squash, and other crops that can be sold as they descend to the plains or when distributors come to their community.

These changes in their ways of living were not considered losses but rather adaptations to their changing environment.

10.6 Changing Relations Within the Community

Because of long-term interactions with the "straight-haired," women and men have gained new roles within the family. Because their families do not solely rely on agriculture as their source of livelihood, some Ayta men as, have been forced to seek livelihood outside the community; meanwhile, women stay in their community to continue agricultural activities, take care of their children, and do household chores. Because they have to sustain more needs, they also need to make more efforts in terms of their labor, time, and sacrifices that consequently erode communal way of life.

The intensified challenges of raising a family put significant pressure on Ayta men. When they fail to meet the family's needs, they often express this sense of failure, leading to tension within Ayta families that was not present before. This tension directly affects wives, who are often subjected to violence by their husbands. Such incidents led to the emergence of cases of domestic violence often against women. This includes physical, verbal, and sexual abuse by their husbands.

According to the Aytas, this kind of violence is alien to their culture. Ayta culture is one that cares, loves, and respects others. Violence in their households and against women has been an influence of outsiders' culture that has infiltrated and gradually eroded the fabric of their own corroding the foundations of their relationships, camaraderie, and way of life.

Because of the interactions between the "straight-haired" and the "curly-haired," the Ayta became receptive to practices that are not natural to them, such as forming gangs and drinking. In one story, a man recklessly spent his money on liquor, and that money would have been used to sustain his family's needs. He would start hitting his wife when she criticized his behavior. These forms of violence that affected Ayta women did not come from the mountains.

10.7 The Role of Women in the Struggles of the Community

Again, the influence of outside culture has to be emphasized. In Ayta culture, women are valued and respected because the Ayta recognizes the significant role that women play in their community.

While community leaders, which are mostly men, take their time in responding to the issues that they face, Ayta women do not stay mum. They have a significant and crucial role in the community. For example, Marivic Capuno, a woman leader, narrated that when there was an attempt to encroach their ancestral landwomen confront those who attempt to trespass. Carrying their children, they form barricades to halt the intrusion of private mining companies. This is alongside and despite the Ayta women's efforts to attend to various responsibilities at home and within their families.

According to Marivic, having household chores does not bar them from participating in community activities, especially when it concerns the need to defend their ancestral domain.

Another role that women play in their community is accompanying their husbands in community meetings because their presence and attention are needed. As they head home, the wives share their observations and feedback from the meeting. This proves that women and men cannot be separated from each other because they work together and cooperate in the majority of the aspects of their lives. There is no big or small role because everyone's roles are important. Therefore, the Aytas clearly treat women and men equally. In addition, women and men have equal rights.

Like what was said earlier, the Ayta are not naturally violent. The Aytas are loving and are advocating a peaceful way of life. In fact, the concepts of rape and swearing are alien to them. But now that there are cases of violence against women, the whole community has recognized changes and realized the need to address these concerns immediately, because to defend women's rights is to defend the community and their culture as well.

Because of these changes, Ayta women are currently strengthening UPKC, a pioneering women's organization. This organization became their foundation in raising the knowledge of women about their rights. It will also serve as an avenue for the return of indigenous practices in the whole community—practices that highly value and respect women. Furthermore, they recognize that strengthening the organization will contribute to the general struggle of the Aytas in defending their interests and their rights especially in terms of their ancestral domain.

In an effort to document these practices and preserve the integrity of their narratives, the researchers have ensured that the voices of the native Ayta narrators remain prominent. For this study, documentation primarily involved voice and video recordings to convey their message in the most accessible way, amplifying the voices of the Ayta Mag-indi of Porac, Pampanga.

10.8 The Alternative Development Practices of the Aytas

The Ayta Mag-indi of Porac, Pampanga, have developed and sustained their community through their unique practices. They have a form of governance informed by their Indigenous Knowledge, Systems and Practices (IKSPs) and manage their ancestral lands according to their traditions. Their natural advocacy for peaceful living fosters harmonious relationships among themselves and a deep care for their environment. The Ayta Mag-indi have showcased a rich culture, knowledge, and indigenous customs shaped by their experiences and history since the time of their ancestors. Today, they preserve these traditions and pass them down to future generations.

Collective action is both a way of life and a means to face and rise above challenges. They recognize that the struggle of one is the struggle of all Aytas. And this recognition led to more systematic organizing that effectively addressed issues that affect their communities.

Clearly, raising women's awareness of their rights is closely tied to strengthening their recognition as integral members of their community. Further, fostering solidarity among women contributes to the community's ability to effectively govern itself. Advancing women's rights goes hand in hand with advancing and defending their culture.

This has been proven true with the experience of the Ayta women in organizing. While the immediate objective of providing support and assistance victims of violence and harassment is met, the organization likewise invigorates the role of women in community governance and helps respond to challenges that their community faces. With their organizing efforts, they have strengthened the community's resolve and capacity to protect and uphold their right to self-determination, self-reliance, and their right to their ancestral domain.

Realizing that organizing is a continuous learning process, the Ayta women make sure that they give time to community discussions, sharing and learning from one another's experiences, further understanding their people's situation and in the context of the larger society.

The Ayta women's organization has also been open to learning from other allowing them to share experiences and draw lessons to enhance their practices. This is evident in their participation in various meetings and workshops organized by numerous partners and networks. Among these efforts are the workshops that Ayta women launched with the help of organizations such as GABRIELA and the faculty of University of the Philippines College of Social Work and Community Development (UP CSWCD). These workshops and training were all done with full respect to and consideration of the Ayta Mag-indi's indigenous practices, culture, and history.

In the continued efforts of their community to show their indigenous practices, they have continued to coordinate with various institutions. A good example is showcasing their knowledge in herbal medicine and featuring their rich history, culture, and even their games. They shared their heritage with professors and researchers from different universities. This activity aimed to make their knowledge of their culture more known, to encourage many people to accept them, and to ensure that they are recorded for their future generations. Some of the first fruits of these efforts are the publication of a book about their native language and studies that describe their indigenous games and unique way of life.

It is clear that for these Ayta Mag-indi women, they are the primary actors in achieving their goals for their community. Indigenous practices sustain them. To them, this is not an alternative model of community governance nor merely an innovation. This is their way of doing, their way of life—collective, communal, and based on social solidarity. This tells us that there is much we can learn from the Women of Ayta Mag-indi Community in governance.

References

de la Cruz, Gwen. 2015. Looking back: When mount pinatubo blew its top. *Rappler*, June 16. https://www.rappler.com/move-ph/issues/disasters/96424-1991-mount-pinatubo-eruption.

Delfin, F.G., Jr., H.G. Villarosa, D.B. Layugan, V.C. Clemente, M.R. Candelaria, and J.R. Ruaya. 1999. Geothermal exploration of the pre-1991 mount Pinatubo hydrothermal system. *United States Geological Survey*, June 10. https://pubs.usgs.gov/pinatubo/delfin/index.html.

Fondevilla, Emma. 1991. *Eruption and exodus: Mt. Pinatubo and the Aytas of Zambales*. Quezon City: Lubos na Alyansa ng mga Katutubong Ayta ng Sambales.

James, Helen, and Douglas Paton, eds. 2016. *The Consequences of Disasters: Demographic, Planning, and Policy Implications*. Springfield: Charles C. Thomas Publisher, Ltd.

Manalili, Angelito. 2012. *Pag-oorganisa ng Pamayanan Tungo sa Kaunlarang mula sa Pilipino para sa Pilipino*. Manila: Kapatiran-Kaunlaran Foundation.

Seitz, Stefan. 2004. *The Aeta at the Mt. Pinatubo, Philippines: A Minority Group Coping with Disaster*. Lungsod City: New Day Publishers.

Shimizu, Hiromu. 2001. *The Orphans of Pinatubo: The Ayta Struggle for Existence*. Manila: Solidaridad Publishing House.

Tayag, Jean, Sheila Insauriga, Anne Ringor, and Mel Belo. 1999. People's response to eruption warning: The Pinatubo experience. *United States Geological Survey*, June 11, 1999. https://pubs.usgs.gov/pinatubo/tayag/.

Tejero, Constantino. 1991. Under the Volcano: Clark on the edge of extinction. *Sunday Inquirer Magazine*.

Ty, Rey. 2010. Indigenous peoples in the Philippines: Continuing struggle. *FOCUS* 62, December. https://www.hurights.or.jp/archives/focus/section2/2010/12/indigenous-peoples-in-the-philippinescontinuing-struggle.html.

Venarica Papa is an Assistant Professor in the Department of Community Development at the University of the Philippines, Diliman. Her research interests span human rights, community governance, and alternative development. She is actively engaged with marginalized groups, including indigenous peoples, persons with disabilities, the urban poor, and women.

Justin Jette Roldan is a development worker at a university in Manila. He is passionate about promoting indigenous knowledge, systems, and practices.

Micah Magaro is passionate about writing, with experience spanning from creative writing to journalism. She particularly enjoys exploring topics related to community development.

Carl Eugene Balbas is a community organizer who has worked with various marginalized sectors. He is currently working for a government project on DRRM focused on establishing Community-based early warning systems for landslides.

Rita Capuno also known as Indun Mayumo, is a respected woman leader and elder in the Ayta Mag-indi community. She plays traditional indigenous musical instruments used in rituals and important community events.

Josie Abuque also known as Indun Saging, is a respected leader and elder in the Ayta Mag-indi community. In addition to her leadership role, she serves as a mangagun (healer) and midwife and has attended numerous trainings and conferences with the objective of upholding the rights and empowerment of her community.

Marivic Capuno also known as Indun Limuel, is a devoted mother, promoter of peace, and a staunch human rights advocate. She is the first female Tribal Chieftain of the Ayta Mag-indi in Katutubo Village, Brgy. Planas, Porac, Pampanga, and a founding member of the local women's organization.

Violeta Valenton also known as Indun Malu, is a prominent female leader and elder in the Ayta Mag-indi community. She is actively involved in advancing women's rights and empowerment within her community.

Chapter 11
Teaching "Pangiyak Ki!": The Lumad School as a Struggle for Land, Life, and Liberation

Jose Monfred C. Sy

Abstract In the Philippines, the alternative schools for the Lumad exemplify how Indigenous peoples can pave their way toward their own modes of development that counter that of the state and the private sector. Responding to the shortfall of state services, the Lumad, with the support of academic, religious, and civil society organizations, established schools that allow students to recognize their potential as shapers of their own lives and future. This case study maps the development of this alternative learning site by situating it in the Lumad people's history and struggle for governance over their yutang kabilin (ancestral land). The paper shall first trace the historical emergence of the Lumad as a political identity of non-Muslim Indigenous groups in Mindanao. It discusses the development of the alternative schools for the Lumad, and later, the mobile Bakwit School, along with the challenges they faced. Finally, the paper delineates the nationalist, pro-people, and scientific pedagogy of the schools that suggests what alternative education could be, one that arises outside the logic of capital and encourages social transformation.

Keywords Alternative education · Ancestral land · Critical pedagogy · Indigenous peoples · Lumad schools

Despite national and international laws, treatises, and initiatives that seek to promote their rights, indigenous peoples (IP) across the world continue to be sequestered to the fringes of society. Abetted by local compradors and the bureaucratic elite, the transnational market—in its search for cheap raw materials and human resources—subject IP communities to land grabbing, resource dispossession, forced removal or relocation, and socio-cultural denial, accompanied by violence by way of threat, intimidation, or even death (IWGIA 2021; Belisario 2019, 77; IBON 2018, 13; UNPFII 2018, 3). A veneer of economic development veils states' abandonment of the lands and resources of its indigenous population to the hands of a destructively technocratic world (Escobar 1995, 44). Threats against these communities' material and cultural

J. M. C. Sy (✉)
University of the Philippines (UP) Diliman, Quezon City, Philippines
e-mail: jose_monfred.sy@up.edu.ph

J. M. C. Sy and A. P. Aban (eds.), *Reimagining Development in Southeast Asia*, Asia in Transition 21, https://doi.org/10.1007/978-981-97-9140-8_11

existence are further exacerbated by the paucity of state support for basic social services such as health, safety, and education in the countryside. The lack of access to education discourages IP communities from scrutinizing and organizing against these attacks, rendering them more vulnerable to corporate business intervention in their ancestral domains (Belisario 2019, 72; Lahoy and Roc 2019, 12).

It would be wrong, however, to frame IP communities as passive actors in the face of these challenges (Canuday 2009, 3). Historically, Indigenous peoples have demonstrated and sustained alternative strategies to pave their way to their own definition of development parallel to or against that of the state and the private sector. The establishment of alternative tribal schools for the Lumad of the Philippines is an exemplar attempt at responding to the incompetence of the state to provide educational services in the farthest reaches of the country. Emerging from the endeavors of religious groups to uplift the high incidence of illiteracy in Mindanao, such schools integrate indigenous knowledges and technologies to what would be the "mainstream" design of formal education homogenously implemented in state schools. As of now, there are around 219 Lumad schools in four regions in Mindanao operated with the efforts of Lumad communities, religious and civil society groups, and activist organizations (Salugpongan n.d.). For the Lumad, these schools become alternative spaces for the development of critical consciousness as their very existence acknowledges (and resists, as I shall show later) the systemic nature of oppression and injustice based on the Lumad's lived experiences as targets of violence (Serrano and Xhafa 2012b, 4).

Unfortunately, over the recent years, the operation of these schools has been disrupted by heavy militarization as part of the Philippine government's counter-insurgency program. As reported by the Save Our Schools Network (2019, 1–2), a campaign network for the right to education of indigenous Lumad children, "the rising number of attacks, child rights violations, and the use of schools for military purposes has effectively denied and continues to deny [almost 4,000] Lumad children [from] their inherent right to education." These attacks forcibly evacuate Lumad communities out of their ancestral domains. In July 2019, Department of Education (DepEd) Secretary Leonor Briones signed the temporary closure of 54 campuses of Salugpongan, a tribal school, "invoking baseless and recycled accusations that [they teach] children to mount ambuscades against the military," affecting 1133 Lumad students (Belisario 2019, 72; SOS 2019a, 1). Most recently, Vice President Sara Duterte who is also the Department of Education Secretary accused teachers and volunteers of the Lumad schools (run by the Salugpungan Ta' Tanu Igkanogon Community Learning Center) as "illegally operating" and "had links with the terrorist groups New People's Army, the Community Party of the Philippines, and the National Democratic Front of the Philippines" (Ombay 2022).

From 2017 and 2018 up until 2021, about a hundred Lumad students, parents, teachers, and volunteers have mounted a Bakwit School in Metro Manila, a makeshift school for displaced indigenous children from Mindanao. Garnering support from allies in the academe, the church, non-government organizations (NGOs), and human rights defenders, the alternative learning venue allows the Lumad evacuees to continue their formal schooling and, at the same time, to hold protest actions around the capital region's centers of political power. It moves from one refuge to another in order to ensure its students, faculty, and staff's security. The Bakwit School stands as a testament to the unyielding struggle of the Lumad not only for education, but also for their ancestral land, the site of their lifeways' flourishing.

This paper seeks to trace the historical and political conditions that necessitated the creation of tribal schools for the Lumad of Mindanao, and later the Bakwit School. After detailing major points in the development of these schools, I shall discuss their pedagogical foundations and objectives to highlight their interrogation of state-sponsored education. This case shows that the schools do not only respond to the shortfall of state services but are also a form of protest. Writing from the perspective of a teacher in the Bakwit School, I shall pin down the challenges that these schools and its supporters have been facing in the recent years, particularly under the Duterte regime's martial rule in Mindanao. Through this, I wish to reveal how such alternative learning spaces aid the Lumad's fight for land, life, and liberation.

11.1 Land and the Lumad: A Short History

The successful establishment of alternative tribal schools is backgrounded by a grim history of land and resource dispossession and military violence in the ancestral domains of IP communities in Mindanao. A short historical review of Mindanaoan political economy positions the nascence of the Lumad identity in conflict and struggle for land. Non-muslim ethnic groups have already settled in Mindanao and several parts of Visayas even before Spain reached our shores in the 1500s. Beginning with Spain in the sixteenth century, colonial powers have attempted to subjugate these groups in the name of socio-political and economic expansion. Spanish settlement and internal migration in the Mindanaoan lowlands drove the tribes that would later become the Lumad to the hinterlands (Tan 2003, 25–26; Zapico et al. 2015, 138).

The nineteenth century witnessed the earnest rise of the United States as an imperial and economic power in the globe (Canuday 2009, 11), and with it the gradual decline of Spanish rule. Armed clashes in Mindanao between the US military and resistant Islamized polities continued to transform and complicate the area's demography and territorialities (Gowing 1977, 21–29). The ability of upland peoples to maintain control over their communal territories in the hinterlands became key to the protection of their lifeways (ICHRP and IBON 2018, 42). The collaborationist and puppet governments that followed the disruption of US power due to the Great Depression in the 1930s and the World Wars in the 1940s saw demographic shifts that minoritized the Indigenous peoples of Mindanao, transforming migrant Filipinos as the new (lowlander) majority in the island (Rodil 2004, 46–55; Tan 2003, 29).

The Marcos regime following the Third Republic marked a decisive turn not only in the Philippine mining industry but also in the lifeworlds of many IP communities. Pressure upon IP communities' land base was compounded by a national economy that became increasingly foreign-dominated and export-oriented (Molintas 2004,

273). Superpower countries targeted nations impoverished by the Wars for social and political intervention (Escobar 1995, 23). International markets and private banks offered cheap interest rate loans for developing countries in pursuit of state-led industrialization (Harvey 2006). At the level of policy, the late dictator Ferdinand Marcos sought a state-led mining regime through an aggressive promotion of domestic firms' mineral exploration in peripheral regions (Camba 2015, 8). Nevertheless, such state-facilitated incentives and large-scale mining activities were propelled by low-interest US commercial loans (Holden 2005). Presidential Decree 1559 of 1978 opened the gates of the countryside for the entry of foreign capital by declaring that cultural minorities and other occupants of public lands may be ejected for the "best use" of a territory in question (ICHRP and IBON 2018, 43). Infrastructure projects in the 1980s such as the Chico Dam, the Bataan Nuclear Power Plant, and state-supported mines expelled entire populations from their settlements (Camba 2015, 10). Through bribery and/or violent coercion, leaders of IP communities were made to surrender their lands to compradors and agents of either or both the national government and foreign corporations. The illiteracy prevalent in some areas rendered such communities more vulnerable; some of them have even been deceived into signing documents in exchange for what should have been given free as relief humanitarian assistance.

National development defined by the fascistic state only provoked and empowered resistance movements. Peoples' organizations and IP groups launched protests against the expansion of mines during Marcos's rule (Thompson 1996). Collectively asserted by several groups belonging to these tribes, the Lumad identity we know now emerged during the martial law regime to affirm and assert governance over their ancestral domains, what they call *yutang kabilin* or land of heritage (Canuday 2009). The autonym "Lumad" is a Cebuano shorthand for *Katawhang Lumad*, literally "people of the land" or "native of the land." This collective political identity is composed of the Ata, Bagobo/Ubo, Banwaon, B'laan, Kalagan, Kaulo, Dibabawon, Higaonon, Mamanwa, Mandaya, Mangguwangan, Manobo, Mansaka, Matigsalog, Subanen, Talaandig, Tiboli, and Tiruray (SOS 2019a, 3). At that time, several IP groups in Mindanao fortified their stance against the Marcos dictatorship to lobby for self-determination over their lands, which are targets of the government's resurrection of a failed mineral regime (Camba 2015, 8). It was on June 26, 1986 during the Lumad Mindanaw People's Federation (LMPF) founding assembly that the identity was affirmed by the same groups.

The downfall of the Marcos dictatorship did not end the Lumad's struggle for land and life. 1997 saw the ratification of the Indigenous Peoples' Rights Act (IPRA), or Republic Act No. 8371, which was touted to echo the "progressive" provisions of the 1987 Philippine Constitution that recognizes customary laws governing property rights and relations and their validity with regard to the ownership and extent of ancestral domains (Molintas 2004, 295). Likewise, IPRA upholds the United Nations Draft Declaration on the Indigenous Peoples, which underscores the collective rights of indigenous communities (ibid., 296). While considered a landmark in legislation for marginalized peoples, IPRA has not been effective in protecting the Indigenous peoples' rights over their land and self-governance, especially given its conflict with other existing laws such as the Philippine Mining Act of 1995. This conflict spells a difference in the conceptions of land ownership and management between IP communities and the Philippine state, allowing the massive land grab of ancestral domains (ibid., 292–94). The formal registration system—indeed a Western concept of land ownership and management (Huggan and Tiffin 2010, 8)—only served to transform communally owned lands to individually titled private lands, easing the process of transferring property (Molintas 2004, 292). The Mining Act of 1995, in the same vein, facilitates the entry of large foreign and local mining corporations into the Indigenous peoples mineral-rich territories, opening up the national mining sector to foreign control (ibid., 294). Such laws and tedious administrative procedures only provide formal legitimacy for the government to declare untitled land to be (sellable) state property and thereby undermine ancestral rights (ICHRP and IBON 2018, 42–43).

Under the iron fist of former President Rodrigo Duterte, mining, logging, and other foreign corporate ventures remain to be a behemoth threat to the Lumad. Today, Mindanao stands as one of the last frontiers of natural resources and fertile soil in the Philippines (SOS 2019a, 3). Ironically, its peoples suffered the highest poverty incidence in the country last 2019 (PSA). In 2021, the BARMM was reported as the poorest region in the country and had the highest poverty incidence in families at 29.8% (Rivas 2022). Living in remote forests and mountains, the Lumad people, which now compromise 61% of around 14–17 million Indigenous peoples in the country, remain one of the poorest with very little access to basic social services (Belisario 2019, 72). Even with IPRA, big plantations and mining corporations still manage to encroach over their *yutang kabilin.* As of 2016, most of the 23 priority mining projects under the government's mining revitalization program in Mindanao lie within or nearby the ancestral domains of the Lumad (Villareal 2016). Some of the largest mining companies in the world are presently operating in Mindanao as well. These include the BHP Billiton, Anglo American Corp., Sumitomo, St. Augustine Gold and Copper, Apex Mining, Toronto Ventures Inc., Indophil, and others. There are also 15 coal operating contracts (COCs) in one of the regions in Mindanao encompassing 42,000 ha documented in 2015. Mindanao has 166 mining tenements as of July 2019 as reported by the Mines and Geosciences Bureau (MGB) (Olea 2011). Aside from mining tenements, about 500,000 ha of land in the five regions of Mindanao are now covered with agro-industrial plantation crops primarily for the export market. This is equivalent to 12% of Mindanao's agricultural land. Another 1,000,000 ha are targeted for oil palm expansion by 2022. Over the years, more than half of the Lumad's ancestral domain have been grabbed by foreign, large-scale mining corporations, with around 700,000 ha converted into cash-crop plantations by the same or related multinational companies (SOS 2019a, 3).

Publicized as a response to the Marawi crisis of 2017, Duterte's imposition of martial law in Mindanao, along with the counter-insurgency scheme Oplan Kapayapaan, has only inflicted unfathomable terror on the Lumad people. With 75% of the Armed Forces of the Philippines (AFP) deployed in Mindanao, the Lumad's ancestral land continues to be denuded, partitioned, and secured for the highest bidders (SOS 2019a, 4). Recent cases of intensifying militarization and the prosecution of Lumad territories across the Pantaron Range and the Andap Valley Complex only exposes how the Duterte government champions its economic interests over the welfare of its people and their lands. Harboring the headwaters of major rivers in Mindanao, the Pantaron Mountain Range straddles across the Davao region up to North Cotabato and Bukidnon. In 1994, the indigenous group Talaingod Manobo successfully fended off the encroachment of the Alcantara and Sons Incorporated (a big logging concessionaire) from ravaging one of the last remaining intact rainforests in the range (Umil 2018a). However, in 2014, Kerlan Fanagel, Secretary General of the Confederation of Lumad Organizations in Southern Mindanao (Pasaka day Salugpungan Kalimudan or PASAKA), learned that the Aquino government plans to build a Php 9 billion (USD 200 million) mining highway that leads up to Pantaron Range, hence the increased military encampment in Lumad communities and the renewed presence of military-organized paramilitary group Alamara to counter resistance against the project (Ayroso 2014). Years after, facilitated by martial rule in Mindanao, the 56th Infantry Battalion of the Philippine Army (IBPA) set up checkpoints that gridlock the paths to Sitio Bulyan, Brgy. Palma Gil, Talaingod, beefing up its surveillance over the community on the premise of countering insurgent or rebellious acts, which includes those against big logging operations (Malayao 2018).

A similar bolstering of military presence in the Caraga region can be explained by the treasure chest that is the Andap Valley Complex. Straddling the towns of San Miguel, San Agustin, Marihatag, Cagwait, Tago, and Lianga, Andap Valley has been targeted for extractive projects by mining corporations such as Benguet Corp., Abacus Coal Exploration and Development Corp., the Chinese-owned Great Wall Mining and Power Corp., ASK Mining and Exploration Corp., and CoalBlack Mining Corp. In 2018, the 75th IBPA built a patrol base beside the community starting with more than 30 soldiers who would later stay in people's houses at daytime and follow farmers on their way to the fields to harvest vegetables (Moderno 2018). Likewise, the Lumad farmers in Opol, Misamis Oriental and Kalabugao, Bukidnon are withheld from tilling their ancestral lands due to the palm oil plantations A Brown Energy and Resources Development Inc. (ABERDI) and subsidiary Nakeen Corporation. In this case, it is the 4th IBPA that protects the corporate plantation by driving away any resistance from the Lumad communities displaced from the area (Miller 2019).

While the global movement for the codification of IP rights at the turn of the century appear to be a boon for the welfare of the Lumad, the conflict-ridden history of the Lumad's emergence and ongoing uphill battle for ownership over their ancestral domains suggests that there is little change between the past and the present. These notable cases of land grabbing and military attacks only prove how the current regime and its martial rule in Mindanao follow the footsteps of Marcos and the precursive colonial regimes in sequestering the indigenous population to the margins in the name of profit and power. Nonetheless, as mentioned earlier, many IP groups have been countering these attacks by (1) engaging with policymakers, (2) launching protests and demonstrations, and (3) acting on the shortfall of state services in the hinterlands. In the case of the Lumad, a successful initiative that encompasses all three strategies is their alternative tribal school.

11.2 Building Schools

That marginalized communities are denied access to basic social services such as education has made the Lumad people all the more vulnerable to the extractive and extortive neoliberal projects of the state. About 9 out of 10 Lumad did not have a proper education, and most, if not all, have no access to adequate health services (SOS 2019c, 3). Children needed to travel miles only to attend classes in town centers. Many students are forced to drop out due to the burden of the high cost of transportation, food, lodging, the discriminatory treatment against Indigenous peoples as second-class citizens, and military surveillance and harassment (Belisario 2019, 72).

These challenges prompted several communities among the Lumad to cultivate a site for educational services with the objective of uplifting the lives of indigenous youth. Starting as outreach missions, literacy and numeracy projects were conducted by religious organizations in Surigao and North Cotabato way back in the 1980s. Through those years, with the help of non-government and peoples' organizations, such programs grew into formalized alternative learning institutions. In 2019, a total of 219 alternative tribal school campuses has been established all over Mindanao with the efforts of the Lumad people, religious institutions, and civil society groups, serving up to 9000 students from indigenous and peasant communities (Salugpongan International n.d.). The long list includes the campuses of Alternative Learning Center for Agricultural and Livelihood Development (ALCADEV, Inc.) and Tribal Filipino Program of Surigao Sur (TRIFPSS) of Caraga; Rural Missionaries of the Philippines' Literacy and Numeracy Schools for Indigenous Children in Northern Mindanao; Center for Lumad Advocacy and Services, Inc. (CLANS) in SOCCSKSARGEN; and Salugpongan Ta Tanu Igkanogon Community Learning Center Inc. (STTICLC), and Mindanao Interfaith Services Foundation, Inc. Academy (MISFI) in Southern Mindanao (Umil 2018b).

Alternative tribal schools for the Lumad respond to the constitutional right of all Filipino citizens to access education. Article XIV, Section 2, Paragraph 4 of the 1987 Constitution states that the "state shall encourage non-formal, informal, and indigenous learning systems, as well as self-learning, independent, and out-of-school study programs particularly those that respond to the community needs." The United Nations Declaration on the Rights of the Indigenous peoples also enshrines the Lumad's right to establish and control their educational systems in a manner appropriate to their lifeways (UN 2007). Most if not all of these alternative tribal schools had also secured permits to operate and have complied with the requirements of DepEd Indigenous Peoples Education Office (IPSEO) for learning institutions. The Salugpongan schools, for instance, had reached recognition status as set by the government. ALCADEV Inc. had likewise been recognized by DepEd's Bureau of Alternative Learning System (BALS) as a duly recognized alternative learning institution. ALCADEV also provides self-learning and out-of-school study programs that are sensitive to the needs of Lumad communities (SOS 2019a, 7).

The program of these schools revolves around the needs of indigenous communities. The schools aim to provide an education relevant and sensitive to the Lumad peoples' culture and needs by offering academic, technical-vocational, and alternative farming skills in coordination with DepEd, local government units, other schools, and civil society organizations (Belisario 2019, 73). Three dimensions of IP survival are triangulated in this design: academics, agriculture, and health. Their academic curricula emphasize basic literacy and numeracy; sustainable agriculture, technology and livelihood education; indigenous languages, Filipino, and English; and the appreciation of indigenous arts and culture (SOS 2019a, 5). Students are segregated based on their grade level, and recently, some tribal schools have adopted the K-12 Basic Education Curriculum by adding a two-year Senior High School component as a requirement for graduation. Primacy is given on sustainable agriculture, the lynchpin of life and livelihood in their ancestral lands, as a subject where other topics could be studies, such as mathematics (weights and measurements, inventory, calendaring), nutrition (selection of crops to be cultivated), communication (reading agricultural texts, interpersonal communication during farming), and values (cooperation and collective labor). These schools' health component involves training in various levels of first aid and the use of indigenous assets for medical purposes. Agricultural practice in schools also sustains the health of the community as it becomes the source of food for teachers, students, and staff. Thus, the use of their ancestral territories to sustain their struggle for education and self-determination also lays claim over the land sought after by mining, logging, and agro-industrial corporations.

For striving to provide an education to the unserved, Lumad schools have been awarded the Most Outstanding Literacy Program in Caraga. ALCADEV in particular has been recognized as the 5th Most Outstanding Literacy Program in the Philippines (de Santos 2018). Despite these accolades, alternative tribal schools for the Lumad have been targeted by harassment, threats, and other forms of attacks through the years, hindering the students' education. During the appointment of Briones as DepEd Secretary, the vilification of alternative tribal schools was adamant. The DepEd, the AFP, and other government institutions have maliciously tagged Salugpongan, ALCADEV, Inc., and CLANS as illegally operating on the basis of a "refusal to comply with the requirements of the Department of Education and of teaching subversive doctrines to their students" (SOS 2019a, 6). The former Davao City Mayor, now Vice President and Secretary of the DepEd Sara Duterte-Carpio previously claimed similar accusations against the Salugpongan, which was countered by the group's Executive Director (San Juan 2019). In the case of CLANS, documents for their operation have long been submitted to DepEd but a permit has yet to be tendered despite successive follow-ups, curtailing classes (SOS 2019a, 7). Government officials also alleged that Lumad schools teach students how to hold and fire a gun and how to sing the "rebel's version" of the Philippine national anthem. While still unproven in any court of law, these allegations echoed by paramilitary elements acting as Lumad leaders have been used to justify the "temporary" closure of schools (Sambalud 2018). The 1333 Lumad students of 54 Salugpongan campuses cited earlier are victims of this black propaganda (SOS 2019a, 1) (Fig. 11.1).

Fig. 11.1 View during one of the classes at the Lumad school (Photo courtesy of the Save Our Schools Network)

Violence inflicted by the AFP and its paramilitary organizations is legitimized by the AFP Directive Letter No. 25 dated July 15, 2013 and through DepEd Memorandum No. 221 dated December 13, 2013. These orders allow the invasion of military elements in schools and communities, infracting Republic Act No. 7610 (Anti-Child Abuse Law), which stipulates that schools must be zones of peace for the youth. On the premise of extinguishing communist insurgency in the Philippine countryside, they have also given way to a number of horrendous cases of attacks including extrajudicial killings of students, teachers, and members of Parent-Teachers-Community Association (PTCA); sexual harassment of students and teachers; threats and harassment; use of students and other civilians as guide and shield during military operations; physical assault and torture of students; fake and forced surrender (as rebels) of PTCA members, teachers and students; arbitrary detention and illegal arrest; forcible evacuation; and the forcible closure of schools that alarmed human rights institutions and the media, both local and international.

Recent cases of attacks this decade have forced students and their communities out of their ancestral land. As early as May 2015, Alamara has already fully occupied multiple villages in Talaingod and in Kapalong, Davao del Norte. The anti-communist group's vilification and indiscriminate firing displaced more than 700 Lumad from their communities, depriving students their right to attend classes in at least 24 primary and secondary schools operated by the Salugpungan Ta Tanu Igkanugon Community Learning Center and Mindanao Interfaith Services Foundation Incorporated Academy. Parents and teachers who opposed the group were threatened to be killed (Manlupig 2015). Until 2017, thousands of residents of Kapalong found refuge in evacuation centers in Davao City due to Alamara's continued harassment (SOS 2019a, 8).

National and international media has also covered the Lianga massacre, what may be one of the most horrendous cases of attacks on schools this decade. Later in 2015, another paramilitary group organized by the AFP known as Magahat-Bagani launched an attack against Manobo communities in Lianga, Surigao del Sur. Magahat-Bagani rounded up the villagers to force them to stop the operations of schools in the area. Community leaders Dionel Campos and Jovillo Sinzo and the director of the local tribal school Emerito Samarca were beaten and shot at close range after resisting the paramilitary's demands. These murders sparked the mass exodus of about 3000 Lumad from the provincial capital of Tandag. The villagers, with the assistance and protection of several humanitarian organizations, believe that the Magahat-Bagani displaced them to prepare the Andap Valley Complex for upcoming mining and logging ventures (Go 2015).

In the three years of the Duterte administration, the Save Our Schools Network has documented 671 cases of attacks on alternative tribal schools, 584 of which were inflicted after the declaration of martial law in Mindanao in 2016. The former president Rodrigo Duterte himself publicly declared in his 2017 State of the Nation Address that "[he] will use the Armed Forces, the Philippine Air Force. *Talagang bobombahin ko 'yung mga ... lahat ng ano ninyo* [I will really bomb all of it]. Because you are operating illegally and you are teaching the children to rebel against government" (Lingao 2017). Indeed, the leader of the Philippines' bombing threat against Lumad children has materialized into barbarity in the Lumad's ancestral domain. The upswing of incidents upon the enforcement of martial law in Mindanao, which endorses and facilitates the bolstering of military presence in much of the Lumad's domain. The most numerous of which are cases of threats, harassment, and intimidation, which in turn sets up conditions for extrajudicial killings, military occupation, indiscriminate strafing, and others. Martial law seems to have expedited the use of aerial bombardment to destroy schools as well. Accompanying the premise of counterinsurgency, many of the students, teachers, and parents who were illegally detained were also forced to confess that they are surrenderees of the New People's Army (NPA), the armed wing of the Communist Party of the Philippines (CPP). Their false confessions have been used to justify more attacks on schools, spinning more fervently a cycle of violence in the Lumad's ancestral domain.

Many of these attacks have happened only a few months after the imposition of martial rule in Mindanao. In February 2017, gunshots were indiscriminately fitted in Sitio Midpanga, Barangay Hinalaan, Sultan Kudarat, forcing more than 60 Dulangan Manobo students and volunteer teachers of CLANS to evacuate. At that time, 30 Lumad schools were closed to protect more than 1745 Lumad children and 80 volunteer teachers. Three of these schools were also victims of a series of air bombings carried out in Kalamansig, Sultan Kudarat (SOS 2019a, 7). Teacher Jolita Tolino, an active campaigner against plantation and mining companies entering the Dulangan Manobo lands, has since then been illegally detained. On June 20, 2017, Rodel Butanlog, CAFGU operative and former Alamara indiscriminately fired gunshots at the school building of Salugpongan in Sito Nasaliban, Talaingod where more than 60 children were holding classes. Classmates and teachers were evacuating the classroom. On July 6, 2017, continuing military operations in Surigao del Sur also disrupted the classes of ALCADEV and TRIFPPS as 2000 Manobos including 545 children were forced to evacuate after elements of the 75th IBPA interrogated villagers and threatened them of launching bomber planes (ibid., 8). Later that year, two Lumad students, aged 16 and 22, were tortured from November 30 to December 2017 in the camp of the 71st IPBA. After being burned alive, the two managed to survive by playing dead.

Attacks against schools have only worsened in the following year. On November 28, 2018, facilitated by the 56th IBPA's encampments and checkpoints nearby, Alamara and elements of CAFGU gathered the communities in Talaingod, Davao del Norte and accused their leaders and children of aiding the insurgent NPA. Community members were coerced to sign petitions testifying to such "crimes" and endorsing the effective closure of nearby Lumad schools. Harassments and violent threats ensued, forcing several families to leave the area (Malayao 2018). The first ever Salugpongan school campus along with 16 others in Talaingod were temporarily shut down for being targets of military encampment (SOS 2019b, 2). This attack began a series of forcible closure of schools that culminated the following year with DepEd's decision to stop the operations of the remaining Salugpongan schools. The link between these closures and the mineral exploration proposals submitted by Metalores Inc., One Compostela Valley Minerals, and Phil. Meng Di Mining Development Corp to the Department of Environment and Natural Resources (DENR) is evident. These explorations would cover 31,180 ha of heavily forested ancestral land of Talaingod Manobo and other Lumad groups in the Pantaron Mountain Range.

In 2019, two years after the Lianga massacre, ALCADEV and TRIFPSS were once again attacked. In the morning of April 28, the 75th IPBA and the 401st Infantry Brigade of the 4th Infantry Division of the AFP, along with Marcos Bocales, leader of Magahat-Bagani, assisted the unwarranted intrusion of the Office of the Presidential Adviser on the Peace Process (OPAPP), Indigenous People's Peace Panel of OPAPP, and the Committee on Indigenous Cultural Communities and Indigenous Peoples (CICCIP) between ongoing classes. Classrooms were forcibly opened, and the unexpected visitors took pictures of students and teachers without consent. Lumad community leaders condemned the lack of notice and absence of local representation in the visitors' team. According to Lumad organization Malahutayong Pakigbisog Alang sa Sumusunod (MAPASU), the reason behind the invasive and pernicious attack is to find proof that the schools serve as training grounds for NPA members. Some children were terrified to recognize Bocales, whose involvement in the 2015 Lianga massacre left a mark on the students of Samarca. Residents have been forcibly evacuated from their homes due to the heavy military occupation of the 4th Infantry Division (Belisario 2019, 71). Just recently, Lumad volunteer teachers Chad Booc and Gelejurain Ngujo II, along with three other civilians were brutally murdered in a "bloody massacre" that transpired last February 24, 2022 in Davao de Oro. They were known Lumad rights advocates and were vilified by the state forces "... to the extent of planting guns and ammunition on the bodies to make it out as if they were combatants" (Save Our Schools Network Cebu 2022).

From 58 schools that were shut down in 2018, there are now 135 Lumad schools forcibly and effectively closed in Mindanao, affecting almost 4000 students and hamleting hundreds of communities (SOS 2019b, 2). Given these attacks, the military has become synonymous to the divestment and destruction of school properties. The now mythic use of counterinsurgency as an excuse to harass Lumad students and handicap their schools further occludes the primary motive behind the communist insurgency in the Philippines: poverty and the effective denial of the poor, oppressed, and marginalized from their democratic rights and state services. Responding to the press, Fanagel clarifies that the presence of elements of the NPA in the Mindanaoan hinterlands is already a given, especially with the historic and geographical significance of Lumad territories in the armed struggle of the CPP. The displacement of Lumad communities, however, covers up the military's human rights violations in commitment to the implementation of the government's counterinsurgency programs (Ayroso 2015). Ensnared by accusations and experiences of armed conflict, the Lumad's successful strides toward affording themselves access to education are eclipsed by hindrances beyond their campaign for governance over their *yutang kabilin* and a share of state services.

11.3 *Og Eskwela A Puron* (To School I Wish)

Dangerous accusations and violent attacks launched against Lumad children, tribal leaders, and their supporters can only propel their resistance further. In light of these challenges, many Lumad communities *bakwit*. Appropriated from the verb "evacuate," to *bakwit* is not simply to cope or to survive. Albeit a dreaded part of the indigenous Lumad life, necessary evacuation in urban centers such as Davao City and Cotabato City saves themselves collectively from military and paramilitary violence in the countryside. It is also a persistent and creative capacity of the displaced to deal with the destructive force of displacement (Canuday 2009, 1). Through *bakwit*, they carve a means of preserving their culture, strengthening their unity, and defending their ancestral homes through a promise of return (Ayroso 2015). It defends the land by leaving. *Bakwets* (the evacuees themselves) bring their campaign from evacuation centers in Mindanao to other parts of the Philippines. In 2014, the Lumad and other IP and Moro communities launched the Manilakbayan (now called Lakbayan), a protest caravan to the heart of Manila in order to bring the capital's attention to minorized challenges like landlessness and military occupation in the countryside.

To sustain the learning of the Lumad youth, *bakwets*, parents, teachers, and supporters mounted the Bakwit School, a refuge classroom that moves from one site to another to facilitate the staging of their protests and to ensure the security of participants (SOS 2019b). More than 3000 Lumad students disenfranchised by martial law's havoc to their schools study in refuges in Metro Manila and other capital cities to expose the situation of alternative tribal schools in Mindanao. Under the Save Our Schools Network, churches, seminaries, universities, and colleges open their doors to the transient Bakwit School (SOS 2019a, 15). From 2017 to 2018, and from April 2019 until today, around 700 Lumad students, teachers, and parents from all over Mindanao have already traveled to Manila, mostly by land, to seek justice for human rights violations in the form of attacks against schools, communities, ancestral land, and resources (Belisario 2019, 78). The evacuation centers in UCCP Haran Center in Davao City and at the Tandag Sports Complex in Surigao del Sur also house Bakwit Schools to let the Lumad continue their studies.

As the fruition of the Lumad's annual Lakbayan, the Bakwit School aims to drum up the demands of the Lumad people to end the military attacks in their schools and communities and to end the national government's all-out-war in the Mindanaoan countryside (SOS 2019a, 15). Hence, the alternative learning venue's thematic campaign is "*og eskwela a puron*," the Manobo phrase for "to school I wish," which embodies the Lumad children's aspiration to attain education. The campaign was developed in response to the first extension of martial law in Mindanao in 2017, which warranted more years of struggle for the Lumad.

The Bakwit School holds classes even with little human and material resources. Given that there are only a handful of Lumad teachers in the School, volunteers are enjoined to teach subjects the students need to take in order to move up to the succeeding level. Volunteers need not be licensed teachers; most of them are undergraduate and graduate students from different disciplines, homemakers, religious people, professionals, and activists. Teachers also undergo educational discussions that relate the Lumad's current situation, culture, and, ultimately, their historic struggle toward emancipation (SOS 2019a, 15). Classroom materials were hardly salvaged from attacked schools, so teachers and students rely on donations by other volunteers and external institutions.

A day in the Bakwit School usually begins with vernacular songs and exercises that the Lumad children would have normally conducted in their original schools, giving students and teachers a semblance of home. In between classes, older students and teachers assigned that day would cook meals to be shared by everyone in the School, including volunteers.

To ensure the maximization of resources, students, teachers, and staff are grouped into committees that are assigned to household duties such as cooking, cleaning, and laundry. Some are relegated to creative tasks, such as in the visual arts and the publicity committees where members craft placards for protests and other artistic materials that may be used in the activities of the Bakwit School. In light of multiple office raids and unwarranted detentions of human rights defenders in the last quarter of 2019 under the whole-of-nation counterinsurgency approach of Duterte's Executive Order 70 of 2018, security measures in the Bakwit School were bolstered by assigning older students, teachers, and volunteers as sentries during the night.

With the Save Our Schools Network, the Bakwit Schools across the Philippines have led successful mobilizations over the years, including the National Children's Day protest every 14th of November. While an emphasis on rescinding martial rule in Mindanao was made during the 2018 protest, the National Children's Day protest in 2019 also called for climate and environmental justice, underscoring the fact that the youth of this generation shall be answering to the ecological impacts of climate change in the years to come. Bakwit School students, together with Moro and other Christian children invited to the event, formed a human tree sizing the University of the Philippines (UP) Diliman's Sunken Garden as part of the protest, catching the attention of both local and international media. Students and teachers of the Bakwit School also join public demonstrations staged by other sectors, such as peasants, workers, women and LGBTQ+, and the urban poor. In every protest, students carry their most fundamental call "Save our schools! Stop Lumad killings!" to urge the government to withdraw military presence in their communities so that they can return to their *yutang kabilin*.

It was on March 29, 2019 at the University of the Philippines Integrated School (UPIS) that the Bakwit School's first milestone was celebrated. With allies from the church, the academe, and civil society organizations, a moving up ceremony was held to recognize the students' one year of struggle for education in the School. It also saw the moving up of Grade 11 students as the first batch of Grade 12 in the Bakwit School. As said by a Grade 10 Lumad student from MISFI Academy in a speech akin to a valedictory address, "*kaya naman, sa kapwa kong mga estudyanteng Lumad, nagtapos man ang isang taon, dapat magpatuloy tayo sa ating mga nasimulan* [that's why, to my fellow Lumad students, one year may have passed, but we must continue what we have started]." Indeed, no amount of threat could break the spirit and determination of Lumad students in asserting their right to education, and through which defend their *yutang kabilin* for the next generation.

11.4 Learn to Struggle, Dare to Win

As alternative learning sites for Indigenous peoples, the pedagogical foundations and design of tribal schools for the Lumad and the Bakwit School address the most immediate concerns of students. While DepEd used to recognize these schools as part of the state's alternative education framework, Lumad schools differ from the objectives and operations of the Alternative Learning System (ALS). The ALS features a ladderized, modular non-formal education program intended for elementary and secondary school dropouts, out-of-school youths, and adults with difficulty in literacy and numeracy (DepEd n.d.). The DepEd specifies as target of the ALS the "group of marginalized learners consists of street children, indigenous peoples, farmers, fisherfolks, women, adolescents, solo parents, children in conflict areas not reached by the formal school system, rebel returnees, and others" (ibid.). Despite this objective, ALS schools remain logistically inaccessible to many Lumad communities across Mindanao.

Conducted either in formal schools, community halls, or private places, ALS classes follow *uniform* lesson modules for all academic subjects. While it is touted by the government as an *alternative* to mainstream education, its focus on livelihood skills training with basic reading and writing still anchors itself on a homogenized set of learning objectives molded by the state. In other words, the ALS program does not seek to empower students in determining and solving challenges faced by their communities. As neoliberal capitalism extends the market to all areas of life, curricular uniformity in mainstream education responds to the needs of the global market for labor by saturating students with skills training and basic literacy and numeracy, which are common qualifications for low-paying, short-term, and unprotected working arrangements (Lanuza 2015, 105–08). With labor development at its core, conventional civic education shepherds the youth toward more productive uses for large profit-making enterprises (ibid., 105; see Giroux 2009).

Community- and civil society-led schools for the Lumad become an alternative modality of education as they arise out of and beyond the logics of capital and profit (Serrano and Xhafa 2012b, 2). As the Save Our Schools Network insinuates, at the core of these alternative schools for the Lumad and the Bakwit School is a "nationalist, pro-people, and scientific pedagogy supplemented by conscious social practice" (SOS 2019a, 5). With a nationalist thrust and a popular orientation, the principles of this curriculum work around learning and exit outcomes that can bolster necessary competencies and eventually specific subjects through agricultural practice, a "skill training" that directly responds to the food security of participating communities and renders the *yutang kabilin* productive not for profit but for the well-being of students, teachers, and staff. Exposed not only to the issues but also the culture, traditions, and lifeways of the Lumad, students are likewise encouraged to serve in local or neighboring communities as a way of giving back to the efforts of everyone involved in establishing and sustaining the schools. This pedagogical foundation fosters the students' holistic development that provides access to life skills that are useful in their daily lives and to their IP communities.

"Nationalist, pro-people, and scientific" may be shorthanded in the liberatory concept of *critical pedagogy*. These alternative schools for the Lumad and the Bakwit School answers Paulo Freire's call for a transformational education that can respond to institutional and ideological domination under capitalism (Freire 2005; Gruenwald 2003). As an alternative and critique of this system, Freirian critical pedagogy is best understood as "an effort to work within educational institutions … to raise questions about inequalities of power, about the false myths of opportunity and merit for many students, and about the way belief systems become internalized to the point where individuals and groups abandon the very aspiration to question or change their lot in life" (Burbles and Berk 1999, 50). This modality of education promotes a materialist consideration of socio-economic position vis-à-vis the ideologies and institutions that gridlock social mobility, including, as mentioned, the uniform and state-defined education and the rationale behind it.

For the schools, their critical pedagogy enables the children to understand social realities and stand up for their ancestral lands and their human rights (SOS 2019a, 5). In the first place, it is impossible not to discuss activism and militancy in the Lumad classroom given that threats to students and teachers' lives linger. More interestingly, this political education even involves their social practice on the ancestral land itself. As connoted by the etymology of the term "Lumad," that land is life is fundamental to the worldview of Indigenous peoples (ECIP 2008, 113–20). The lifeworld of IP is moored to their ancestral land where their lifeways form and is formed by the endemic environment. Equipped with indigenous knowledge of their immediate ecologies, students demonstrate agricultural stewardship over land, interrogating the commercialized development aggression that looms over their *yutang kabilin*. Their productive usage of lands that the state would construe as "idle" in order to render them convertible (and thus sellable) meshes with the Lumad's decades-long struggle for their ancestral land. Students are thus intimated to learn critical consciousness, the ability to understand and analyze the systemic nature of day-to-day problems and the interconnectedness of various struggles (Serrano and Xhafa 2012b, 26).

Agriculture as a social practice and source of critical consciousness seems inapplicable to the circumstances of the Bakwit School. However, as explained earlier, *bakwit* articulates the Lumad's deep personal-political attachment to land. Defending the land by leaving, the political act of *bakwit* endeavors to bring the social realities of Mindanaoan Indigenous peoples to the fore and closely engage with policy and humanitarian institutions, from the government to the academe, in city centers. Parallelly, the critical pedagogy of the Bakwit School lies in its mobility, which ensures the safety of the *bakwet* and, by extension, their access to education. The School is a key venue for the internalization of the consequences of oppression (Freire 2005), such as landlessness and homelessness. In a way, learning in the Bakwit School is the inexorable learning of the oppressive structures that necessitate the Lumad struggle.

The critical pedagogy of tribal schools for the Lumad and the Bakwit School demonstrate what alternative learning could be. Even though it seeks to serve marginalized communities, the DepEd can only subsume students to the state's mold of civic education through "alternative" non-formal classroom spaces. Alternative schools for the Lumad give DepEd a good run for its money as they locate education in the very political struggle of the Lumad people, transforming students in the process, and thus providing an alternative to the capitalist logic of the formal school system (Serrano and Xhafa 2012a, 289). Education defined by the Lumad people hammers the "understanding of the need to resist the underlying causes of peoples' oppression and exploitation" (Serrano and Xhafa 2012b, 4).

As demonstrated by the case of these schools, the struggle for education for the Lumad people firmly roots to the Indigenous peoples' struggle for self-determination over their *yutang kabilin*, the right to independence, in the political sense, from the "oppressor nation" (Longid 2010, 3). The long history of the Lumad struggle for land and liberation from colonial powers at first and the authoritarian state today keys in students, teachers, and supporters' everyday resistance of operating these schools. The mobile Bakwit School further testifies to how Indigenous peoples forge alternative strategies to pave their way to their own vision of development despite the challenges posed by mining, logging, and other foreign corporate ventures and the state itself. Knowledge of reality is attained in these tribal schools through common reflection in the classroom and constant action in the form of protests and civic engagement, helping students be critically conscious of their potential as the "re-creators" of their lives and history (Freire 2005, 73). Ultimately, what these schools teach the next generation of the defenders of their *yutang kabilin* is *pangiyak ki*, to shout out in the Manobo language, and in their case, to reclaim a future withheld from them.

References

Ayroso, Dee. 2014. The Lumad school in Pantaron Range. *Bulatlat*, December 1. https://www.bulatlat.com/2014/12/01/the-lumad-school-on-pantaron-range/

Ayroso, Dee. 2015. 'Bakwet' | Defending the land by leaving. *Bulatlat*, October 25. https://www.bulatlat.com/2015/10/25/bakwet-defending-the-land-by-leaving/.

Belisario, Paul. 2019. Lumad schools: A struggle for land and learning. In *Repression and Impunity: Continuing Attacks on People's Rights in the Philippines*, ed. Jennifer del Rosario-Malonzo, 71–80. Quezon City: IBON International.

Burbles, Nicholas C., and Rupert Berk. 1999. Critical thinking and critical pedagogy: Relations, differences, and limits. In *Critical Theories in Education*, ed. Thomas S. Popkewitz and Lynn Fendler, 45–66. New York/London: Routledge.

Camba, Alvin Almendrala. 2015. From colonialism to neoliberalism: Critical reflections on Philippine mining in the 'Long Twentieth Century.' *The Extractive Industries and Society* 2: 287–301. https://doi.org/10.1016/j.exis.2015.02.006

Canuday, Jose Jowel. 2009. *Bakwit: The Power of the Displaced*. Quezon City: Ateneo de Manila University Press.

de Santos, Jonathan. 2018. For Lumad schools, even holding class is a struggle. *Philstar.com*, July 11. https://www.philstar.com/headlines/2018/07/11/1831716/lumad-schools-even-holding-class-struggle

DepEd (Department of Education). n.d. *ALS Frequently Asked Questions*. Department of Education. https://www.deped.gov.ph/k-to-12/inclusive-education/als-frequently-asked-questions/

ECIP (Episcopal Commission on Indigenous Peoples). 2008. Indigenous peoples education: From alienation to rootedness. *Human Rights Education in Asian Schools* 11: 113–125.

Escobar, Arturo. 1995. *Encountering Development: The Making and Unmaking of the Third World*. Princeton, New Jersey: Princeton University Press.

Freire, Paulo. 2005. *Pedagogy of the Oppressed*, trans. Myra B. Ramos. New York: Continuum.

Giroux, Henry A. 2009. Hard lessons: Neoliberalism, education, and the politics of disposability. *Policy Futures in Education* 7 (5): 570–573.

Go, Vincent. 2015. Displaced Lumads of Mindanao. *Al Jazeera*, October 9. https://www.aljazeera.com/indepth/inpictures/2015/09/displaced-lumads-mindanao-150929074732377.html

Gowing, Peter. 1977. *Mandate in Moroland: The American Government of Muslim Filipinos, 1899–1920*. Quezon City: Philippine Center of Advanced Studies.

Gruenwald, David A. 2003. The best of both worlds: A critical pedagogy of place. *Educational Researcher* 32 (4): 3–12. https://doi.org/10.3102/0013189X032004003.

Harvey, David. 2006. *The Limits to Capital*. New York: Verso.

Holden, William N. 2005. Civil society opposition to nonferrous metals mining in the Philippines. *Voluntas: International Journal of Voluntary and Nonprofit Organizations* 16 (3): 223–249. https://doi.org/10.1007/s11266-005-7723-1

Huggan, Graham, and Helen Tiffin. 2010. *Postcolonial Ecocriticism: Literature, Animals, Environment*. New York: Routledge.

IBON (IBON Institute for International Development). 2018. *Simmering Crisis, Rising Inequalities Amid Corporate Concentration: International Situation*. Quezon City: IBON International.

ICHRP (International Coalition for Human Rights in the Philippines) and IBON (IBON International). 2018. *Duterte Killings Continue: State Terror & Human Rights in the Philippines*. Quezon City: IBON International.

IWGIA. 2021. *The Indigenous World: Philippines*. International Work Group for Indigenous Affairs (IWGIA), March 18. https://www.iwgia.org/en/philippines/4243-iw-2021-philippines.html

Lahoy, Rodolfo Jr., and Renz Roc. 2019. A creeping reign of repression, impunity in the Philippines. In *Repression and Impunity: Continuing Attacks on People's Rights in the Philippines*, ed. Jennifer del Rosario-Malonzo, 1–16. Quezon City: IBON International.

Lanuza, Gerry. 2015. Reactivating youth activism in the age of slacktivism. *Plaridel: A Philippine Journal of Communication, Media, and Society* 12 (1): 97–122.

Lingao, Amanda. 2017. Duterte threatens to bomb Lumad schools. *CNN Philippines*, July 25. https://cnnphilippines.com/news/2017/07/25/Duterte-threatens-to-bomb-Lumad-schools.html

Longid, Beverly. 2010. Human rights framework: Self-determination and liberation. Reference paper by the *KATRIBU Partylist at the International Conference on Indigenous Peoples Rights, Alternatives and Solutions to the Climate Crisis*, Sagada, Mountain Province, Philippines, November 6–9.

Malayao, Pya Macliing. 2018. Testimonya sa Pangyayari sa Talaingod. *Bulatlat*, December 6. https://www.bulatlat.com/2018/12/06/testimonya-sa-pangyayari-sa-talaingod/

Manlupig, Karlos. 2015. Timeline: Attacks on the Lumad of Mindanao. *Rappler*, September 16, 2015. https://www.rappler.com/nation/105847-timeline-attacks-lumad-mindanao

Miller, Brad. 2019. 'It is Open Season Right Now' martial law intensifies in the Philippines. *Mongabay*, March 14. https://news.mongabay.com/2019/03/it-is-open-season-right-now-martial-law-intensifies-in-the-philippines/

Moderno, H. Marcos C. 2018. Group says mining interests behind military presence in Lumad lands. *Mindanews.com*, July 24. https://www.mindanews.com/top-stories/2018/07/group-says-mining-interests-behind-military-presence-in-lumad-lands/

Molintas, Jose Mencio. 2004. The Philippine indigenous peoples' struggle for land and life: Challenging legal texts. *Arizona Journal of International & Comparative Law* 21 (1): 269–306.

Olea, Ronalyn V. 2011. Philippine mining industry: A source of profit for TNCs, a curse to communities. *Bulatlat*, April 11. http://bulatlat.com/main/2011/04/11/philippine-mining-industry-a-source-of-profits-for-tncs-a-curse-to-communities/

Ombay, Giselle. 2022. Sara, Makabayan bloc trade barbs over closing of Lumad schools in Davao. *GMA News Online*, August 1. https://www.gmanetwork.com/news/topstories/nation/840116/sara-makabayan-bloc-trade-barbs-over-closing-of-lumad-schools-in-davao/story/

PSA (Philippine Statistics Authority). 2019. Proportion of poor Filipinos registered at 21.0 percent in the first semester of 2018. *Philippine Statistics Authority*. http://www.psa.gov.ph/poverty-press-releases/nid/138411

Rivas, Ralph. 2022. Philippines' poverty incidence rises to 18.1% in 2021. *Rappler*, August 15. https://www.rappler.com/business/poverty-statistics-philippines-january-december-2021/

Rodil, B.R. 2004. *The Minoritization of the Indigenous Communities of Mindanao and the Sulu Archipelago* (rev. Philippine ed.). Davao City: Alternate Forum for Research in Mindanao, Inc.

Salugpongan International. n.d. Save our schools. *Salupongan*. http://salupongan.org/save-our-schools-1

Sambalud, Mart. 2018. DepEd Davao Norte chief says no closure order issued vs Lumad schools. *Mindanews*, December 4. https://www.mindanews.com/top-stories/2018/12/deped-davao-norte-chief-says-no-closure-order-issued-vs-lumad-schools/

San Juan, Ratziel. 2019. Lumad groups slam Sara Duterte's order to shut down school. *PhilStar*, October 2. https://www.philstar.com/headlines/2019/10/02/1956821/lumad-groups-slam-sara-dutertes-order-shut-down-school

Save Our Schools Network Cebu. 2022. Press release. *Facebook*, February 25. https://www.facebook.com/SaveOurSchoolsCebu/posts/2754246331538125

SOS (Save Our Schools Network). 2019a. *Intensifying the People's Struggle Amidst State Fascism*. Primer.

SOS (Save Our Schools Network). 2019b. *Lumad Bakwit School*. Primer.

SOS (Save Our Schools Network). 2019c. *Save Our Schools: History and Background*. Primer.

Serrano, Melisa, and Edlira Xhafa. 2012a. Grounding thinking utopia. *Labor History* 53 (2): 284–291. https://doi.org/10.1080/0023656X.2012.679405.

Serrano, Melisa, and Edlira Xhafa, eds. 2012b. *The Pursuit of Alternatives: Stories of Peoples' Economic and Political Struggles Around the World*. München, Mering: Rainer Hampp Verlag.

Tan, Samuel K. 2003. *Internationalization of the Bangsamoro Struggle*. Quezon City: UP Center for Integrative and Development Studies.

Thompson, Mark R. 1996. *The Anti-Marcos Struggle: Personalistic Rule and Democratic Transition in the Philippines*. New Haven: Yale University Press.

Umil, Anne Marxze. 2018a. Alliance for Lumad schools holds first national confab. *Bulatlat*, September 18. https://www.bulatlat.com/2017/09/18/alliance-lumad-schools-holds-first-national-confab

Umil, Anne Marxze. 2018b. Context of Talaingod incident I The decades-old struggle of Lumad in Pantaron Mountain Range for ancestral land, right to self-determination. *Bulatlat*, December 1. https://www.bulatlat.com/2018/12/01/context-of-talaingod-incident-the-decades-old-struggle-of-lumad-in-pantaron-mountain-range-for-ancestral-land-right-to-self-determination/

United Nations. 2007. *United Nations Declaration on the Rights of Indigenous Peoples.* New York: United Nations General Assembly.

UNPFII (United Nations Permanent Forum on Indigenous Issues). 2018. Discussion on the theme: Indigenous peoples' collective rights to lands, territories, and resources. In *Proceedings from the 17th Session of the Permanent Forum on Indigenous Issues*, April 17. https://www.un.org/development/desa/indigenouspeoples/wp-content/uploads/sites/19/2018/04/Concept-note-theme-discussion-UNPFII17-final.pdf

Villareal, Gilbert G., Jr. 2016. The Lumads of the Philippines: Struggling from conflict toward peace. *globalEcco*. https://globalecco.org/the-lumads

Zapico, Florence Lasalita, Catherine Hazel Aguilar, Angelie Abistano, Josephine Carino Turner, and Lolymar Jacinto Reyes. 2015. Biocultural diversity of Sarangani province, Philippines: An ethno-ecological analysis. *Rice Science* 22 (3): 138–146. https://doi.org/10.1016/j.rsci.2015.05.018.

Jose Monfred C. Sy is an assistant professor of Philippine studies at the Department of Filipino and Philippine Literature, University of the Philippines (UP) Diliman. He received his Master of Arts (Philippine Studies) and Bachelor of Arts (Comparative Literature), summa cum laude, from the same university. He is currently a research fellow of the Program on Alternative Development of the UP Center for Integrative and Development Studies, where he is a co-editor of the multi-volume monograph series *Alternative Practices across Southeast Asia* (2020, 2023). He is interested in social movements, anticolonialism and decoloniality, environmental humanities, and children's literature. His research appears in journals such as *Kritika Kultura, Humanities Diliman, Social Science Diliman, Daluyan, Suvannabhumi, AlterNative*, and *Philippine Studies*. He is a member of the Congress of Teachers/Educators for Nationalism and Democracy (CONTEND) and the Alliance of Concerned Teachers (ACT). He is also the author of several children's books.

Chapter 12
The Rural Internship Training Program in Sarawak, Malaysia

Andrew Aeria

Abstract The Rural Internship Training Program (RITP) began as a response to the overly market-oriented teaching programs of Malaysian universities which presently focus on preparing students of all academic disciplines to serve the market and to be servile to market values. Started in 2015, the RITP aims to expose social science students to the diversity of livelihoods, cultures, and value systems in rural Sarawak as an alternative to the market. It also trains students to be good social science researchers in the real world even as it aims to develop a sense of empathy, solidarity, and belonging between the students and the rural communities.

Keywords Rural internship · Teaching social science · Sarawak communities

12.1 Introduction

The Rural Internship Training Program (RITP) originated from a growing concern at the overall direction of university training in Malaysia.[1] Over the last 30 years, not unlike other universities around the world and very much in line with the rise of neoliberalism globally since the Thatcher–Reagan years, the social ethos and values of Malaysian universities have undergone a major transformation.

Once dynamic, socially conscious institutions filled with intellectual ferment and student activism, Malaysian universities have transitioned today into academically laidback institutions that prize and promote individualism, corporate identification, and wealth accumulation. The revolution in digital technologies has not helped either. While access to information and knowledge has exploded, digital technology has

[1] I am deeply grateful to all the students from the Faculty of Social Sciences, Universiti Malaysia Sarawak, who had the courage and tenacity to join this RITP. It was they who made the RITP. But arguably, the RITP also made them. This article therefore is their article. I write it on their behalf and in dedication to them all.

A. Aeria (✉)
Persatuan Pemangkin Kesedaran Sosial, Kuching, Malaysia
e-mail: andrewaeria@gmail.com

J. M. C. Sy and A. P. Aban (eds.), *Reimagining Development in Southeast Asia*, Asia in Transition 21, https://doi.org/10.1007/978-981-97-9140-8_12

generally also promoted intellectual laziness, academic shortcuts, plagiarism, and self-centered narcissism instead of promoting a reading culture, academic rigor, critical awareness, and social responsibility. Mediocrity is pervasive among academic staff and students in Malaysian universities. Intellectual curiosity and personal initiative are rare. Social analysis is rarer still and a desire to "remake and transform the world into something better" nigh absent. Indeed, the rise of our digital society has also seen the focus of much social science research move from the real world into doing virtual research in the digital realm.

This dumbing down of universities and graduates has led to a serious mismatch of graduate skills with the needs of industry and that of our larger society. Students today generally are apathetic, timid, and eschew asserting themselves let alone championing any social cause. The influx of Saudi Arabian funding into supporting the development of conservative Islam has also spawned a very narrow (read: Wahhabi) form of Islam in the country which prioritizes form over substance, rhetoric over thoughtful speech (Hamid 2009).

Complicating matters, the present emphasis of university training in Malaysia has focused on meeting the needs of the corporate world with little effort spent on meeting society's needs. Consequently, internship training in Malaysian universities has largely concentrated on preparing graduates for corporate careers; the internship being a period to be spent in an air-conditioned office (preferably) to expose and introduce students to the world of work. Given the way universities support and are supine to the corporate agenda, one could almost be made to believe that no work occurred outside the corporate office.

The tragedy of such an approach within the social sciences has led students to being more inclined to seek well-paying jobs in a cushy, temperature-controlled environment instead of them being attuned to complex social issues (like poverty, discrimination, marginalization, crime, social neglect, etc.) or made aware of interesting social phenomena that give communities their specific identity (like social histories, language, belief systems, gender relations, land, tangible and intangible cultural heritage, etc.) or even equipping students with the rudiments of good social science training (like theoretical perspective, research methods, questionnaire preparation, observation skills, people handling, data collection and interpretation, report writing, etc.).

It was within this context that after having been an academic in university for 20 years, I came to conclude in 2014 that the present-day system of university education and internship training was failing our social science students from becoming competent and analytically sound social researchers (Fig. 12.1).

I realized that we were not graduating students with the necessary skill sets, policy perspectives, training, and ethics needed to ensure the continuation of good social science thinking and analysis that would complement physical and other socio-economic developments. Instead of developing theoretically grounded, analytical, perceptive, and socially aware students, we were graduating students with a sense of bland acceptance of society and even a listless anomie about their own life paths and roles in society.

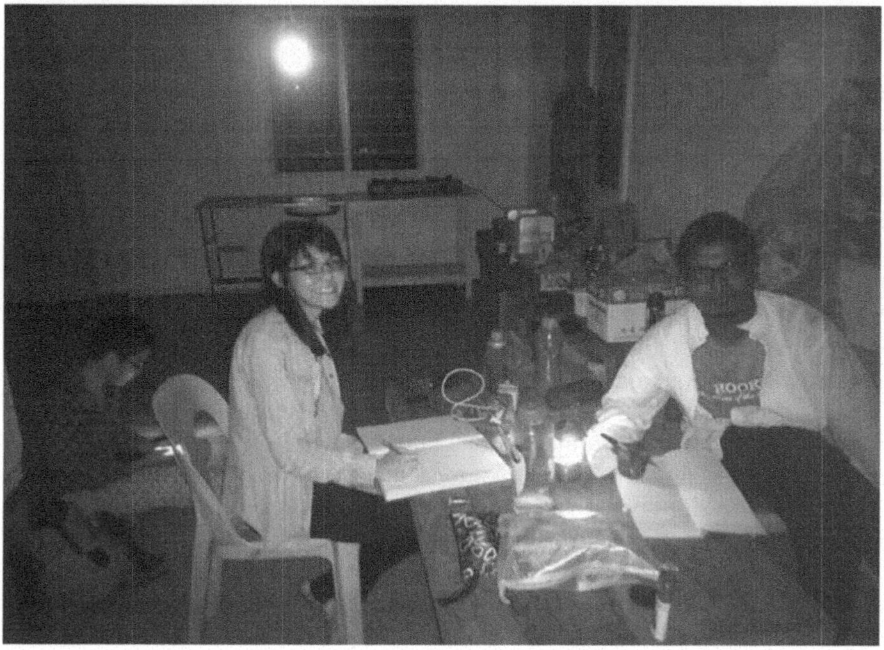

Fig. 12.1 Writing in student journals under solar light (Photo courtesy of the author)

This forced me to stop and question myself. We are training social science students. But what are we training our social science students *for*? Was university social science training designed to expose students to the uniformity of the market or to life in all its multiple diversities? Was university social science training designed to mold students into unquestioning narcissistic robots filled with a hunger for wealth accumulation or to push them into realizing their own compassionate humanity with a love for society, our environment, and hunger for knowledge? Was it to form social science students who fear and kowtow to the authoritarian state OR to courageously challenge the authoritarian state to realize a more democratic and compassionate society respectful of human rights? I felt that there must be a better way to nurture competent and thoughtful social science students.

Consequently, with support from a few equally concerned faculty members, an alternative internship training program was developed that aimed at reviving the whole range of traditional social science training presently not emphasized in campus. As well, I hoped that the alternative internship experience would expose students to and also generate a sense of empathy among them for other (marginalized) communities, which they would not otherwise come into contact within their university training, let alone their lifetimes. The RITP was thus mooted in 2015.

12.2 The Rural Internship Training Program

In designing the RITP, I was inspired by my own experiences as an undergraduate who underwent social exposure-immersion programs and social analysis training when I was an active member of the Malaysian Catholic student movement and the International Movement of Catholic Students (IMCS). As well, I was inspired by the 1961 Cuban Literacy Program (Supko 1998; Lakhani 2010) except that this time I wanted the rural folk to teach the students their indigenous language and share their indigenous knowledge, not vice-versa. After all, did not the Cuban literacy Program "raise the social consciousness of Cuba's youth?" And did not the Cuban students sent to the "most remote corners of Cuba to teach literacy" also end up being "educated by the experience?" (DuRand 2005).

I thus decided that the RITP must focus on language training and be research-based to occupy the student-intern's time in the village (for 8–10 weeks). It would involve copious note-taking in the traditional manner of pen and notebook. And the tougher the internship conditions in the remote villages, the better for student-interns. Indeed, the RITP was promoted among students as "The Internship of Doom!" with catchphrases like "Do you like to read novels? Do you consider yourself to be a modern Indiana Jones? Are you seeking adventure? Do you like to watch the stars at night?" It was my panacea to present-day university students who are often addicted and glued to their handphones instead of being focused on the colorful diversity of society, the environment and their relationship to it.

The RITP had the following four key objectives. Firstly, it aimed to expose and immerse social science students into the diversity of non-market livelihoods, rural cultures, value systems, and living environments; secondly to train students to be good social science researchers in the real world, having a sense of empathy, understanding, solidarity, and belonging with marginalized communities; thirdly to channel students toward non-corporate careers working with and among marginalized peoples; and lastly to help students get in touch with themselves and to realize different "alternative" values.

From 2015 to 2019, we ran five cohorts of the RITP for 25 student-interns in ten different rural communities.[2] All these villages were rural, remote, and marginalized communities on account of various factors like geographic distance, vindictive politics, lack of public amenities and infrastructure, poverty, neglect, deforestation, dam inundation, etc. Of these interns, 19 were women and only six were men. This dearth of men was due to most male applicants declining the internship offer as it did not pay any monthly stipend! Of the 10 villages, three were Penan; three were Kenyah; two Iban; and two Bidayuh.

[2] The first four cohorts comprised 19 interns. In 2019, as this paper was being written, six more interns were placed in three more villages.

12.3 The RITP Research 'Tools'

Being research-based, the RITP is designed around a set of focus areas and research tools. Interns are required to undertake a comprehensive village survey (covering areas like demography, occupation/employment, household income and expenditure, educational levels, physical household assets, health status and illnesses, food intake, voter registration status, public amenities, social issues, environmental resources, work tools and ancestral artifacts, etc.). The interns also are required to be attentive to intangible cultural heritage like oral history, traditional music and dance, weaving and painting arts, culture, and religious beliefs. They also are required to document genealogies of various families within the village. They were also required to produce a hand-drawn physical and environmental resource map of the village which they would share with the villagers so as to allow the villagers a good visual representation of their village as well as their common economic and environmental resources.

The interns spent a week to 10 days preparing to collect data prior to entering their village. During this week or so, they were briefed about the key objectives of their internship. They also learned about their host's ethnic/cultural background, their community's historical migratory movements, how to draw genealogical charts, how to draw rivers via fishbone representations, how to operate and take Global Positioning System (GPS) readings, basic first aid, behavior in the village and how to relate to villagers, how to take notes, sex and relationships, etc. Appendix is a general program schedule for the interns in their first week of the RITP. They were also introduced to four key tools/skills for their internship experience, namely language, paper notebook, personal journal, and databook to help them succeed in their work in the village. The fifth tool/skill of report writing comes after their village internships. These tools have proved invaluable to the RITP.[3] Additionally, to ensure that these tools really worked, the interns were barred from using their mobile phones to refrain from accessing WhatsApp messaging, Facebook, Instagram, or other digital platforms.[4] This was to ensure minimal distraction by social media during the internship training. At the end of the first week of RITP, the interns would be paired into teams of two individuals. Each intern team was then placed in a remote rural village.

The RITP required the interns to immerse themselves socially into their respective communities. To do so, they had to learn the local village language as the primary research tool to help them communicate with villagers. Villagers were thus requested to speak to the interns in their local language as much as possible. Although it seemed difficult, the results of previous internships indicated that the interns ended up crash learning the local language, often with the aid of their own homemade language phrasebook that listed hundreds of words.

[3] For the purpose of this paper, I use the words "tools" and "skills" interchangeably since the skill is also the tool of investigation and vice versa.

[4] In many cases, the villages the interns were placed in had no telephone or internet connectivity on account of being very isolated. Where available, mobile phones were for emergency use only. In all other cases, handwritten notes and letters were the norm and strongly encouraged.

They were able to speak and hold basic conversations with villagers within a month. Via the local language, the interns were able to research comfortably and to communicate with all as adopted members of the village community. Basic command of the local language by the interns allowed villagers to communicate reasonably well with the interns on their own terms without having to use any English or Malay, which many villagers (especially women) were uncomfortable with. Such a perspective on language as a tool of research is not unheard by anthropologists and social science scholars of yesteryears. Unfortunately, the present widespread use of English and Malay as the sole language of teaching, learning, and research in many market-oriented university faculties under-emphasize the importance of local languages. This has inadequately prepared social science students to transform them as serious independent researchers and scholars who speak from an immersed perspective and with an ability to see through the eyes of their host communities. Additionally, present-day social science practitioners often resort to the ubiquitous lengthy questionnaire and even the "hit-and-run" focus group discussion methodology for data gathering (that often results in "superficial" data) which does not adequately reflect the depth and complexity of social reality (Fig. 12.2).

The second core training tool of the RITP was that interns were required to jot down onto a paper notebook "everything" they encountered in the village throughout their research. This was insisted upon since old-fashioned paper notebooks filled with

Fig. 12.2 Visit from the Flying Doctor service (Photo courtesy of the author)

pencil and ink-scribbled notes, doodles, and sketches are far more practical tools for research in rural areas that have little access to any reliable electricity supply. Paper notebooks are also more versatile with interns able to stuff them into their pockets and go anywhere throughout farms, forests, and rivers. Handwritten notes allow for better mental recall of details of activities, meetings, discussions, travel, observations, thoughts, ideas, samples, etc., especially when interns are drawing up their data sheets and/or writing up their reports. Notebooks seemed less intimidating to villagers compared to the use of devices such as mobile phones, voice recorders or computers. At times, mobile phones were utilized as handy cameras or video recorders that in turn entertained villagers. Written documentation from the paper notebook served as "raw, messy, unfiltered, and unvarnished" account of their research. These notes, doodles, and sketches shall later be used by the interns as references for factual basis needed in substantiating their academic research (Fig. 12.3).

The third core training tool imparted to the interns was writing their personal journal. Although it had to be handwritten, the personal journal was different from the notebook since it detailed the daily activities, thoughts, reflections, and feelings of the interns from the beginning right until the end of their internship. Interns were free to pour out their deepest feelings, thoughts, ideas, reflections, suggestions, their feelings of joy, and even their frustrations. The journals served as their repository for sketches of the forest, trees, houses, animals, and people-things that they encountered during the period of their internship. The journals also contained a compilation of mundane things such as strange leaves, sticks, seeds and plants, insects, and even tiny

Fig. 12.3 Interns walking a bamboo bridge (Photo courtesy of the author)

soil samples that they had collected to remind them of their research endeavors. This material served as a rich resource for their research data and a personal memento that kept their valuable experience among the rural community.

The fourth core training tool of this RITP was the intern team's databook. In this, all the data collected by the intern teams were systematically logged and stored in two general sections. Firstly, all statistical data collected over two months on individual households were compiled comprising detailed information about the population and number of households in the village, the household's gender breakdown, age, occupation, income, health status and problems, voter registration status, household utensil and amenities, daily food intake, births and deaths, farmland size, agricultural crops and handicrafts produced within each household. Secondly, the intern team's databook would have detailed information about the village in general, namely its forests and environmental resources, its oral history, traditions, folklore, belief systems, musical instruments and dances, and the genealogy of various families or even the origin story of the whole community. All these tangible and intangible cultural resources were collected either by regular observation or via interviews with key elders within the community, as they were the ones who were the keepers of individual and community memory. A list of various issues/problems faced by the village would also be documented. A central component of this databook would be the village map that the teams would draw by hand with inputs from the villagers. This map would show the physical location of each house, workshop, school/clinic (if any), farmland, river/stream, fishpond, animal husbandry hut, forest, and other environmental resources of the village.

The last core training tool of the RITP, (but will be known after they have completed their community immersion), would be their report writing skills. Here, interns were oriented how to organize their data thematically to conclude their final report over a period of one to two weeks. All final team reports would contain an Executive Summary, while the full report would be organized into subtopics (with relevant photos). A concluding section would detail suggestions for further research. Copies of all photos, language phrasebook, notebooks, databook, and personal journals would also be submitted to CIVICA as an essential component of and complement to their final report. Although copies of all materials were submitted to the CIVICA for archiving, the interns retained ownership to all materials. Publication as monograph of these field notes is currently being discussed.

12.4 RITP Views: In the Interns' Own Words

A short questionnaire was circulated in the middle of 2018 to 12 RITP interns in the purpose of feedbacking about their experience.[5] Ten responses were gathered with the following results—students signed up for the RITP out of curiosity and interest in an adventurous and challenging internship. Some stated that previous interns expressed that the RITP internship were interesting and experientially diverse, definitely not just sitting in an office and photocopying newspapers or answering front office queries of customers. Others were taken in by the RITP marketing slogans of "*Do you like to read novels? Do you consider yourself to be an Indiana Jones?*" One artistically inclined student joined the RITP because she was "attracted to the 'fear of the unknown' and the 'thrills of the uncommon.' She had heard rumors about the RITP and how it would be 'dangerous' for students to be left alone in an unknown rural village." These expectations did not disappoint albeit posing a challenge to RITP interns.

All the interns responded that they learned new research skills in unfamiliar circumstances that radically challenged their bookish view of research. They also learned about very different rural realities from that of their urban lifestyles and from what they learned on campus. They learned about hardship, poverty, physical risks, isolation, boredom, fear, and painful decision-making. Many also learned that they often took their lives for granted. Being urbanites, they often emphasized on the importance of money. But in the village, they discovered that money didn't solve more urgent issues like the lack of access to proper roads, regular electricity supply, clean water, and mobile phone signals. They realized how difficult it was for villagers to access schools and clinics, many of which were miles away on logging roads or through dangerous river journeys. They also discovered that many of their assumptions were misconceived. They never realized how important forests and rivers were to rural villagers. They discovered that wherever there was a forest, there would always be food like meat, vegetables, fruits, and even clean water from bamboo. They discovered that villagers did not worry so much about money. Instead, villagers worried more about the continuous degradation of forests (due to logging and oil palm plantations) and river pollution and the impending insecurity their children will face once their forests, rivers, and lands were gone.

They also realized the importance of rural development and the contribution of smallholder agriculture to the local and regional economics. They discovered that many villages only had old people and young children staying in them. The working adults mainly worked in distant towns and only returned to the village during weekends and long holidays. They realized that village life was quiet, humdrum, and boring.

---

[5] Only 12 interns from Cohort 1 to Cohort 3 were canvassed for their views. There were seven interns from Cohort 4 in 2018 in the villages at the time this questionnaire was circulated. They were therefore not canvassed. Cohort 5 only began their RITP in late June 2019, which ended in early September 2019.

Although many felt daunted at first, all learned a new language. The students were certainly proud of this especially when fluent native speakers expressed their admiration of their new-found language skills. Everyone expressed their learnings about the native customary land issues, the problems that arise due to oil palm plantation expansion and its effects such as conflicts and extensive loss of native lands. As one intern noted, "How many peoples' homes and livelihoods were sacrificed for this plantation? I just find that greed is really very, very scary."

Others appreciated the simple lifestyle of villagers. As urbanites who never understood why people in the village seemed to enjoy their boring routines, they emerged more appreciative of village life, "not that there is not much to do, it's just that you need to find your own activity." Quoting one intern, the RITP was a "unique life experience journey that really opened my eyes regarding the hardship yet simple lifestyle of the villagers. It made me appreciate life itself."

Some learned how to take GPS readings and map key locations in and around the village.[6] Others learned how to catch fish and birds like villagers do. Others discovered beauty and meaning behind cultural and religious rituals. Farmlands, rivers, and waterfalls once only given a cursory glance or passing thought in tutorial classes were now interesting places for them to explore. Each location had its own story.

Others learned "what true vulnerability looks like … When fear struck me at my most vulnerable time, I learnt to accept it and leaned on my partner's shoulders. From that too, I learnt that I was able to work together professionally to a common goal despite personal issues and differences." Working in a partnership was a lesson many took to heart; helping, caring, and supporting each other through thick and thin during the weeks in isolation from their usual support systems. Many thus learned about themselves and improved their interpersonal skills. Some learned how to talk and significantly, how to be curious. "Yes people laughed at me, and I felt embarrassed for constantly asking 'stupid' questions. But only through asking, I got to know what I wanted to know!" Others realized that the villagers genuinely cared for them, showered them with love and kindness that made them feel valued—something seldom experienced by them in an urban setting. As another intern noted, "I've learnt that I actually have value, and have the ability to do something for someone, as long as I am willing to learn. I am also a human, tempted by materials and entertainment. Yet, I've learnt to not put those as priorities in life after this internship. Using the best phone and laptop or being 'pretty' doesn't really interest me anymore. I feel okay with my dark skin and scarred legs after I came out from my village. Totally worth it!". The RITP changed many of the interns. "It changed me to be my better self, to be aware of what's happening around me, and to cherish what I have at all times. I began to explore more about the indigenous culture of Sarawak and to learn their languages (if possible), as their knowledge can never be learnt from a book."

Many interns also were shocked at the deeply patriarchal way of living in rural villages. Many of the women interns got really annoyed. They realized that in the

[6] In some years, this was not possible when we lost some GPS sets while others were spoiled, and we did not have the funds to buy new sets. So, two cohorts lost out.

village, men often were the decision-makers and who gave out information to others. Women (especially the younger ones) played more domestic roles of serving tea and cooking. They also discovered that villagers tended to engage in serious ethnic stereotyping of certain urban ethnic communities as "rich" and "greedy." Drunken behavior, the scourge of alcohol, and sexual harassment in villages also made deeply intrusive and stressful impressions on some interns.

In other words, many emerged more mature, independent, and alive after their RITP experience. Many were emotionally tougher. They were able to look at different perspectives for most issues, especially about development. They had learned about doing research and had discovered friendships and surrogate families in their villages. But most importantly, they had also learned a lot about themselves, about their own resilience and courage in the face of rural adversity and daily humdrum boredom (Fig. 12.4).

Fig. 12.4 Into the forest with study materials (Photo courtesy of the author)

12.5 Conclusion

What the RITP suggests is that what present-day social science students learn in university is but a shadow of social reality. It remains too theoretical and impractical. They do not know enough of what is happening in real life. They certainly need to get out and dirty their hands and feet. And it does bear fruit. Of the 19 interns who have participated in the RITP since 2015, one works with her mother in the family business. Four work in the commercial sector. Three have just graduated and are working as part-time researchers while they seek permanent work. One just completed his master's degree and is also seeking university work. But ten are now working with various issue-based civil society organizations, mainly in Kuala Lumpur and in various towns in Sarawak. The numbers speak for themselves. And it can be attributed to the interns becoming self-aware of their identity and role in society.

Indeed, for many interns, two months in the RITP was not enough. It was a bitter-sweet experience which was too short. Still, it made them realize that as "privileged persons, they should do something more to help the villagers. Or at least to bring their voices out of the woods, to let others know who they are, acknowledge their rights to land, to the forests, and above all, to the basic amenities that we all have in towns."

As one intern noted, "What is development to us if what we take is less than what we give?".

Appendix

Schedule of Activities

Time	June 17 (Monday)	June 18 (Tuesday)	June 19 (Wednesday)	June 20 (Thursday)	June 21 (Friday)
9.00 a.m.–10.30 a.m.	• Introductions and expectations • RITP overview • Photos and social media addresses	Health clinic check-up	• Ethnic groups of Sarawak • Culture and rituals	Briefing on personal accident (PA) insurance and safety; family/emergency contact; RITP agreement; rabies; medicines; any other business	Field methods in a rural community

(continued)

(continued)

Time	June 17 (Monday)	June 18 (Tuesday)	June 19 (Wednesday)	June 20 (Thursday)	June 21 (Friday)
10.30 a.m.–11.00 a.m.	Tea break		Tea break	Tea break	Tea break
11.00 a.m.–12.30 p.m.	Researching the village I: your research focus and tools		• Penan resources • Drawing rivers and maps	• Former intern's sharing • Challenges you will face!	Field methods (continuation)
12.30 p.m.–2.00 p.m.	Lunch	Lunch	Lunch	Lunch	Lunch
2.00 p.m.–3.15 p.m.	Researching the village II: your research focus and tools	Researching the village III: spider network analysis; samples of notebook, personal journal, databook, maps; handphone as camera only	Doing genealogy	Shopping for stuff	Rivers, forests, and communities
3.15 p.m.–3.45 p.m.	Tea break		Tea break/ Malaria pills		Tea break
3.45 p.m.–5.00 p.m.	Medicines needed (first aid kit and vitamins, mosquito repellent, tiger balm, etc.)		• Introduction to villages • Personal stuff needed • Sex and relationships • Dealing with boredom • Planting native trees		GPS basics

Time	June 22 (Saturday)	June 23 (Sunday)	June 24 (Monday)	June 25 (Tuesday)
9.00 a.m.–12.30 p.m.	Overnight field trip	• Forming the teams! • Travel back to CIVICA	Rest day/Final preparations/ Last-minute shopping	Depart for villages
12.30 p.m.–2.00 p.m.	Lunch	Lunch		
2.00 p.m.–5.00 p.m.	GPS field practicum	Inputting GPS data into Google Earth Pro		
5.00 p.m.–7.00 p.m.	Rest and swim in river			
7.00 p.m.–onward	BBQ			

Students' Research Focus

Village demographics, socioeconomic situation, environmental resource, health and social issues, census:

- Demography: Full census of village population
- Educational achievement
- Occupation and income
- Housing and village amenities
- Health issues
- Daily food and nutrition intake (all individuals within families)
- Village natural resource inventory (via GPS mapping)
- Work (agriculture and animal husbandry; daily work log of men and women)
- Traditions, cultural practices, and taboos
- Genealogy and oral history of the village (interview with knowledgeable elders)
- Folklore, origin myths, and legends (interview with shaman)
- Handicrafts and traditional foods
- Music, art, and religion
- Social issues and environmental challenges facing children and adults (men and women)
- Hand drawing of village map
- Gender roles and social status (men vs. women)
- Hopes and dreams for the future.

Students' Research Tools

- Your brain!
- Five senses
- Pen
- Notebook
- Databook
- Personal journal
- Camera
- GPS set
- Basic questions (the five Ws: *who*, *what*, *when*, *where*, *why*).

References

Hamid, Ahmad Fauzi Abdul. 2009. The new challenges of political Islam in Malaysia. *Working Paper No. 154, June*. Perth: Asia Research Centre, Murdoch University.

DuRand, Cliff. 2005. Cuba today: A nation becoming a university. *Monthly Review Online*, October 13. https://mronline.org/2005/10/13/cuba-today-a-nation-becoming-a-university/

Lakhani, Nina. 2010. Latin lessons: What can we learn from the world's most ambitious literacy campaign? *The Independent,* November 7. https://www.independent.co.uk/news/world/americas/latin-lessons-what-can-we-learn-from-the-worldrsquos-most-ambitious-literacy-campaign-2124433.html

Supko, Ruth A. 1998. Perspectives on the Cuban national literacy campaign. Paper presentation, 1998 Meeting of the Latin American Studies Association, Palmer House Hilton Hotel, Chicago, Illinois, September 24–26.

Andrew Aeria, Ph.D. was formerly Associate Professor in the Department of Politics and International Relations, Faculty of Social Sciences, University Malaysia Sarawak. He is currently President of *Persatuan Pemangkin Kesedaran Sosial Kuching—PEMANGKIN* (Kuching Society for Social Awareness), a Civil Society Organization (CSO) based in Kuching. Established in 2020 with the support of an alliance of civil society groups, PEMANGKIN undertakes research, training, and advocacy activities to support other CSOs promote sustainable development in Sabah and Sarawak.

Andrew is also Chairperson of Non-Timber Forest Products Exchange Programme Malaysia (NTFP-EP Malaysia), a CSO based in Miri working to conserve and rehabilitate forests, strengthen social forestry policy and practices, and so improve the livelihoods of indigenous communities. Andrew actively promotes reforestation with native tree species and the development of non-timber forest products (NTFPs) among indigenous communities. Andrew was also recently elected into the Board of Trustees of the Non-Timber Forest Products Exchange Programme Asia (NTFP-EP Asia) for 2023–2026.

Part V
Women-Led Development and Communities of Care

Chapter 13
Vietnam's Peace House Shelter as a Safe Space for Women to Reclaim Their Rights

Ananeza P. Aban

Abstract Cases of gender-based violence (GBV) that is on the rise across Vietnam, that traverses national boundaries, have propelled the women's movement in this country to create their own initiative that actively responds to the needs of women in this very critical situation. On the other hand, Vietnam has a state obligation to promote and protect the rights of women as signatory to international human rights treaties and conventions. In 2007, the women's movement in Vietnam promptly responded to the call to help these women during the time when there was no safe space yet for them to take shelter and get comprehensive assistance. The Peace House Shelter was established when there was also a significant gap for providing holistic support for the trafficked women.

Keywords Gender-based violence · Peace shelter · Women's movement · Human trafficking

On the rise across Vietnam, cases of gender-based violence (GBV) that traverse national boundaries have propelled the women's movement in the country to create their own initiative that actively responds to the needs of women amid the critical situation. Here, gender-based violence (GBV) refers to "forms of violence that are directed at individuals because of their gender and which are predicated on gender inequalities. It also maintains inequalities between men and women and reinforces traditional gender roles" (United Nations Viet Nam 2014).

The Center for Women and Development, (CWD) as a national women's organization under the umbrella of the Vietnam Women's Union, (VWU) promptly responded to the call to help these women during the time when there was no safe space yet to shelter them and get comprehensive assistance. That was in 2007. The Peace House Shelter was established when there was also a significant gap for providing holistic support for the trafficked women. At the time of its inception, there was no similar center or house in the country.

A. P. Aban (✉)
Ghent University, Ghent, Belgium
e-mail: ananeza.aban@ugent.be

J. M. C. Sy and A. P. Aban (eds.), *Reimagining Development in Southeast Asia*, Asia in Transition 21, https://doi.org/10.1007/978-981-97-9140-8_13

Vietnam has a state obligation to promote and protect the rights of women as signatory to the following international human rights treaties and conventions: the Convention on the Elimination of all Forms of Discrimination Against Women (CEDAW), International Covenant on Economic, Social and Cultural Rights, the Convention on the Rights of the Child, 57th Commission on the Status of Women in 2013, the UN Convention Against Transnational Organized Crime and its supplement, and the Palermo Protocol[1] (United Nations Viet Nam 2014). The raison d'être of the Shelter is to contribute to the realization of these international commitments.

At the national policy level, the social protection instruments in Vietnam have included the gender component within the social equity measures (Jones and Van Anh Tran 2012). Within this frame, Vietnam has the following national laws as legal reference in protecting women from violence: the 2006 Gender Equality Law; the 2007 Domestic Violence Prevention and Control Law, which demonstrates another perspective among Vietnamese society that recognizes the concept of domestic violence as a violation of women's rights and therefore a human rights violation (Kwiatkowski 2011); and later, the Law on Human Trafficking Prevention and Combat 2012 (Jones and Van Anh Tran 2012; Kwiatkowski 2011). Particularly, Vietnam has a National Plan of Action on Anti-Human Trafficking and the Plan of Action on Domestic Violence Prevention and Control that aims to strengthen the implementation of these laws (Jones and Van Anh Tran 2012).

The Shelter's activities are in accordance with these government policies and programs, specifically addressing issues of human trafficking, domestic violence, and other forms of abuse, and to support the victims. CWD is mindful of institutionalizing international standards in responding to GBV, as well as implementing CEDAW which the Vietnam State Council ratified in 1982 (Fig. 13.1).

13.1 Building a Peace Shelter for Women

The Shelter provides comprehensive and long-term support for the rehabilitation and reintegration of women and children who were victims of gender-based violence (GBV), mainly domestic violence and human trafficking.

Led primarily by women, the Shelter is under the supervision of a Shelter Manager. An average of one (1) social worker is assigned per five (5) cases. These social workers are assigned to handle the domestic violence cases separately from those of human trafficking cases, given that the plight of these women are context-specific. Weekly updating among social workers on the cases are done in the Shelter. There are also housekeepers and security personnel as support workforce in the Shelter. All of them are salaried staff. The staff coordinates with the VWU, hospital, police, and legal aid practitioners. Researchers of the UP CIDS Alternative Development

---

[1] The Palermo Protocol is otherwise known as the protocol to prevent, suppress, and punish trafficking in persons, especially women and children, supplementing the United Nations Convention Against Transnational Organized Crime.

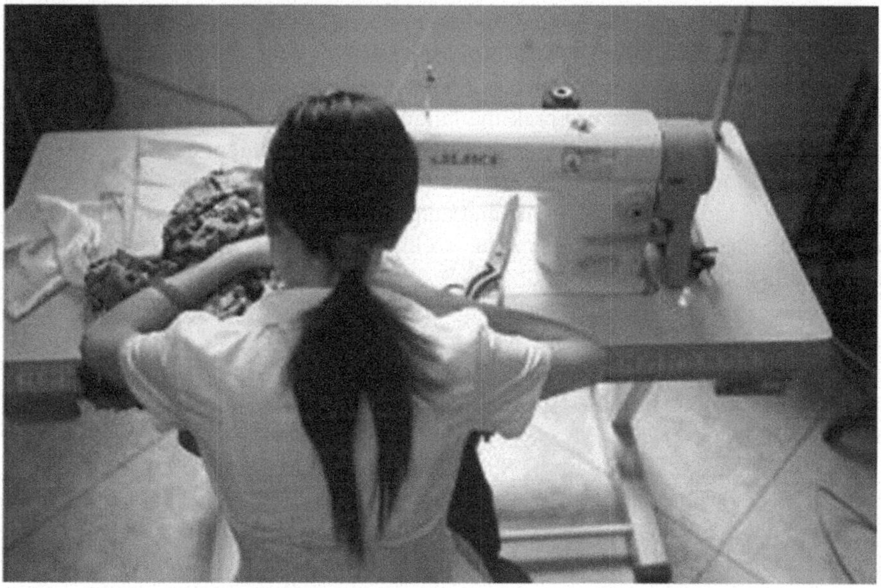

Fig. 13.1 Residents can engage different productive activities (*Source* Peace House Shelter)

Program were able to visit this undisclosed Shelter in Hanoi and met some of its women residents with permission from CWD and VWU. Interviews were done with CWD and the Shelter manager and staff.

The Shelter provides safe accommodation for six (6) months for women who experienced domestic violence and 18 months for trafficked women and girls. The longest stay recorded was 22 months and the shortest was one (1) month. Since counseling and intervention differ among these women, the Shelter clusters those who fled from domestic violence to one floor and to another those who escaped human trafficking.

While these women are in the Shelter's custody, they receive medical treatment, psychological care and counseling, and are engaged in indoor and outdoor activities. Social workers responsible in the counseling use diverse and creative approaches in order to raise awareness about gender equality, reproductive health, and human rights.

There are instances when the Shelter staff needs to inform the local VWU and visit the family members in the community because they are unaware of the violence that has happened. In some cases, perpetrators of violence such as the husbands are also given separate counseling. However, CWD also recognizes the cycle of violence as a reality, and therefore stressed that it is not an assurance that husbands will cease to abuse their wives after undergoing counseling.

Capacity building through education is provided both for the women and their young children who live with them in the Shelter. CWD sees the need for literacy and numeracy sessions since most of the women in the Shelter have very limited education and are even incapable of speaking and comprehending the Vietnamese language, as some of them come from ethnic minorities (such as the H'Mong and Kho Mu peoples). CWD gets volunteer educators to teach these women. In fact, some of them speak the ethnic languages to break the language barrier. Thereafter, these women are referred to vocational centers to improve their development capacity, personal skills, and practical capabilities. Primary education (kindergarten) is provided to their accompanying kids. Job placements and career orientation are provided to prepare them to start their new lives. Otherwise, livelihood development is provided to help them establish their own business.

After the women decide to return to their communities, reintegration support and monitoring until two (2) years are also provided in collaboration with VWU and its local organization to ensure local community support. As a part of this, the women are also asked to participate in poverty alleviation activities of the VWU and the local government.

One important component of the Shelter is providing legal support to their women residents in coordination with legal aid centers and local authorities. As divorce is legal in Vietnam, victims of domestic violence can file a divorce against their husbands. On the other hand, a different case is filed against the perpetrators of human trafficking. The legal processes will have to follow the law that penalizes perpetrators of human trafficking.

For the domestic violence cases, after a woman has registered to the Shelter, her case is brought to her community, and CWD contacts the local authorities to make sure they receive the information. The woman, on the other hand, is informed of the necessary steps in filing her case. A health test is administered to her, and if the results are below the acceptable rate of 11%, CWD calls police attention. An agreement is made between the woman and her perpetrator in the presence of local authorities. If the woman decides to file a divorce, CWD will coordinate with the legal authorities for the processing of the court case. The policy department of VWU will also coordinate with the court.

Another service includes temporary residence registration in coordination with the police. Some of these trafficked women lost their legal identification documents such as birth certificates during the traumatic situation they went through. In order to strengthen the chain-of-referral system, both border guards (Vietnam–China border) and police receive training courses from CWD on how to relate and work with the victims.

CWD maintains its communication with these women after leaving the Shelter. Meetings are held with the local women's union and community. Sometimes, the CWD staff performs field visits. Nevertheless, there is a handful who return to the Shelter as it provides them a safer space than their community.

13.2 Understanding Domestic Violence and Human Trafficking in Vietnam

Scrutinizing a broader landscape of Vietnam's domestic violence and human trafficking problems is of utmost importance to be able to locate the value and emerging challenges of the Shelter.

13.2.1 Domestic Violence

In 2006 findings of the National Study on the Family, it was discovered that 21% of the couples had experienced at least one type of GBV. That same year, the Vietnam Multiple Indicator Cluster Survey found out that 64% of women aged 15–49 years old accepted violence of husbands as 'normal.' And in a 2010 study of the General Statistics Office, 31.5% of women have experienced physical violence, while more than 50% admitted to have undergone emotional violence (Jones and Van Anh Tran 2012). This 2010 National Study on Domestic Violence against Women significantly cited that incidents of physical, sexual, and psychological violence were higher in the countryside and among the less educated people (United Nations Viet Nam 2014).

Domestic violence is acknowledged as a serious public health and widespread social problem in Vietnam (Vu et al. 2014) as it also haunts many ethnic groups in the country with diverse socio-cultural and environmental context (Kwiatkowski 2013). It is deeply rooted in the dominant mindset in Vietnamese society that perceives women as the bearer of family harmony and protector of the 'happy family' ideology (Vu et al. 2014). This notion can be traced back from its ancient Confucian values of authority, familial loyalty, and community (Le 1996; Kwiatkowski 2013) where the male possesses dominant gender roles over the female and therefore, the husband has the authority or control over his household (Vu et al. 2014). The culture, the local political system including the power relations within, and the legal system support play key roles in advancing (or not) the rights of women and empowering women to seek justice for the violation of their rights.

The prevalence of gender stereotyping, however, has discouraged many abused women from seeking support who often opt to endure the life of suffering within an abusive marriage, in the name of the family (Vu et al. 2014). The UN Vietnam (2014) also identified this dilemma when stating that this 'happy family' ideology reflected in social attitudes and norms hinders the enforcement of the Domestic Violence Law because it perpetuates gender power inequalities and stigmatizes women for reporting violence in the family.

Kwiatkowski (2011) noted the rise of divorce rates in years, most likely due to changing public opinion or increasing social acceptability of divorce. It doubled from 1991 to 1998, from 22,000 to 44,000 and further surged to 59,551 in 2004 (Phi 2007 cited in Kwiatkowski 2011). The Shelter, on the other hand, has facilitated around

80% of court cases of divorce in response to domestic violence, as a result of their counseling and processing with the women residents.

However, the data do not say that divorce is becoming a viable option to confront domestic violence, unless a comprehensive state policy formation and law enforcement will be strongly centered on protecting women from violence and inequality. Kwiatkowski (2011) noted that this does not necessarily translate to a decreasing trend of abuse of husband to wife citing that the pervasive culture of reconciliation within the family through the state-mandated reconciliation processes is well in place. She argued that this demonstrates the power of social institutions to prolong the agony of abused women by advocating this mediation scheme. The state rather perpetuates domestic violence as, in practice, it tolerates a husband's violent actions toward his wife and reconciliation committees composed of community leaders, who include women, prioritize the preservation of the family over women's rights to freedom from violent marital relations, thereby pushing women to a cycle of abuse (Kwiatkowski 2011).

The cycle of violence can be better understood in this manner: unequal gender norms and attitudes of masculinity and femininity bring about the depreciating human value of women and girls while imperially appraising hegemonic masculinity. This results in disempowerment of the female gender in both public and private spheres that increases their vulnerability to violence and the violation of their human rights. In this chain reaction, the normalization of violence against women and girls becomes commonplace in society that enables non-action of duty bearers and authorities. And to finally complete the cycle, this normalization fortifies the structural context which is the unequal gender norms and attitudes (United Nations Viet Nam 2014).

Vu et al. (2014) explained that the principle behind this reconciliation process focuses more on the violation of the 'harmony' within the family that can affect the community, instead of the emphasis on the safety of women from a violent relationship in the home and community. It has been observed that these groups that deal with domestic violence before it is brought to court work more on preventing divorce rather than protecting women from violence (Schuler et al. 2011 and UNFPA 2007 cited in Vu et al. 2014).

Whereas women already feel reluctant to file a divorce in the initial stage, even in cases of extreme violence, the court will also attempt to entertain reconciliation. The divorce process is so complicated, which includes several reconciliations that somehow discourage women to proceed with this legal procedure (Vu et al. 2014).

In their continuing intervention, CWD recognizes the culture of silence that remains ingrained among women victims of domestic violence. There are cases when the victims understandably refuse to share their story because of the trauma or mental stress they have been through, aside from social pressures to maintain 'family harmony.' Unfortunately, this, in serious cases, led to extending their unsafe condition. Economic dependency on their abusive male partners (or husbands) remains an issue. With no sources of income as fallback, having solely devoted to unpaid domestic work, women find it difficult to escape this abusive relationship.

CWD also noted that while the divorce process is extended, and pending court decisions, the husband or his family seeks every measure to legitimize property division in favor of the husband in order to disenfranchise the woman of her right over these properties. Divorce is perceived to bring sorrow and embarrassment thereby reflecting lack of support from many Vietnamese families, or nothing. Eventually, this access to property issue stands in the way of abused women from seeking child custody (Vu et al. 2014).

On the other hand, the existing quandary can be interpreted through a political economic lens. While the state seriously champions women's rights and human rights in line with the global goals to achieve gender equality, the steps toward a socialist-oriented market economy have repositioned the household to serve as the foundation for the market economy and integration into the global capitalist economy. Bound by the 'happy family' ideals, the family should therefore not be disintegrated as it is important in the pursuit of the development of both domestic and national economic systems (Kwiatkowski 2011). Vietnam's transition to commodity economy has been acknowledged to achieve some 'remarkable successes.' However, along with this system change produces many 'social evils,' including the escalation of domestic violence (Le 1996).

Vu et al. (2014) enumerated the following recommendations to help address the problem of GBV: the government should adopt rights-based policies that strengthen women's property rights; provide legal support for domestic violence survivors; educate lawyers about the issue of domestic violence; strengthen law enforcement and build the capacity of people involved in the process of 'reconciliation' and divorce; and though this is challenging but essential, there is a need to engage men in gender equality and issues of violence against women.

13.2.2 Human Trafficking

The globally accepted definition of human trafficking will be that of the UN Convention Against Transnational Organized Crime: "the recruitment, transportation, transfer, harboring or receipt of persons, by means of the threat or use of force or other forms of coercion, of abduction, of fraud, of deception, of the abuse of power or of a position of vulnerability or of the giving or receiving of payments or benefits to achieve the consent of a person having control over another person, for the purpose of exploitation." The term exploitation shall include the prostitution of other and other forms of sexual exploitation, forced labor, slavery, or similar practices as stated in the Palermo Protocol. Article 6 of the Palermo Protocol enumerates the assistance the State Party should provide for the protection of victims of human trafficking, such as measures to provide for the physical, psychological, and social recovery of victims of trafficking (United Nations 2000).

Vietnam is a source and destination country for trafficking. Between 2004 and 2012, as many as 6,184 Vietnamese women and children were reported to have been trafficked across the Vietnam borders. According to the United Nations Economic and Social Council 2017 report (as cited in Hutt 2022), "about 18,000 people from Vietnam to Europe [is estimated] annually, an illegal trade worth around €300 million

($315.6 million)." This trend also reflects the proliferation of traffickers. The low education and social economic status of these women, some coming from broken and violent homes, are the common factors why human traffickers prey on these women for exploitation, according to CWD.

Based on CWD's records of victims supported by the Shelter, more than 80% of these trafficked young girls are within the age range of 18–25 years old, while 37% are in their adolescence. In their juvenile age, some of these girls have already been pregnant or bore a child during the trafficking period.

They are trafficked to Cambodia, China, Thailand, Hong Kong, Malaysia, Taiwan, Korea, and as far as England and Czech Republic for purposes of sexual exploitation, forced marriage, and labor exploitation. The Northern provinces of Ha Giang, Lao Cai, Lang Son, and Quang Ninh are the hotspots where 60% of the deals are located. The increasing trend in the number of human trafficking cases are located mostly in remote and rural poor areas. Kidnap cases were also recorded, which include those from isolated residents and children traveling on deserted roads.

China is the most common destination for Vietnamese women trafficked for forced marriage or sex work, using deception as a common recruitment tactic among traffickers, although there were some cases of abduction (Stöckl et al. 2017). Based on the report by the USAID from the 2013–2019 implementation of Decree No. 09/2013/ND-CP (Decree 09), 98% of the victims that received assistance were mostly female and were rescued at "on the border with China" (Luu 2020 in USAID 2021). The cross-border bride trade has been an increasingly alarming phenomenon as a result of a social norm that treats women as commodities, that mirrors China's patriarchal structures as manifested in its one-child policy with preference for sons and that later created demographic imbalances (United Nations Viet Nam 2014; Stöckl et al. 2017). This is also due to the uneven socio-economic condition between China and Vietnam that makes crossing state boundaries and migration to China lucrative, given that the passage will only require a border pass than any other legal travel documents such as a passport (Stöckl et al. 2017).

While in the trafficking situation, the study of Stöckl et al. (2017) reveals that these women experienced extremely high levels of violence (90%) including sexual violence, mainly by their husbands and traffickers. On top of their physical pain and illnesses, their mental health has been gravely at stake as they experience strong symptoms of depression, post-traumatic stress disorder, and anxiety. Complementing these particular results of this study, CWD also confirmed that these women, including the girl children who returned from the border countries are often traumatized. Many of them have been referred by the border guards and provincial organizations.

CWD said that there is a proportion of women from ethnic minority groups who are unable to communicate in the Vietnamese language and thus it becomes difficult for supporters to extend immediate help. These women also have difficulties in approaching vocational training, loan programs for livelihood development, or finding a sustainable job once they return home. Albeit the Shelter addresses these needs, CWD recognizes its limited capacity to only cater to those who are residents. There remain countless trafficked women out there badly needing support.

A challenging factor that CWD also identified is the police, many of whom still do not recognize GBV as a serious social problem. Another troubling fact is the change of transport route of the perpetrators for trafficked women. In order to evade border guards, perpetrators have explored the province-to-province route before finally reaching the border. The sea border has also been utilized.

Confronted with this alarming and complex trend, Giang Pham Huong, the director of CWD explained that they have intensified their collaboration with the border guards, especially in the China border, and the police ministry, in order to curb the trend and find the perpetrators.

The Shelter, through CWD and VWU, has also broadened its policy advocacy on anti-trafficking and other forms of GVB through policy dialogue, public awareness raising, and by maximizing the mainstream media. This policy dialogue with the government aims to strengthen national policy to protect women and children, and the chain-of-referral system that is a necessary component of the Shelter (Fig. 13.2).

Fig. 13.2 Residents collectively engage in artmaking activities (*Source* Peace House Shelter)

13.3 Sustaining and Strengthening the Shelter

During the early years of the Shelter, funds were sourced from AECID/Spanish Embassy, Australia Aid, and UNICEF (Burchill and Le 2015). Looking into the need for sustainability of the Shelter and the necessity to call government accountability on the plight of women, the leadership of VWU and CWD lobbied to the Vietnam government to recognize state responsibility over this program. They submitted a proposal to the Ministry of Finance to be able to access state funding that can sustain the Shelter. As of this writing, the Shelter gets funds from the Vietnam government budget, as a 'secular' support. Government subsidy is for the period 2013–2027.

As of latest count, there are 1300 women plus their children, who have been provided support by the Shelter (70% are those who experienced domestic violence, while about 30% were rescued from human trafficking). The CWD records showed that they come from 50 provinces and cities across Vietnam and from 16 ethnic minorities. The youngest girl recorded is between 12 and 13 years old. Around 80% of those who have been referred by the local Women's Union to the Shelter are related to domestic violence.

CWD admitted that there are still provinces where local authorities are unwilling to cooperate with the Shelter to support the victims. This is highly needed for the prompt response in dealing with GBV cases right after being detected and reported. The referral system which concerns coordination and information sharing between multi-agencies of government and support services providers remains inadequate. There is still a need to widely disseminate the information about the Shelter's support services for the victims. In addition, CWD also advocates to change the gender inequality mindset.

Jones and Van Anh Tran (2012) emphasized that social protection and gender equality are cross-cutting policy issues, thus, the importance of inter-sectoral coordination. Example of such is capacitating the local level or the communes to integrate gender in their poverty reduction plans. Highlighting inclusive accountability, they stressed that women's issues are not just a concern of the Women's Union but a professional responsibility of government agencies.

13.4 Outcomes of the Shelter

The culture of patriarchy that cements male dominance over women and thereafter creates this violent attitude and behavior over women is heavily entrenched on a global scale where Vietnam is not spared. Victim blaming and discrimination exist not only in Vietnam but elsewhere in the world.

But for the women of CWD, the Shelter that is being operated by the women themselves is already a big step forward in order to advance the interest of women, particularly from the marginalized sector, and their long-term agenda for social change with a gender equality lens.

Despite the compelling externalities earlier mentioned, CWD and VWU continue the campaign for women's rights at the community level and in training the media on how to handle these women stories that must be anchored on human rights, on women's rights. As it has been and continues to be so, CWD has effectively contributed to Vietnam's GBV policy development and engagement in GBV-related networks at the national, regional, and global level. This direct engagement with the government got favorable results when they observed the changes in the policy ministry system. With their constant prodding, one department has been established to deal with GBV. This department has requested CWD and VWU to teach and orient them of their current work through the Shelter and the pressing need of protecting women's rights.

Upon further reflection, CWD and VWU gauge the relevance and responsiveness of their comprehensive intervention when they already see the women residents in their empowering situation, in such a way that they finally recognize their rights and assert them, and have thought of or planned how to rebuild their lives thereafter.

The Shelter, under the leadership of women and the women's movement in Vietnam, although still facing complex issues in trying to implement policies that benefit the women survivors of violence, has asserted its space in the fight against human trafficking and domestic violence.

This multi-level coordination involving various stakeholders from government agencies to the grassroots, being managed by women yet independent from that of a government agency responsible for social welfare, or charity, becomes a strong point of the Shelter. The funding support both from government and external donors has significantly contributed to institutionalize and even improve this comprehensive approach to GBV.

These stages, from the moment a woman registers to the Shelter until the time of her reintegration (including the underpinning legal interventions), reinforce the capacity of the women sector as a collective to fight for women's rights and work toward women empowerment. The experience of establishing this kind of shelter or rather safe space for women in Vietnam has also proven women's efficiency as a collective to reach, not only the policy makers and implementers, but also the basic social structures—the commune and family unit which are responsible in ensuring the safe reintegration of these women survivors (Fig. 13.3).

While the Shelter still strives to become more efficient and to expand its services, its resilient identity among the police, border guards, and provincial authorities as the core in the chain-of-referral system, manifests the big responsibility of the women's movement in Vietnam and their huge potential to promote and protect women and children's rights, and even substantially contribute to the fulfillment and improvement of international laws on women's rights and the overall status of women.

Fig. 13.3 Campaign and advocacy work are crucial to spreading awareness about domestic violence (*Source* Peace House Shelter)

References

Burchill, Anggia A., and Thi Binh Le. 2015. *Peace House Shelter Project: Supporting Victims of Trafficking in Persons (Final Evaluation)*. Hanoi: Embassy of Spain in Vietnam, Spanish Agency for International Development Co-operation, and Center for Women and Development.

Hutt, David. 2022. How EU police disrupted Vietnamese human trafficking rings. *DW,* December 7. https://www.dw.com/en/how-eu-police-disrupted-vietnamese-human-trafficking-rings/a-64013468

Jones, Nicola, and Thi Van Anh Tran. 2012. The politics of gender and social protection in Viet Nam: Opportunities and challenges for a transformative approach. *Background Note*, May 2012. London: Overseas Development Institute.

Kwiatkowski, Lynn. 2011. Domestic violence and the 'Happy Family' in Northern Vietnam. *Anthropology Now* 3 (3): 20–28. https://doi.org/10.1080/19492901.2011.11728331

Kwiatkowski, Lynn. 2013. Domestic violence, ethnic diversity, and political and economic change in an upland community in Vietnam. *Alterstice* 3 (2): 51–65.

Le, T. Q. 1996. Domestic violence in Vietnam and efforts to curb it. In *Vietnam's Women in Transition*, ed. Kathleen Barry, 263–274. Hampshire/London: Macmillan Press Ltd.

Stöckl, Heidi, Ligia Kiss, Jobst Koehler, Thuy Dong Dung, and Cathy Zimmerman. 2017. Trafficking of Vietnamese women and girls for marriage in China. *Global Health Research and Policy* 2 (28): 1–9. https://doi.org/10.1186/s41256-017-0049-4

United Nations. 2000. United Nations convention against transnational organized crime and the protocols thereto. *United Nations Office on Drugs and Crime*. https://www.unodc.org/unodc/en/organized-crime/intro/UNTOC.html

United Nations Viet Nam. 2014. From domestic violence to gender-based violence: Connecting the dots in Viet Nam. *UN Discussion Paper*. Ha Noi: United Nations Viet Nam.

USAID. 2021. *Assessment of Countering Trafficking in Persons Activities in Vietnam: Situational Analysis Report*, October. https://www.usaid.gov/sites/default/files/2022-05/C-TIP_Assessment_report.pdf

Vu, Ha Song, Sidney Schuler, Tu Anh Hoang, and Trang Quach. 2014. Divorce in the context of domestic violence against women in Vietnam. *Culture, Health & Sexuality* 16 (6): 634–647. https://doi.org/10.1080/13691058.2014.896948

Ananeza P. Aban is currently a Ph.D. researcher of the Department of Conflict and Development Studies and a recipient of the Special Research Fund doctoral scholarship of Ghent University in Belgium. She also serves as a senior research analyst of UP Center for Integrative and Development Studies Program on Alternative Development. Among her major tasks are coordinating the field research across Southeast Asia, co-editing the Alternative Practices monograph series (2020, 2023), and ensuring the international linkages of the program. She finished her Master of Community Development degree from the University of the Philippines in Diliman.

Chapter 14
Women's Empowerment Through Weaving: The Case of the Women Weavers Cooperative of Lepo Lorun

Angeli Fleur G. Nuque

Abstract This case study documented how the initiative to preserve ikat weaving eventually became means to empower women weavers across villages of the Sikka Regency and an estimate of over 1200 women weavers. The Lepo Lorun Women Weavers Cooperative has confronted the challenge of preserving their cultural heritage and tradition and how to make it a sustainable occupation for women weavers in the face of commercial textile production. Moreover, the case study documented how weaving has become a "socialization" activity and community building.

Keywords Ikat weaving · Textiles · Women weavers · Traditional knowledge · Commercialization

The Lepo Lorun Women Weavers Cooperative was founded in 2003 (but has begun in as early as 1998) by Alfonsa Horeng at Nita Village, Sikka Regency in Flores Island, East Nusa Tenggara (NTT), Indonesia. What began with small households is now a cooperative spread across villages in the Sikka Regency which engages over 1200 women weavers and replicated in multiple towns across Flores (Potter and Wijanarko 2015). The term *lepo lorun*, which means "weaving house" in Sikka language, is also a name of a homestay accommodation and workshop/outdoor studio that shares the weavers' weaving culture with guests. The researchers of UP CIDS Program on Alternative Development, were able to stay in Lepo Lorun and had an exceptional opportunity to watch the women master weavers demonstrate by hand the tedious work starting from the yarn spinning process out of natural cotton fiber, the dyeing of threads, preparation of the warp and frame, until the backstrap weaving phase. The most important part was the series of conversation with these women master weavers—specifically *Du'a* (nickname for a woman who came from Sikka Regency) Siska, *Du'a* Kompiana, *Du'a* Antonia, *Du'a* Beth, and *Du'a* Gout—and the founder of the collective, Alfonsa Horeng.

A. F. G. Nuque (✉)
University of the Philippines Center for Integrative and Development Studies (UP CIDS), Quezon City, Philippines
e-mail: afg.nuque@gmail.com

J. M. C. Sy and A. P. Aban (eds.), *Reimagining Development in Southeast Asia*, Asia in Transition 21, https://doi.org/10.1007/978-981-97-9140-8_14

Alfonsa Horeng was driven to set up a weaving cooperative to preserve the heritage and tradition of *ikat* weaving and traditional agricultural technology, provide additional income for women weavers, and most importantly, "raise the social, cultural, and economic value of the magnificent *Tenun Ikat*—the traditional weavings—that embody the stories of Flores" (Weaving Community and Culture Together, n.d.).

Textiles in general hold both practical and symbolic or religious meanings (Allerton 2007; Livingston 1994). While most literature and anthropological studies on textiles place emphasis on the ritualistic and ceremonial meaning of textiles, Allerton (2007) focused on the utilitarian function or materiality of textiles—which she called the "everyday life"—that straddle the sensual and intimate life of everyday garments. She gave *sarong* as an example and its numerous purposes as a wrap worn in various locations and events in life (in the house; outside while working or out in the fields; while bathing or sleeping; at birth, marriage, or death for protection; and for hiding objects or bodily conditions; for accentuating properties of the skin; for carrying babies or more menial items like beans or small amounts of coffee; or for concealment of money, pregnancy, and secret medicines) (ibid., 23). Allerton further added that focusing on the everyday meaning of textiles has re-oriented discussions of textiles as two-dimensional panels "displayed in museums or printed in the pages of glossy art books" to three-dimensional objects to be worn, "animated by the shape and movements of human bodies" instead (ibid., 24).

Looking at the everyday, private, and utilitarian purpose of textiles is an important assertion in looking at the longevity of (material) culture such as textile simply because of its everyday function—it is used every day and therefore will continue to exist because of its everyday function. During our stay in Lepo Lorun, the women weavers exhibited the so-called "everyday life" of their sarongs where they separate the ones they wore to church or for special occasions, from those for their everyday activities like farming, doing household activities, or weaving. Alfonsa had even let us borrow sarongs to wear for the duration of our stay and had explained the meaning of each *motif* (which refers to the designs and patterns), the distinctive character of every hand-woven fabric, and from which regency it belonged to or represented. We were also lucky to have touched a piece of heirloom which Alfonsa got from another regency for preservation and historical reference. She and the women weavers plan to build an *ikat* museum in Lepo Lorun, as a legacy for the next generation.

Other scholars focus on the symbolic and religious meanings of textiles and argue that textiles are never simply worn as clothing but are deeply emblematic. Motifs locate the textile owners' complex social and cultural matrix. The motifs "on the cloth tell stories and hold symbolic and archival significance for Flores people" (Simanjuntak 2015).

Allerton (2007, 24) also noted that textiles have a public life as an "exchange (sic) valuable" or "commodities for sale in the market." In an interview with Alfonsa, she had mentioned that woven textiles were presented as *belis* (dowry) to the woman's future family while men present ivory; textiles were also seen as heirlooms with significant exchange value (ibid.).

Ikat woven fabric holds a "special place because of the intricacy of the tying, dyeing, and weaving techniques essential to their production" (Livingston 1994,

155). The practice of *Tenun Ikat*, or the traditional practice of weaving, is not simply a craft but an art and an archive of ritual, stories, and traditions (Potter and Wijanarko, 2015). *Ikat* comes from the word *mengikat* which means "to tie or bind" (Livingston 1994).

Ikat weaving is almost an exclusively female function in Indonesia (Livingston 1994) and even signifies the entry from girlhood to womanhood. In fact, some cultures in Indonesia prescribe a certain level of skill in weaving and to weave a certain number of textiles before they are considered to be marriageable. A girl or a woman's weaving capability, especially if highly skilled, may also mean that she is more sought after as a wife and is then seen as an opportunity for economic mobility (because of possible marriage to a husband with a higher social class). Scholars who look at the more gendered aspect of weaving see it as within the female realm that is "parallel to men's ritual speech" (Allerton 2007, 24) and a "highly gendered craft" associated strongly with women and "demonstrates readiness for marriage on the part of a young woman" (O'Brian 1999, 31).

Weavers of Lepo Lorun explained that they learned the art of weaving traditional textiles and clothes from their mothers. The labor-intensive weaving process has taught them to be more patient, strong, and independent. Because of the intricacy of the technique, it takes months for a weaver to complete her work, depending on her other reproductive roles (such as house chores and child rearing) that she needs to comply with. During our stay in Lepo Lorun, we met young daughters of weavers being taught to weave using a backstrap loom, a weaving technique where a strap is worn around a weaver's waist. Alfonsa herself shared that she learned the weaving process at an early age and that prior to formally taking on the loom, she was already helping and watching her mother, aunt, grandmother, and sometimes women neighbors who weave in their home. Although weaving is not entirely a female function or occupation in some societies [e.g., in India where *ikat* weavers are mostly men (Livingston 1994)], it was observed that weavers in Lepo Lorun were mostly women and they serve a significant role in the preservation of the weaving tradition. Striking a different perspective from the common notion of weaving for trade and other commercial production purposes, Alfonsa is resolute in calling the women weavers of Lepo Lorun "profesoras" or professors of their pieces, and that ikat woven cloth is more than a commodity for sale (Fig. 14.1). In an interview with *Jakarta Post* (2015), she said: "[We, weavers] are not craftspeople; we are not menial workers. We are [maestros, *profesoras*]. We don't weave for money; we do it because of our belief in tradition. The cloth is our heritage." This is a reason why weavers in Flores society are considered a source and preservers of traditional knowledge, further proving that they also "function equally as visual reminders of the past" (Duggan n.d., 77, as cited in Heringa 2003, 629).

Although the women of Lepo Lorun get extra income from selling their handwoven fabrics at a fair price, their main purpose is to enliven a tradition that has been an integral part of their domestic and social life.

Alfonsa cited the preservation of tradition, knowledge, and culture as one of the primary movers for forming the weaving cooperative in Lepo Lorun. There is an unwritten science behind weaving that is passed down from one generation to another:

Fig. 14.1 Showcase of various textiles woven by members of the Cooperative (Photo courtesy of the author)

from how dye is processed, to making patterns, to body positionality. Livingston (1994, 155), for example, noted the "mathematical certainty required to know in where to tie or bind unwoven threads together in order to achieve the desired pattern after dyeing is complete." Body tension also affects the weaving process. In making textiles using a backstrap loom, one's body tension affects how correctly the handle is used (Andon 2014).

Minturn (1996, as cited in O'Brian 1999) in an earlier work looked into women and the role of technological complexity in weaving. She argued against Murdock and Provost's (1973, 212, as cited in O'Brian 1999, 33) suggestion that the invention of a new artifact or process supplants an older and simpler one where both the activity of which it is a part and other "closely related activities tend [to be] more strongly to be assigned to males." Further adding that women are exposed to such complex techniques: "spinning and weaving technologies are in themselves complex and constitute a case in which a technologically complex activity remains strictly within the control of women" (Minturn 1996, as cited in O'Brian 1999, 33).

The weaving process observed in Lepo Lorun was a confluence of traditional knowledge and practices with science. During our stay, we were able to see the ingredients being used to make natural dye, especially the indigo color (Fig. 14.2). Before seeing it, however, we were asked whether or not we were menstruating. Part of their traditional belief, we were told that women who are menstruating cannot

Fig. 14.2 Gathering raw materials from immediate environments (Photo courtesy of the author)

process or observe the dyeing of textiles. But more than that, women possess the indigenous wisdom of successfully selecting the best time to pick materials for dyes (for example, bright colors are usually processed when the weather is cooler). As mentioned previously, each woven cloth follows a motif. There is a philosophy behind each pattern that explains various themes like the harmonization of families, a regency's origin story, and male and female relationships. Each motif also has a distinctive style and color scheme that represents their region or regency it belongs to.

Weaving as an occupation provided an excellent opportunity to examine the cultural attributes of an occupation. Weaving is a commonly occurring occupation across cultures and has been described as "an ancient human activity that is related to some of the touchstone concerns of modern art" (Perl 2000, 32, as cited in Bonder 2001, 311). According to Perl, this occupation can be described as an art, a craft, or can be a pragmatic and practical task leading to production of a needed product (ibid.).

One of the challenges being faced by weavers not only in Indonesia is the commercialization of textiles and the shift to using synthetic colors, primarily because of commercial production. It also followed that textiles using synthetic or artificial colors have become cheaper and readily available to the market that textiles using natural dyes appear to be considerably more expensive than the synthetic ones. O'Brian (1999, 32) cited Maya women as examples where while they maintained small, home-based knitting and weaving businesses, the "increased availability of cheap, mass-produced goods has driven prices up and has, for example, made traditional clothing prohibitively expensive." It cannot be denied however that while more expensive, *ikat* woven cloth boasts of high-quality production because of the long and laborious process of weaving it entails. Erb (in Hamilton 1994) reported that in southern Manggarai, cloths woven for sale in the market are almost always of an inferior quality, bought by town-dwellers or by those from areas of Manggarai with historical taboos against weaving.

This presents two problems to the women weaver's cooperative both culturally and economically: how do they tread the fine line between weaving as a way to preserve tradition and heritage and make it a sustainable occupation for the women weavers who have financial responsibilities to themselves and their families? The response to this dilemma was setting up and strengthening a weaver's cooperative that would address both issues.

A similar effort was done by the *matmee* weavers in the Laotian border of Thailand (also known as the "forgotten North East' of Thailand") where young women from the slums who were drawn into prostitution in the capital were organized by the Good Shepherd Sisters (Wilshaw 1991). The "Village Weavers" focused on economic poverty alleviation and offered support to women through vocational training programs in hairdressing, secretarial and business skills, and sewing; maternity units and day-care centers were also available for their children (ibid.). The "Village Weavers" was also a response to the lack of infrastructure and development projects in Northeast Thailand as it was generally untouched by the growing economic and developmental progress in the rest of Thailand. Through the "Village Weavers" initiative, "wells have been sunk, irrigation schemes started, and rice and buffalo banks set up. [Above] all, they have looked for ways to keep women and girls in the villages" (Wilshaw 1991, 204). More than the economic success and security of the young women in the "Village Weavers" in successfully providing income to more than 300 weaving families, the tradition of *matmee* has been sustained and passed on to young girls in the village.

Wilshaw (1991, 205) also reported a previous attempt in Indonesia during the 1990s to make *ikat* (which she described as a traditionally made cloth "using simple equipment in an extremely labor-intensive way") for an income-generating project. A foundation motivated weavers to form thirty self-help groups with over a thousand members and they set up a company to buy and market the textile and cloth products. She further added that the company paid twice "the exploitative rates normally paid by the traders and guaranteed to buy all the weavers' products, as long as they met set standards of quality control, irrespective of whether any orders have been received from them" (ibid.). When we asked why weavers would sell their textiles at a cheaper price, their answer was practical: the money is instant, most especially when there are school fees or household needs to pay. Although the weavers of Lepo Lorun understand the cultural value and appraised market value of their textiles, immediate and practical needs force them to sell cloths at a cheaper price to compete with the prices of commercial textiles available in the market. The Women Weavers Cooperative of Lepo Lorun tries to carefully straddle the fine line between maintaining that textiles are not just a commodity and giving opportunities for economic empowerment to the weavers.

Alfonsa noted that prior to joining the collective, the women weavers themselves initially expressed reservations in pursuing weaving as an occupation when more lucrative jobs were available. Eventually, the weavers realized that the textiles they wove could be a source of income and, in our interviews, they took note of the positive economic impact Lepo Lorun had to their families and how they were able to fulfill their role as providers for their families during times of difficulties. *Du'a* Elizabeth shared that Lepo Lorun not only gave her a steady income from her work in the homestay, she also found a sense of belongingness in the collective and work that makes her feel happy. She was also able to discover activities she enjoyed doing in Lepo Lorun. Other weavers shared that because of more opportunities in Lepo Lorun—weaving and weaving workshops, homestay, venue and clothes rental, singing, and dancing—they were able to save up to build houses, send their kids to school, and start small businesses. The impact of Lepo Lorun should not only be measured by how it has helped the women weavers economically. Lepo Lorun was able to help the weavers explore new avenues to hone their talents and skills. This was echoed by the women weavers we had interviewed, saying they felt a sense of fulfillment in weaving and being able to provide for their families when they previously could not. They also felt a sense of dignity, an increased self-esteem, and a new found community to belong to because of weaving.

Initially, some of the weavers' husbands did not support their activities in Lepo Lorun. But through the persistence of the women weavers to continue working at Lepo Lorun and showing the positive economic impacts of their work, they began having a better impression of the collective. Lepo Lorun was able to extend the economic opportunity of families where husbands were hired to build the traditional houses in the homestay.

Women Weavers Cooperative of Lepo Lorun also place a strong emphasis on *sama-sama* or *socialisasi* (socialization), coming together as a group to do a certain activity together. The weavers we have talked to said that weaving is also a social activity—they talk together while they weave and they teach their daughters how to weave—which is why they are able to form stronger relationships with each other (Fig. 14.3). During one afternoon in Lepo Lorun, we were able to sit with the weavers and saw how they stayed in one area of the homestay where some continued to weave their unfinished textiles; some observed their daughters' progress in weaving; and some were singing and practicing playing traditional instruments. They made sure to make us feel welcome in their *social* circle and told us stories about their past and, at one point, encouraged their best singers to sing a traditional song for us. *Socialisasi* is also done while sharing and sourcing raw materials—wherein information and technical knowledge is shared. We were fortunate to experience this practice when we were invited to go to a nearby farm to pick the vegetables we would eat for lunch and dinner. Through *socialisasi*, women weavers consciously try to avoid gossiping as it strains their relationship with one another and rather learn how to deal with conflict by constructive conversation. They said that while there

Fig. 14.3 Women of Lepo Lorun weaving and working together (Photo courtesy of the author)

are times when they get angry or frustrated with each other, weaving has been an instrument to learn how to resolve it and laugh together afterwards.

Dodi Wijanarko filmed a documentary on the case study of the Lepo Lorun Weavers Collective in 2015 informed by the concepts of participatory media and positive deviance—using such approaches highlight the capacity of individuals to participate in a "variety of activities of self-empowerment" (Au Lorun: I Am Weaving n.d.). The specific focus on the weaver's collective model was to showcase how the women weavers have become economic providers for their families and re-focus the existing scholarship on *tenun ikat* from technology, motifs, or trade to how women weaving collectives have been a key mechanism to reinvigorating the art of *tenun ikat* and its emergence (Au Lorun: I Am Weaving n.d.). Moreover, Wijanarko said that he decided to focus on the collective and the stories of the weavers because he was impressed by the important role women held in both their families and in upholding the traditional rituals of the community (Au Lorun: I Am Weaving n.d.).

Alfonsa mentioned that there is a lack of support for weavers from the local government of farther islands in Indonesia, especially in the case of Flores, and how their government supports artists from bigger and more popular cities like Bali. She added that national government officials from other countries have been more welcoming of her and the weavers of rural islands. Despite this, the Women Weavers Cooperative of Lepo Lorun remains steadfast in its mission to preserve traditional *ikat* weaving techniques against the rampant commercialization of textiles and the preference of younger generations to pursue other professions and wearing western clothing; promote economic and social empowerment for women weavers; and share the traditional weaving heritage and culture to the Indonesian and global community.

References

Allerton Catherine. 2007. The secret life of sarongs: Manggarai textiles as super skins. *Journal of Material Culture* 12(1): 22–46. https://doi.org/10.1177/1359183507074560.

Andon, Nelly. 2014. Meeting Alfonsa Horeng, a devoted weaver from flores. *Wastra Indonesia,* April 23. https://wastraindonesia.uk/alfonsahoreng.

Au Lorun: I am weaving. n.d. DRO, Deakin University. http://dro.deakin.edu.au/view/DU:30108758.

Bonder, Bette. 2001. Culture and occupation: A comparison of weaving in two traditions. *Canadian Journal of Occupational Therapy,* December 68(5): 310–19. https://doi.org/10.1177/000841740106800507.

Hamilton, Roy W., ed. 1994. *Gift of the cotton maiden: Textiles of flores and the solor islands.* Los Angeles, CA: Fowler Museum of Cultural History, University of California.

Heringa, Rens. 2003. Review of Ikats of Savu: Women weaving history in Eastern Indonesia by G. Duggan. *Bijdragen tot de Taal-, Land- en Volkenkunde,* January 159(4): 627–29. https://doi.org/10.1163/22134379-90003744.

Livingston, Judith. 1994. Ikat Weaves of Indonesia and India: A comparative study. *India International Centre Quarterly* 21(1): 152–74 (Spring).

O'Brian, Robin. 1999. Who weaves and why? Weaving, loom complexity, and trade. *Cross-Cultural Research* 33 (1): 30–42. https://doi.org/10.1177/106939719903300103.

Potter, Martin, and Wijanarko, Dodid. 2015. Au Lorun: I am weaving. ResearchOnline@JCU, James Cook University Australia. https://researchonline.jcu.edu.au/51785/.

Simanjuntak, Tertiani. 2015. Flores women weave a safety net for tradition. *The Jakarta Post,* June 6. https://www.thejakartapost.com/news/2015/06/06/flores-women-weave-a-safety-net-tradition.html.

Tadem, Eduardo. 2017. Preliminary draft concept note on alternative practices in Southeast Asia: Towards a new model of peoples' regional integration. Unpublished document.

Weaving community and culture together. n.d. Big stories, small towns. http://bigstories.com.au/page/flores-indonesia.

Wilshaw, Rachel. 1991. The role of marketing in income generation: Two case studies. *Development in Practice* 1 (3): 204–206. https://doi.org/10.1080/096145249100076391.

Angeli Fleur G. Nuque was Junior Research Analyst at the University of the Philippines Center for Integrative and Development Studies (UP CIDS) Program on Alternative Development (AltDev). She is an alumna of the Department of Women and Development Studies of the UP College of Social Work and Community Development. Her graduate thesis focused on feminist gerontology, aging, and sexuality. This study was undertaken in cooperation with the Women Weavers Cooperative of Lepo Lorun and the researchers and the Program extends its thanks to Shantoy Hades, who facilitated the field visits and interviews for this chapter. The field visits in NTT were conducted with Ananeza P. Aban.

Chapter 15
Women Household Heads in Indonesia and Their Role in Addressing Gender Inequality

Nathaniel Punongbayan Candelaria

Abstract This chapter utilizes a feminist lens in studying the *Pemberdayaan Perempuan Kepala Keluarga's* (PEKKA's) case. It also tackles gender inequality issues experienced by women in Indonesia. To gain a deeper understanding of PEKKA's activities to promote women's advocacies, in light of gender inequality issues in Indonesia, an interview with the PEKKA secretariat and a focus group discussion (FGD) with the PEKKA union members were conducted by a research team from the University of the Philippines Center for Integrative and Development Studies (UP CIDS) to document their experiences, including the actions they took to promote the rights of women-headed households in Indonesia. Given the extent of PEKKA's activities, this chapter not only considers its example as a feminist practice, but also as an alternative practice.

Keywords Gender inequality · Women-headed households · Marriage · Women's rights · Feminism

15.1 Introduction

Attempts to promote the rights of women are covered through the conventions, declarations, and agreements signed at the international and regional levels. For instance, the United Nations spearheaded the efforts to have an international convention on the rights of women ratified. And at the level of the Association of Southeast Asian Nations (ASEAN), declarations were signed and institutions were created at the regional level as far as its promotion is concerned.

Despite these noble intentions and goals, the reality is that gender inequalities remain. As mentioned by Ortiz-Ospina and Roser (2018), women earn less than men, are underrepresented in managerial positions, are overrepresented in low-paying jobs,

N. P. Candelaria (✉)
Department of Political Science, College of Social Sciences and Philosophy, University of the Philippines Diliman, Quezon City, Philippines
e-mail: npcandelaria@up.edu.ph

J. M. C. Sy and A. P. Aban (eds.), *Reimagining Development in Southeast Asia*, Asia in Transition 21, https://doi.org/10.1007/978-981-97-9140-8_15

have less control over their income, and have limited decision-making capabilities in households.

Even if improvements were noticed as far as women's rights are concerned, challenges still remain. In Southeast Asia, for instance, women's role in agriculture and food security are not recognized despite their big role (Devasahayam 2018). As Indonesia is one of the largest states in the region, issues on women and gender are significant. In terms of the World Economic Forum's Global Gender Gap Report 2023, Indonesia ranks 87th out of 146 countries surveyed (World Economic Forum 2023). In other rubrics, Indonesia is also located in the lower half of the strata.[1] These issues and concerns have implications for future generations—especially regarding women and their children's access to services (PEKKA Secretariat 2018; Duff et al. 2016; Sumner 2015).

In Indonesia, there are attempts from different civil society organizations to bridge the gap and address this concern. One of these organizations is the *Pemberdayaan Perempuan Kepala Keluarga* (PEKKA). This paper will attempt to document the experiences of PEKKA and what they have already achieved as far as the fight toward the promotion of women's rights is concerned.

For this reason, the chapter will employ a feminist lens in looking at the issues of gender inequality and the actions of PEKKA. As pointed out by Randall (2010), feminist approaches can help probe the private–public divide, which is prominent in Indonesian society and the advocacy of PEKKA. Aside from this paper's being a feminist analysis of PEKKA, it is also a contribution to documenting cases of alternative practices in the Southeast Asian region. Documenting these endeavors is one of the projects of the UP CIDS Program on Alternative Development. Based on the *Integrating Alternative Development Efforts in Asia* Report, alternative means that it is a reaction against the dominant trend of development; however, it does not mean that even if there are aspects of the dominant being practiced by different organizations, it automatically adheres to the dominant idea (Raina 1997). Moreover, for an idea to be considered alternative, the processes undertaken must be implemented and integrated for it to challenge the dominant paradigm (Raina 1997).

15.2 Women and Gender Issues at the State Level: The Situation in Indonesia

As one of the founding members of ASEAN and as the country with the largest population in the region, Indonesia is significant. While it has no official state religion, with the exception of the belief in one God in Pancasila, Indonesia has the largest Islamic population. And while Indonesia was heavily influenced by the military after

[1] Based on the data of the World Economic Forum, Indonesia ranks 96th out of 149 countries in terms of economic participation; 107th in terms of educational attainment; 79th in terms of health; and 60th in terms of political participation.

the New Order, Megawati Sukarnoputri rose to power as Indonesia's president, even when women heads of state or government are rare even within the ASEAN region itself (Blackburn 2004).[2] However, even though Indonesia already has a woman head of state, the stark reality of women remains. This part of the chapter will therefore look into the development of the Indonesian state and connect the discussion with the issues of women across different timeframes.

15.2.1 Women and Gender Issues vis-à-vis the Development of the Indonesian State

Before declaring its independence in 1945, Indonesia was a Dutch colony; however, it was briefly controlled by the Japanese forces during the Second World War (Blackburn 2004). Since gaining its independence from its colonizers, Sukarno's Guided Democracy focused primarily on the identity-building of the Indonesian state, where women issues are treated secondarily to national interests (Blackburn 2004). Under the Guided Democracy era, Blackburn (2004) noticed that women found themselves having difficulty to air their opinions and concerns; nevertheless, women's groups in Indonesia such as the *Gerwani* were allowed to exist by the Sukarno government. Moreover, women's organizations and politicians flourished under the Sukarno government (Firdaus 2019).

When Sukarno's administration fell under the coup of Suharto, the administration also changed from Guided Democracy to the New Order. As part of the changes made in the Indonesian society, women's role has also changed substantively. The New Order era was geared toward the promotion of the family such as birth control and taking care of the household (Rinaldo 2011). From being active in Indonesian politics during the Sukarno government, women were renegades back to their respective household duties (Firdaus 2019). This change has been reflected heavily in the Indonesian Marriage Law of 1974, where men were recognized as heads of households and women as managers, with the aim of making marriages harmonious (Parker, Riyani, and Nolan 2016). This is known as *kodrat* (natural destiny), where the role of women as *ibus* (mothers) are emphasized vis-à-vis the role of men as *bapak* (fathers) (Blackburn 2004). Aside from these structural changes, a Ministry for Women's Roles under the Suharto's regime was established; however, it also disallowed the existence of women's organization with the example of *Gerwani* as a group included in the Communist purge in Indonesia (Blackburn 2004).

The *reformasi* period followed after the fall of Suharto's government in 1998. Under the *reformasi* period, women in Indonesia are divided as to how they will deal with issues on women; some of them are liberal, and some want to promote more conservative values on women (Blackburn 2004). In terms of policies, the state under the *reformasi* period implemented Presidential Decree No. 9/2000, which proposes to

[2] Megawati Sukarnoputri is the daughter of Indonesia's first president, Sukarno. She became the President of Indonesia after the resignation of her predecessor, Abdurrahman Wahid, in 2001.

include gendered perspectives on different government policies (Blackburn 2004). However, conservative policies restricting women's rights have also been implemented under this period. For instance, policies on women clothing is incorporated in Indonesia's Anti-Pornography Act of 2008, thereby restricting how women should act in public (Parker et al. 2016).

In the Indonesian society, there seems to be a divide between global values and local values. As discussed by Kusujiarti et al. (2015), there are differences on how objectification of women is perceived in terms of western values vis-à-vis the teachings of Islam. While western values see the case of Indonesia as restriction to women's rights, for the locals, adherence to the teachings of their religion meant freedom from point of view as well (Kusujiarti et al. (2015). In a case study of Acehnese women, it was observed that local norms help with the mobility of women in Aceh, which is different from how western scholars usually depict women mobility in the context of global norms (ibid.). The apparent clash between the universalizability of human rights values v. Indonesia's local culture and traditions affects how legal services are accessed by women (Curnow 2015).

15.2.2 Contemporary Issues of Women in Indonesia

As pointed out in the discussion of Indonesia's history, a division between conservative and liberal forces were identified under the Reformasi period. The debated points of contention between these forces cover issues affecting women such as, but not limited to polygamy, pornography, divorce, and relations of women within the family. This is important to be discussed since despite the existence of global norms to protect the rights of women, issues at the national and local levels remain. In fact, the global–local divide on women's rights remains in Indonesia as argued by both Rinaldo (2011) and Curnow (2015).

In contemporary Indonesian society, it was documented that women are disadvantaged in terms of their marital status. It is especially difficult for women to be divorced and/or to be widowed, since the society will categorize these women as *janda* (Parker et al. 2016; Zulminarni et al. 2018). This is because when a person is being referred to as *janda*, that person is accused of sexual promiscuity; and therefore, women are viewed to be immoral (ibid.). In order for women to de-stigmatize this situation, they remarry to avoid society's ostracism; thus, contributing to higher remarriage rates in Indonesia (Van Bruinessen in Parker et al. 2016). And in terms of pornography and sexual promiscuity, Indonesia made pornography illegal; however, the formulation of the law negatively affects women as it dictates the proper attire for women in Indonesia (Parker et al. 2016).

Aside from sexual promiscuity, polygamy is also a big issue for women in Indonesia. Despite the presence of a law regulating marriages in Indonesia, people tend to follow the traditional way; thus, making the process not valid under Indonesian laws (Nurmila 2016). This also happens when a man seeks to marry another woman (ibid.). Since some of the women are married illegally, there are also implications for the children. Because of the lack of legal basis of the marriage, children are not registered when born (Zulminarni et al. 2018; Sumner 2015; Duff et al. 2016). Moreover, the poor are much more vulnerable because they are more likely not to be able to register due to accessibility issues; the lack of registration also leads to the stigmatization of both the mother and child (ibid.).

Women issues do not spare even single women. This is usually reflected in terms of their goals to migrate in order to find better opportunities elsewhere (Silvey 2000). When women migrate, they adjust their familiar roles and encounter challenges both at the local and international contexts, thus changing their conceptions of space and home (Williams 2005). There are many reasons as to why women chose to leave their homes in search of opportunities elsewhere. In a study conducted by Williams (2005), women shared the reasons why they migrate, i.e., to experience a different community and escape from constraints and at the same time to help their household and fulfill their role as "dutiful daughters." However, women are hindered by the views of the state on women vis-à-vis men (Silvey 2000). Because of these negative views, they do not help women escape the cycle of violence and disadvantages against women in the society in general (ibid.).

While women are autonomous and have their own goals and ideas in the first place (Williams 2005), the abovementioned issues make women vulnerable in Indonesian society. Nevertheless, women themselves are finding ways on how to change the system. One of the ways to deal with issues of women in Indonesia is through participatory action of women youth in Indonesia (Fergulio et al. 2017). The paper pointed out how Indonesian women find it difficult to deal with issues and how the group helped these women to empower themselves (ibid.).

15.3 The Case Study Method

In spite of the international and regional norms promoted for the advancement of women's rights, Indonesia still has a long way to go to promote them. One of the active women groups in Indonesia is PEKKA, a group organized in 2001. While there are many women organizations in Indonesia, one of their members stressed that PEKKA was the first to organize women in the society. Given these reasons, the research will document the case of PEKKA as a women's organization in Indonesia. Thus, a case study research design was adopted for this particular research. As discussed in political methodology books, case studies are important in dealing with the intricacies of different cases and how their experiences lead to another one (Flick 2009; Babbie 2013).

To conduct this research, a team of researchers from the UP CIDS visited the PEKKA Office in Jakarta, Indonesia, and conducted interviews and focus group discussions (FGDs) with the members of PEKKA Union. As discussed by Zulminarni et al. (2018), the PEKKA as an organization became federated in 2014, thus decentralizing the operations of the organizations within a national secretariat and a local group as well. This is the reason why the team was able to conduct a data gathering process from both sectors of PEKKA. In this regard, interviews with the members of the PEKKA Secretariat and FGDs from the PEKKA Union members are used in order to gather the viewpoints and to expound these ideas better. Aside from these interviews, desktop review of materials and other information provided by PEKKA were also used in the research.

15.4 PEKKA Amidst Gender Issues in Indonesia

As explained in the methodology section of the chapter, the discussion will focus on the experiences of PEKKA toward the advancement of women's rights in Indonesia. For this section of the paper, a discussion on the historical roots of PEKKA will be highlighted. This will be followed by the discussion of the interventions they implemented to promote the welfare of women in Indonesian society.

15.4.1 PEKKA's Origins as a Women and Women-Headed Organization

In 2001, PEKKA was established by Nani Zulminarni. PEKKA, as an organization for women-headed families, traced its roots from the project called *komnas perempuan* (widow's project), a project that documented the experiences of widows in Aceh (Zulminarni et al. 2018). The organization aims to provide help and intervention on "women-headed households in Indonesia" because of the gender imbalances in Indonesia and the stigma of being *janda* felt by women under Indonesian society (Zulminarni et al. 2018). *Janda* are regarded to be a disgrace in the society because they are treated to be sexually promiscuous (Parker et al. 2016; Zulminarni et al. 2018). Because of this stark reality, women find it difficult to access basic social services such as subsidies, health insurance, and cash transfer programs (Sumner 2011 in Zulminarni et al. 2018).

15.4.2 Advocacies for Women in Indonesia

PEKKA, as an organization, adheres to feminist thoughts on how the organization does things (PEKKA Secretariat 2018; Zulminarni et al. 2018). For PEKKA, feminism helps in empowering women in an imbalanced society. Moreover, as pointed out by Randall (2010), feminism as a lens helps discuss the intricacies of the private–public discourse. That is also reflected in the gendered analysis of food security in Southeast Asia by Devasahayam (2018). Under the logic of feminist analysis, the experiences of PEKKA will be discussed thoroughly, especially with regard to their interventions, as well as challenges that this women organization has faced as far as the promotion of their rights is concerned.

PEKKA promotes the welfare of the women-headed households in Indonesia, which the organization defines as the "widows, divorced, abandoned, disability husband, under polygamous married, single, unmarried but have a child, having irresponsible husband" (PEKKA Secretariat 2018). As of 2018, PEKKA has membership of 31,447 members across 2559 villages in Indonesia (Zulminarni et al. 2018). As discussed during the interview of the members of the PEKKA Secretariat (2018) in Jakarta, they mentioned that PEKKA is present nationally, except in Guinea primarily due to cost-related concerns. Figure 15.1 shows the map of operations of PEKKA.

In a nutshell, PEKKA intervenes in order to promote the welfare of women, particularly women abandoned by their respective husbands, in Indonesian society. In order for PEKKA to promote women's welfare in the society, the members have organized women by establishing cooperatives; and the organized women were taught and trained by PEKKA members extensively (Zulminarni et al. 2018). PEKKA has been training women in order to promote change in their society. Figure 15.2 shows how the organization envisions change for Indonesian women and the society in general.

As explained during the interview by members of the PEKKA Secretariat, the process starts from women themselves. Through the community, PEKKA was able to intervene and help them train. However, the members of PEKKA in the interview nevertheless recognize how powerful institutions are. That is the reason why whatever interventions they envision for women, society has to change also to fully promote women's rights. Even if international norms exist as far as women's rights

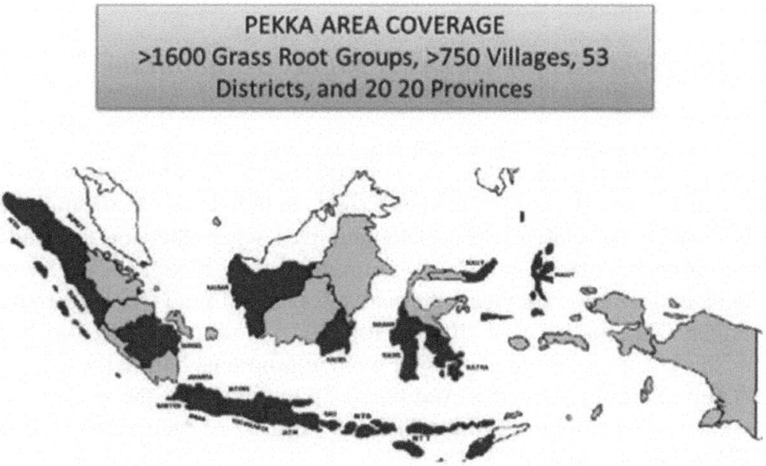

Fig. 15.1 Map of operations of PEKKA in Indonesia. *Sources* PEKKA; Zulminarni et al. (2018, 8)

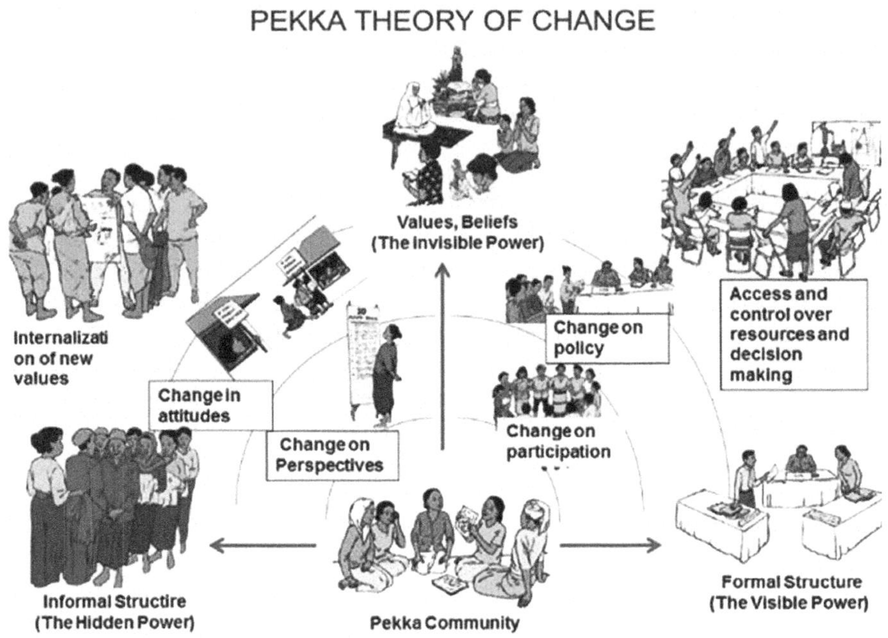

Fig. 15.2 PEKKA's theory of change. *Sources* PEKKA; Zulminarni et al. (2018, 11)

are concerned, the member of the secretariat discussed that PEKKA operates more on changing Indonesian society from the below.

15.4.3 Interventions for Women's Rights in Indonesia

To concretize these objectives and goals, PEKKA has several interventions to promote women's welfare in Indonesian society. One of the interventions promoted by PEKKA to improve the state of women's rights in Indonesia was the mobilization of women through the establishment of cooperatives, where they must go and introduce the program to women one-by-one (Zulminarni et al. 2018; PEKKA Secretariat 2018). These cooperatives provided alternative sources of funds for women through the movement (Zulminarni et al. 2018). However, the initiative received backlash from loan sharks and local elites running women programs before the existence of PEKKA (Zulminarni et al. 2018). And based on the inputs of the members of the PEKKA union, they primarily joined PEKKA because of their services and their help (PEKKA Union Members 2018).

As part of promoting women's rights in Indonesia, PEKKA provides leadership training for women in the communities. Called *Paradigta* Academy, the program seeks to capacitate women "for community engagement, public deliberation, and decision-making" (Zulminarni et al. 2018, 20).[3] The program's modules help women from basic skills such as reading and writing to managing one's village (Zulminarni et al. 2018). The training school spearheaded by PEKKA was first established in Nusa Tenggara Timur (East Nusa Tenggara) and spread across different areas of Indonesia such as Sulawesi, Aceh, and Jakarta in Java Island; other schools were also created in Nusa Tenggara Barat (West Nusa Tenggara), Moluccas, Central Java, and West Kalimantan (PEKKA Secretariat 2018). Based on the views of the PEKKA Union members, the leadership training gave them confidence to pursue the needs of women in their society.

Furthermore, PEKKA's documentation of the socioeconomic conditions of their members shows that it is also impossible for these women to afford the education their children need. For instance, the cost to PEKKA members of sending their children to school ranges from 51 to 178% of a family's income (ibid.).[4] Poverty therefore does not help women-headed households to send their children to schools. Because of this, PEKKA provides education for the children of these parents, as discussed by members of the PEKKA Secretariat in the interview.

But leadership training and education is not sufficient to address the situation of women in Indonesia. As one of their initiatives to promote women's rights, the organization has gathered pertinent data to highlight how widespread women's issues are (Zulminarni et al. 2018). As explained during the interview, government data are not enough to highlight the reality of the situation of women in Indonesia, thus PEKKA gathered their own data to help the government or to show to the government the implications of their inaction on the issues (PEKKA Secretariat 2018). Moreover, PEKKA has also been advocating for the government to recognize and to update definitions of heads of household to be gender-neutral to reflect the reality on the ground (Zulminarni et al. 2018). Based on what PEKKA gathered on the ground and shared by PEKKA to the research team, 49% of PEKKA members are considered to be poorest of the poor, and 57% of women heads of household are illiterate (PEKKA Secretariat 2018). In this regard, PEKKA was able to objectively assess what interventions are needed to promote women's welfare in the society.

Another reality that PEKKA faced in the course of their service is that women lack pertinent legal services, which are crucial for the actualization of their rights both as women and as citizens of Indonesia. Many of their members, as discussed during the interview, lack pertinent documents such as divorce papers, since many of the marriages are religious in nature (PEKKA Union Members 2018). As discussed in the literature earlier and as pointed out by PEKKA Secretariat members, religious marriages and divorce processes, unless registered under the processes of the state, are not recognized to begin with. These have implications for the well-being of women and their family, including the lack of a birth certificate for their children.

[3] According to Zulminarni et al. (2018, 20), "Paradigta comes from an ancient Javanese word meaning 'a woman who stands tall and strong'."

[4] Based on PEKKA's data, 51% of a family's income is needed to finish elementary school, 140% for junior high school, and 178% for high school.

Legal services are highly needed as the lack of children's documents such as birth certificates enables young children to get married young (PEKKA Secretariat 2018).

Providing legal services for women and their family was made possible as PEKKA was able to successfully argue their point at the Indonesian Supreme Court; thus, it paved the way for the government to allot funding for women to access their pertinent documents (Curnow 2015; PEKKA Secretariat 2018). Aside from the Supreme Court, PEKKA was also able to establish linkages with religious courts to provide legal services for women (Zulminarni et al. 2018). Mobile clinics for legal processes were initiated by PEKKA to make legal access for women much easier (ibid.). As shared by one of the members of the PEKKA Union, while it was difficult for them to secure the certificate of their children, PEKKA made their access toward these services easier through their mobile clinics (PEKKA Union Members 2018).

15.5 On PEKKA and the Promotion of Women's Rights and Welfare

Since PEKKA's formation in 2001, it has now been able to reach almost all of the areas of Indonesia. Many activities have been implemented by PEKKA to promote women's welfare across different areas of the state, especially education and legal services—which are vital for Indonesian women to be able to minimize the gap, as reported in the World Economic Forum data. The discussion below will tackle the viewpoints of women in terms of their issues, as well as their viewpoints with regard to the services rendered by PEKKA toward them, and how their lives have changed due to PEKKA.

Given the activities mentioned by PEKKA, the participants of the study have mentioned that the presence of PEKKA in their lives is very positive. One of the members, for instance, has pointed out that through PEKKA, they were able to gain confidence. In spite of the positive impact of PEKKA to their lives, they were able to realize that the concerns are actually structural in nature. Table 15.1 will summarize the responses of both the members of the PEKKA Secretariat and the members of their union.

Views on Women and Their Situation: Based on the discussion, PEKKA was able to characterize the imbalances in terms of gender in Indonesia, and they were able to support it by providing data. At the local level, PEKKA Union members discussed that being head of a household is perceived as a negative situation. They further pointed out that members of the society judged women for heading households. Moreover, they shared that they were called out by the members of their community for joining PEKKA and for pursuing activities for the improvement of women's status in their society (PEKKA Union Members 2018). For PEKKA, this is basically caused by structural problems in Indonesia as Indonesia still has restrictive laws (PEKKA Secretariat 2018). This translates to economic problems and problems

Table 15.1 Thematic coding of the issues and impact of PEKKA as mentioned by the women of PEKKA across different levels[5]

	How women are viewed by society?	Specific problems affecting women	Actions	Changes
PEKKA Secretariat[6]	Women's issues are structural as policies are codified defining the roles of women in the society. Because of this legal process, women-headed households are actually a taboo under Indonesian society	Women and gender imbalances are felt at the societal level. Women lack the opportunities to support their families. They also lack access to basic social services such as education and access to pertinent documents	PEKKA intervened in terms of organizing women, providing leadership training, gathering data, and providing legal services for women and their families	Members of the PEKKA Secretariat are happy as they help women. They also felt improvement in terms of their skills in community organizing
PEKKA Union members[7]	Women-headed households are viewed negatively, and are judged by the Indonesian society, as experienced by some of the members of the PEKKA Union	There are immediate issues and concerns for women at the village level such as lack of access for services and education. They also lack access to financial services	Members of the PEKKA Union joined the organization and mobilized themselves. They also gather data from the community and help address issues at the community-level	The impact of PEKKA is positive as they were able to participate in the political practices at the village level. Trust between the local government and PEKKA Union members was established

in accessing basic social services from the Indonesian government, based on the interviews conducted by the team (ibid.).

While PEKKA gathers data both at the national and local levels, one participant from PEKKA Union discussed that in her experience, the local government authorities questioned the validity of the data they gathered (PEKKA Union Members 2018). Nevertheless, the participant from West Kalimantan mentioned that the local government keeps working with them. In this regard, both the secretariat and the members of the union recognized the cross-sectionality of issues felt at the societal and individual levels in Indonesia.

[5] Thematic coding summarized by the author based on the interviews and focus group discussions conducted by the research team in Jakarta, Indonesia on October 12, 2018.

[6] The Deputy Director of PEKKA and two staff members of the secretariat were interviewed by the team on October 12, 2018 at PEKKA office in Jakarta, Indonesia.

[7] Two PEKKA Union leaders from Kalimantan and Nusa Tenggara Barat (West Nusa Tenggara) were interviewed by the team on October 12, 2018 at the PEKKA Union Office in Jakarta, Indonesia. The interview was assisted by Romlawati Kamad, Deputy Director of PEKKA.

Interventions: Given this, PEKKA was able to promote the welfare of women and was able to intervene in the problems. PEKKA has acted for the welfare of women in terms of education and leadership training, as well as gathering and using data to promote the welfare of Indonesian women (Zulminarni et al. 2018). Aside from these activities, PEKKA was also able to successfully lobby for women inclusion in the Indonesian Supreme Court and religious courts, which reacted in their favor (PEKKA Secretariat 2018; Zulminarni et al. 2018). Because of the change of this policy, the secretariat intervened in order to democratize access to pertinent social and legal services, thereby making them more available to women.

And for the PEKKA Union, one member from West Kalimantan shared that the data they gathered helped the local government promote further inclusion of women through the provision of health insurance covering women-headed households in the area (PEKKA Union Members 2018). The member from West Kalimantan further elaborated that the inclusion to the local government's health insurance was the result of PEKKA Union members filing a petition to the local government regarding the non-inclusion of women-headed households in the insurance scheme by a medical practitioner overseeing the project (ibid.). And from the experience of the member of the PEKKA Union from Nusa Tenggara Barat (West Nusa Tenggara), she was able to get the certificate of her children, which were very difficult to secure because of red tape in the bureaucracy before PEKKA's intervention (ibid.).

Impact: One of the positive impacts PEKKA has brought to the lives of these women is that the organization was able to change the worldviews of these women. For instance, one of the participants from PEKKA Union has mentioned that she views single-parenthood as a curse and thinks of their faults as reasons why their husbands left them in the first place. She further shared that through PEKKA's intervention, she was able to realize that it was not her fault that her husband left her; it's her husband's problem why he chose to be disloyal in the first place (ibid.). Moreover, PEKKA was able to help them achieve more in the society, as they were trained to be leaders from their respective communities (ibid.).

Because of PEKKA's intervention, one of the members of the PEKKA Union remarked that they were able now to help other women in the society and are now running for village positions to help them advance the cause of PEKKA further (ibid.). In fact, one of the members of PEKKA ran in the local elections and won despite opposition from more conservative people in their community (Zulminarni et al. 2018). These women were able to rise again despite the horrible experience that occurred. For instance, one member from West Nusa Tenggara shared that she was abandoned by her husband in a foreign land, which led to her incarceration before being sent back to Indonesia (PEKKA Union Members 2018).

And for the PEKKA Secretariat (2018), one of their members shared that it makes her happy that she was able to promote women's welfare. She also shared that through her experience with PEKKA, she was able to feel empowerment as she was able to go to the field and interact with members of the PEKKA Union during field activities (ibid.). Aside from this, the deputy director of PEKKA mentioned that the secretariat team was also able to help members of the PEKKA Union to share these negative experiences by helping them publish their stories (ibid.). By documenting these cases, this will help people further understand how prevalent these cases are.

In spite of these successes and the positive impact that PEKKA has brought for these women, the secretariat has recognized that they still have a long way to go in order to mainstream women's rights in Indonesia. They were able to provide these services, but nevertheless recognize that the system has to change in order to fully emancipate women from their respective issues (ibid.). And for PEKKA Union members (2018), they shared that they were still judged by conservative women for the services they rendered for women's welfare. And for the organization itself, it was mentioned that conservative forces are calling out PEKKA for their initiatives (PEKKA Secretariat 2018). To secure the gains that PEKKA had through the years, the key takeaways from their experience discussed that there is a need to address social inclusion, help build foundation for participation, document evidence to highlight cases of inequality, and establish linkages with allies to help promote these advocacies (Zulminarni et al. 2018).

15.6 PEKKA as Both a Feminist and as an Alternative Practice

As stated explicitly, PEKKA adheres to feminist values in the promotion of women's welfare (Zulminarni et al. 2018). And because of the following framework, it was able to uncover that the problem of women is structural, and gender inequality permeates at the level of individuals and private spheres. As pointed out by Devasahayam (2018), feminist analysis helps break the dichotomy between public and private spheres, which are normally left out when it comes to political analysis. Since these issues are felt by women, PEKKA organized women in order to uplift their conditions in the society and promoted activities and initiatives that cater to the needs of women (Zulminarni et al. 2018).

Aside from PEKKA's case being feminist, it is also an example of a case of alternative development. Based on the criteria set by Raina (1997), PEKKA's actions were able to promote women empowerment by building its base through cooperatives. Moreover, PEKKA's actions were able to provide services such as education and legal integration, despite their being part of the mainstream processes. However, what made PEKKA's case compelling as an alternative practice is the reality of Indonesian society where women are recognized to be managers of the household under the law, which created the inequality Indonesia is facing today as far as gender issues are concerned. Furthermore, while political analysis pointed out the unilateral role of the state in terms of decision-making, PEKKA's case was able to challenge this.[8] As pointed out by Blackburn (2004), the state should not be considered as an all-powerful actor, and that women groups are important to help shape the discourse of women in the society.

[8] Katznelson and Milner (2002), in an attempt to discuss the parameters of political science as a field of study, had focused on the discussion of the state as the primary locus of the field.

Fig. 15.3 PEKKA member weaving together with the PEKKA Team and the UP CIDS AltDev researchers (Photo courtesy of the author)

Overall, the experience of PEKKA was able to point out how the organization was successful in promoting the welfare of women in Indonesian society despite the challenges they faced. PEKKA's case was able to exemplify how their experience is a feminist endeavor, as enshrined in their principles. It was also able to exemplify how their experience can be considered as an alternative practice as it seeks to challenge the dominant paradigm on gender inequality in Indonesian society in the present (Fig. 15.3).

15.7 Conclusion

While international agreements promote women's rights across the globe, its implementation at the local level varies. In Indonesia, for instance, the literature has pointed out that global and local forces are still contradicting each other as far as women's rights are concerned. Given this, Indonesian women themselves are mobilizing in order to ensure their rights. PEKKA, as a national and women-led organization, seeks to uphold women's rights across different levels.

PEKKA has been instrumental in providing necessary services toward the actualization of women's rights through leadership training, education, and legal services. As pointed in the discussion earlier, these issues do not only affect women but also their children. Moreover, poor women in Indonesia are in a much more vulnerable situation as they have less resources to access essential services to promote the welfare of the household.

With emphasis on the experiences of women both at the public and private levels, the case of PEKKA was able to highlight that it is a feminist study. While normal discussions of power focus on the role of the state as discussed in Katznelson and Milner (2002), PEKKA's case helps to locate politics across different levels, including the private sphere. Thus, feminism as an analytical tool helps demystify the dichotomy between private and public affairs, as argued by scholars such as Devasahayam (2018).

Aside from PEKKA's case's being feminist, it is also an example of an alternative practice. As pointed out by Raina (1997), it should be able to integrate itself against the dominant paradigm, for it to be recognized as an alternative, which PEKKA was successfully able to do so.

Indonesia still has a long way to go in promoting the well-being of women in its society. However, it does not mean that these will not improve over time. The case of PEKKA should make society realize that it is important to discuss women's rights. And given that alternative practices are usually located at the marginal levels of society, states and other organizations can learn from PEKKA's experience, especially with respect to women's mobilization toward the actualization of women's rights. Moreover, PEKKA's case should also help states realize that they should not be complacent with their current objectives and should do more to fully incorporate women's rights in the society.

References

Babbie, Earl. 2013. *The practice of social research*, 13th ed. Belmont, CA: Wadsworth/Cengage Learning.

Blackburn, Susan. 2004. *Women and the state in modern indonesia*. Cambridge/New York: Cambridge University Press.

Curnow, Jayne. 2015. Legal support structures and the realisation of Muslim women's rights in Indonesia. *Asian Studies Review* 39 (2): 213–228. https://doi.org/10.1080/10357823.2015.1025695.

Devasahayam, Theresa W., ed. 2018. *Ensuring a square meal: Women and food security in Southeast Asia*. Singapore: World Scientific Publishing Co., Pte. Ltd.

Duff, Putu, Santi Kusumaningrum, and Lindsay Stark. 2016. Barriers to birth registration in Indonesia. *The Lancet* 4 (April): 234–235.

Feruglio, Francesca, Niken Lestari, Emma Bell, and Karen Brock. 2017. Building safe spaces to support young women's participation in Local Governance in Indonesia. Practice paper (June). Brighton: Institute of Development Studies.

Firdaus, Febriana. 2019. The long struggle of the women's movement in Indonesian politics. *The Interpreter*, March 8. https://www.lowyinstitute.org/the-interpreter/long-struggle-women-s-movement-indonesian-politics.

Flick, Uwe. 2009. *An introduction to qualitative research*, 4th ed. London/Thousand Oaks, CA: SAGE Publications.

Katznelson, Ira, and Helen V. Milner. 2002. *Political science: The state of the discipline*. New York/London: W. W. Norton.

Kusujiarti, Siti, Elizabeth W. Miano, Annie L. Pryor, and Breanna R. Ryan. 2015. Unveiling the mysteries of aceh, Indonesia: Local and global intersections of women's agency. *Journal of International Women's Studies* 16(3): 186–202. http://vc.bridgew.edu/jiws/vol16/iss3/13.

Nurmila, Nina. 2016. Polygamous marriages in Indonesia and their impacts on women's access to income and property Al-Jami'ah. *Journal of Islamic Studies* 54 (2): 427–446. https://doi.org/10.14421/ajis.2016.542.

Ortiz-Ospina, Esteban, and Max Roser. 2018. Economic inequality by gender. *Our World in Data,* March. https://ourworldindata.org/economic-inequality-by-gender#differences-in-pay.

Parker, Lyn, Irma Riyani, and Brooke Nolan. 2016. The stigmatisation of widows and divorcees (*Janda*) in Indonesia, and the possibilities for agency. *Indonesia and the Malay World* 44 (128): 27–46. https://doi.org/10.1080/13639811.2016.1111677.

PEKKA Secretariat. 2018. Interview with Karl Hapal, Venarica Papa and Nathaniel P. Candelaria. Personal Interview. Jakarta, Indonesia, October 12.

PEKKA Union Members. 2018. Interview with Karl Hapal, Venarica Papa and Nathaniel P. Candelaria. Personal Interview. Jakarta, Indonesia, 12 October 2018.

Raina, Vinod. 1997. The idea behind IADEA. In *Integrating alternative development efforts in asia: Report on the workshop at Palakkad, Kerala. India, March 1–4, 1996*, edited by Dessa Quesada, 13–18. Tokyo, Japan: PP21 RUA, JCNC, Asian Regional Exchange for New Alternatives and Kerala Sastra Sahitya Parishad.

Randall, Vicky. 2010. Feminism. In *Theory and methods in political science*, 3rd ed., ed. David Marsh and Gerry Stoker, 114–135. Basingstoke, Hampshire: Palgrave Macmillan.

Rinaldo, Rachel. 2011. Muslim women, moral visions: Globalization and gender controversies in Indonesia. *Qualitative Sociology* 34: 539–560. https://doi.org/10.1007/s11133-011-9204-2.

Silvey, Rachel M. 2000. Stigmatized spaces: gender and mobility under crisis in South Sulawesi, Indonesia. *Gender, Place and Culture: A Journal of Feminist Geography* 7 (2): 143–61. https://doi.org/10.1080/713668869.

Sumner, Cate. 2011. Increasing access to justice for women, the poor, and those living in remote areas: An Indonesian case study. *World Bank Justice for the Poor Briefing Note* 6(2). https://openknowledge.worldbank.org/server/api/core/bitstreams/36aaee50-c19b-550f-8bbe-f8988880f855/content

Sumner, Cate. 2015. Indonesia's missing millions: Erasing discrimination in birth certification in Indonesia. Center for Global Development Policy Paper 064, June 2015. Washington, D.C.: Center for Global Development.

Williams, Catharina Purwani. 2005. Knowing one's place': Gender, mobility and shifting subjectivity in Eastern Indonesia. *Global Networks,* October 5(4): 401–17. https://doi.org/10.1111/j.1471-0374.2005.00126.x.

Zulminarni, Nani, Valerie Miller, Alexa Bradley, Angela Bailey, and Jonathan Fox. 2018. Movement-building for accountability: Learning from Indonesian Women's Organizing. Accountability Research Center, Working Paper 3, October 2018. Washington, D.C.: Accountability Research Center.

World Economic Forum. 2023. *Global Gender Gap Report 2023.* Geneva, Switzerland: World Economic Forum.

Nathaniel Punongbayan Candelaria is currently an assistant professor at the Department of Political Science, UP Diliman, where he also earned his Bachelor of Arts (Political Science) and Master of Arts (Political Science) degrees. He also previously worked as Teaching Associate at the UP Department of Political Science, as Senior Project Assistant at the UP Center for Integrative and Development Studies (UP CIDS), and as Research Assistant at the Institute for Strategic and Development Studies (ISDS Philippines) and at the Foundation for Integrative and Development Studies (FIDS). His research interests are non-traditional security issues, International Relations, religion and politics, and Southeast Asian politics.

Chapter 16
Women's Access to Justice Through Legal Aid Clinics and Paralegal Formations

Angeli Fleur G. Nuque and Ananeza P. Aban

Abstract This case study looks at the local initiative of the Association for Development of Women and Legal Education (ADWLE) in handling violence against women (VAW) cases. It interrogates how the status of women in Lao society, which is deeply rooted in traditional social norms, impacts legal empowerment and access to justice. ADWLE Legal Clinics, which specialize in legal aid for women and paralegals, enable women to gain greater access to justice not only with regards to violence, but also across a range of other issues, such as family and property law cases. ADWLE, which utilizes women and community legal empowerment as an alternative socio-political practice, strives to address patriarchal and cultural gender norms and the subsequent recurrence of VAW cases. To critically respond to cultural attitudes towards women or the "cultural tolerance" of VAW, and the low awareness of women's rights in rural villages and peri-urban areas in Laos (officially named Lao People's Democratic Republic), ADWLE developed program streams for education and training, empowerment of women's access to justice, and "advocacy for change" in select target villages.

Keywords Gender-based violence · Access to justice · Women's rights · Legal aid clinics · Women empowerment · Legal systems

---

A. F. G. Nuque
University of the Philippines Center for Integrative and Development Studies (UP CIDS), Quezon City, Philippines

A. P. Aban (✉)
Ghent University, Ghent, Belgium
e-mail: ananeza.aban@ugent.be

J. M. C. Sy and A. P. Aban (eds.), *Reimagining Development in Southeast Asia*, Asia in Transition 21, https://doi.org/10.1007/978-981-97-9140-8_16

16.1 Introduction

The issue of violence against women (VAW) appears to remain "culturally tolerated" in Lao People's Democratic Republic (Lao PDR) despite the growing global and domestic recognition of VAW as a "serious public health problem and a violation of women's rights" (National Commission for the Advancement of Women 2015, 2).

A *Multiple Indicator Cluster Survey* conducted in 2006 revealed that 81% of women aged 15–49 years old in Lao PDR believe that the husband is justified in exacting physical violence against a wife or a partner when she does not conform to traditional gender norms and expectations like "[going] out without telling him; [neglecting] the children; [arguing] with him; [refusing] sexual relations with him; and [burning] the food" (Asian Development Bank and World Bank 2012, 58). The *Lao National Survey on Women's Health and Life Experiences,*[1] another survey conducted in 2014, echoed similar findings and also identified the "cultural tolerance" of VAW as one of the underlying factors for the persistence of violence against women and intimate partner violence. But patriarchal, traditional gender norms and stereotypes not only impact societal attitudes and perceptions towards women; it also results in women being conditioned to internalize said cultural and gender norms and stereotypes. Such internalization was reflected in the 2014 survey with 58% of women reporting that "VAW was justified if women did not adhere to traditional gender norms" (National Commission for the Advancement of Women 2015, p. 2). Alarmingly, the survey revealed that girls and boys aged 13–17 years old echoed the same belief that a woman "should tolerate violence to keep the family together" and that it is "acceptable" under certain circumstances, especially when reproductive and care roles are not fulfilled (National Commission for the Advancement of Women, Mothers and Children 2018, 5). Girls and boys regurgitating such beliefs reflect how ingrained the construction of the "cultural tolerance" of VAW is vis-à-vis the regulation and operation of gender roles. Girls internalize women's roles and identities as wholly connected to their ability to fulfil reproductive and care roles within their families.

Lao traditional and societal norms also severely undermine women's leadership and control of decision making in the public sphere, community, and home (United Nations Entity for Gender Equality and the Empowerment of Women [UN Women] 2016, 11). Challenges faced by ethnic women due to these norms are compounded by the lack of access to education and health services, limited economic development opportunities, and fewer resources for developing themselves and their families (UN Women 2016, 11).

Faming (2018), however, warns that policies anchored on gender equality in Lao PDR have been largely market oriented and focused mostly on women's livelihood, motherhood, and economic participation. This has ushered in "a new form of poverty within the market and money-based economy" (3). While policies address women's economic participation, women's access to political power and public participation

---

[1] The *Lao National Survey on Women's Health and Life Experiences* is the first national survey on violence against women in Lao PDR.

have been largely unaddressed. Faming (2018) views this as the *phu nyai*[2] concept still largely at play, which continues to subordinate and marginalize women and ethnic minorities in Lao society. Policies hinged on "big man" politics has created a "socio-cultural hierarchy among women based on differences in sex, ethnicity, and social class in Lao society" (Faming 2018, 2).

While these complexities continue to confront women in Lao society, including those among the ethnic communities, cases of VAW alarmingly remain underreported in Lao PDR. The main findings in the *Lao National Survey on Women's Health and Life Experiences* reported that 71.4% of women did not seek help from formal networks or agencies and persons in authority. Lao women turned to informal and immediate networks such as family and friends instead. Of the women who encountered physical and/or sexual violence, "only 28.6% [...] sought help from local authorities, 18.9% from local leaders and 11.7% from a Village Mediation Unit (VMU)," and only a handful of women approached women's organization (4.3%), the police (3.8%) or healthcare facilities (2.6%)" (National Commission for the Advancement of Women 2015, 88–89). The reasons for seeking help outlined in the survey were being unable to endure anymore (at 64.2%) and being encouraged by family or friends (at 39.7%); some cited being badly injured or threatened with death by their husbands or partners (National Commission for the Advancement of Women 2015, 88). Several women did not seek help because of embarrassment or shame (at 36.6%), trivialization of the situation and thinking of it as not serious (at 35.3%), fear of ending the relationship with their partner or being told by family and community leaders to be patient and resolve the issues at home (National Commission for the Advancement of Women 2015, 88).

The lack of knowledge regarding the legal system was also cited as a barrier to reporting VAW cases (National Commission for the Advancement of Women 2015). Only 16.9% of women in urban areas, 5.4% of women in rural areas, and 1.5% of women in rural areas without road access reported knowledge of current laws concerning VAW or women's rights. Underreporting may also be due to having no systematic reporting system for monitoring domestic violence and limited services for gender-based violence in Lao PDR (Asian Development Bank and The World Bank 2012, 57).

[2] Faming (2018) uses two translations for the word phu nyai: "big man" and "big person" (literal) (4). When explaining the "social world" of phu nyai and *phu noi* (literally "small person"), Faming referred to power relations and social hierarchy between men and women, and women and women. The social world of phu nyai and phu noi is "based on a patron–client relationship in a male dominant context, in combination with the kinship-like hierarchical relationship among women" (4). Faming also describes phu nyai as based on a "political order filled by individuals who are subordinates and followers" and a "hierarchical socio-cultural" system which places women and ethnic minorities in a "rigid caste-like system" (8). Emphasis was placed on the power dynamics present in such relationship, with a phu nyai not being exclusive to a man or woman or a person with economic power or wealth but a person "who should be given respect" (i.e., adults vs. children, teachers vs. students, seniors vs. freshmen, older siblings vs. younger sibling) (8).

16.2 Defining Violence Against Women in Lao PDR and Strengthening the Recognition of Women's Rights: Domestic Legal Frameworks

Understanding the VAW phenomenon in Lao PDR, as revealed in the surveys, also underscores the need to comprehend how violence and women's rights are locally understood and addressed. In this country, violence against women and children is defined as "any behavior that results in or is likely to result in danger; harm, physical, psychological, sexual, property or economic suffering to women and children." The *Law on Development and Protection of Women* (LDPW) defined domestic violence against women and children as "an act or omission committed by someone in the family which causes physical [or] mental impact on[,] or [which impacts on the] assets of[,] women and children in the family." The LDPW is one (1) out of the two (2) identified major laws in Lao PDR in relation to violence against women.

Enacted in 2004, the LDPW is considered a more comprehensive response to domestic violence and various forms of abuse faced by children. Although considered landmark legislation in its own right because of how it has defined "physical and mental impact on assets" (Article 29 to 32), outlined the rights of victims (Article 33, 38), and the responsibilities of the State and the society for the development and protection of women (Article 39 to 42), CARE International still flagged several concerns and limitations of the LDPW such as the (1) exclusion of unmarried couples and non-partner relationships (LDPW only covers violence between married couples) and (2) how it does not specify marital rape in the context of sexual violence (National Commission for the Advancement of Women 2015, 8). Additionally, at the time of the report, "Lao PDR does not have pre-litigation measures to immediately stop violence, such as intervention orders to protect survivors or refer them to shelters/ safe places as in the case in other ASEAN countries. The only existing pre-litigation action is resolution through VMUs. Fourth, the LDPW categorized levels of violence into 'serious' and 'non-serious', with non-serious cases encouraged to go through VMUs and be settled through mediation (Article 35), which is likely to 'cause a diversion from the court or criminal procedures and may remove the recognition of domestic violence as a human rights violation' (UNIFEM, cited in CARE 2009)" (National Commission for the Advancement of Women 2015, 10).

The distinction between "serious" and "less harmful acts of violence" has been noted with concern as it could "socially legitimize violence and foster a culture of silence and impunity" (National Commission for the Advancement of Women 2015, 10). The recognition of women's rights in Lao PDR can also be seen in the adoption of the *National Action Plan on the Protection and Elimination of Violence against Women and Children.*

16.3 International Obligations and Forming the *Association for Development of Women and Legal Empowerment*

At the global level, Lao PDR has an obligation to protect and promote women's rights as a signatory to international human rights treaties and conventions. By virtue of ratifying the *Convention on the Elimination of All Forms of Discrimination against Women* (CEDAW) in 1981, Lao PDR is obligated to submit a national report every four years to account for their compliance with CEDAW principles and regulations.

Attuned to these commitments, the Association for Development of Women and Legal Education (ADWLE) conducted a capacity assessment of the implementation of CEDAW in Lao PDR. ADWLE observed a low awareness at the grassroots level, among leaders of international non-governmental organizations and non-profit associations, and levels of the government of CEDAW (Association for Development of Women and Legal Education 2016, 7). Other key constraints identified were insufficient human resources, weak governance, limited technical expertise, and limited funds allocated by the government to effectively implement the law (Association for Development of Women and Legal Education 2016, 7–8).

The findings of ADWLE mirror that of the *Concluding observations on the combined eighth and ninth periodic reports of the Lao People's Democratic Republic* on CEDAW released on 12 November 2018. The Committee on the Elimination of Discrimination against Women raised concerns over women's access to justice, citing "persistent barriers, including stigma, fear of retribution, deep-rooted discriminatory gender stereotypes and limited legal literacy, that deter women and girls from registering their complaints regarding gender-based discrimination and violence, including domestic violence, marital rape and sexual harassment" (UN Committee on the Elimination of Discrimination against Women 2018, 4–5). The Committee has likewise taken cognizance of how the persistence of discriminatory gender stereotypes regarding the roles and responsibilities of women and men in the family and in society continue to contribute to the "perpetuation of harmful practices, such as child marriage, and the high levels of gender-based violence against women and girls, including domestic violence" (UN Committee on the Elimination of Discrimination against Women 2018, 7).

ADWLE finds its history very much entwined with CEDAW, with founding members belonging to the CEDAW Resource Pool. The organization was formalized as a non-profit association under the Ministry of Home Affairs in 2012 and was re-registered in 2017. ADWLE has emphasized the ratification of CEDAW and the corresponding obligation of the State to ensure that all standards are applied throughout the country, including customary law systems and discriminatory cultural practices.

ADWLE's Founding Director Inthana Bouphasavanh in an interview with World Bank said: "I [have] been in the legal sector since 2001, when I started working at the National University of Laos Faculty of Law and Political Sciences. We were helping build the capacity of law teachers, and I volunteered to train the lecturers and the students on gender equality and [CEDAW]. During this work I came to understand that women in Laos lacked knowledge about the law and also had limited access to the justice system. This gave me the inspiration to set up [ADWLE with its] formal registration completed in 2012" (World Bank 2022).

16.4 Lao Women and Legal Systems: Challenges in Access to Justice

With the presence of a pluralistic legal system in Laos that incorporates state and non-state systems and having the informal justice system heavily influenced by customary rules (UN Women 2016), this notion of access to justice together with its underlying principles increasingly becomes an important element in helping women exercise their right, especially in relation to the issue of VAW cases in this country.

Access to justice is defined as "the ability of people to seek and obtain a remedy through formal or informal institutions of justice and in conformity with human rights standards" (UNDP 2011). A primary concept of access to justice "rests on the individuals' equal capacity to find remedy regardless of the type of justice system accessed," which includes finding remedies within formal justice systems or informal justice systems (UNDP 2011). This implies that assessments of access to justice look at both formal and informal means of justice to "ensure that poor, disadvantaged and marginalized populations gain understanding, knowledge, confidence, a voice, and the physical access to appropriate and effective means of meeting their justice needs and furthering their right" (UNDP 2011, 19).

Lao PDR has three (3) overlapping systems in practice: the formal justice system (the state-sponsored formal justice system), a semi-formal justice system (comprised of informal practices recognized by the state), and an informal justice system (based on customary laws and social norms, particularly prevalent in ethnic minority communities).

In documenting Lao women's access to justice in the context of customary rules and laws in ethnic communities, the United Nations Entity for Gender Equality and the Empowerment of Women (UN Women 2016, 9) reports that women are "often victimized by the justice system which has been designed to uphold the values and customs of a given society". Specific to the experience of rural women and women belonging to ethnic groups, traditional problem-solving methods remain the dominant conflict resolution mechanism in homes and communities because of the lack of access to information or services (UN Women 2016). However, the failure to include women in the planning, information collection or analysis and men's significant influence over ethnic groups' traditional rules, there remained significant gaps in understanding women's access to justice (UN Women 2016).

Laos is still predominantly a rural population which has about 85% of the people living in remote areas and exhibit significant ethnic diversity. The men are traditionally recognized as the "stronger" beings and therefore entitled to leadership roles and decision-making in both the public and private (i.e., home) spheres (UN Women

2016, 11). The ethnic minority women as a result, face "high levels of economic poverty, limited livelihood options or capacities, excessively high workloads, very low voice in decision-making at any level, poor maternal health and nutrition status and the risk of gender-based violence and exploitation" (UN Women 2016, 31). They remain largely marginalized in legal systems and in accessing justice. Socio-economic factors and lack of resources such as money and time make them unable to access formal justice systems while traditional, patriarchal social norms result in unequal access to informal justice systems (UN Women 2016).

The burden of unpaid care work has also been cited as disproportionately affecting women in many ways. Deviating from expected social norms makes them vulnerable to violence, with men using violence to "punish" women if work is not completed or if meals are not prepared on time (UN Women 2016). Women's multiple burden brought about by unpaid care work also affect their access to services, education, and justice mechanisms. As succinctly described by UN Women (2016), "women's lack of free time, coupled with men's control of time, prevents women from considering justice issues or seeking redress in any form" (2016, 41).

Ethnic minority women's access to information also remains a challenge due to women's low literacy rates and language barriers.[3] Although printed information has been said to be available, it is mostly in common Lao language; written documentation and file claims with formal courts also require the use of Lao language.

16.5 Program Streams of the *Association for Development of Women and Legal Empowerment*

The birth of ADWLE took significant notice of all these structural and systemic problems of marginalized Lao women which prompted Ms. Bouphasavanh and her colleagues in ADWLE to design program streams bearing in mind structural and socio-cultural barriers in accessing formal, semi-formal, and informal justice systems. ADWLE aimed to address the low awareness of CEDAW in both the grass-roots and levels of the government through (a) legal education on gender and applicable national and international women's rights laws, (b) capacity building activities with villages and communities to promote women's rights and community behavioral change towards women and girls, and the (c) provision of legal aid services to survivors of gender-based violence. It continues to envision a society where there is gender equality, including equal opportunities, contributions to decision-making and access to legal system.

ADWLE further elaborated that although Lao PDR has ratified CEDAW, Lao women have "limited ability to assert them [...] for a number of reasons including illiteracy, lack of knowledge of above-cited laws, cultural norms, limited legal aid or lawyer's services' (Association for Development of Women and Legal Education

[3] Women from ethnic minorities are limited to speaking their ethnic language and not Lao language, the language used in courts.

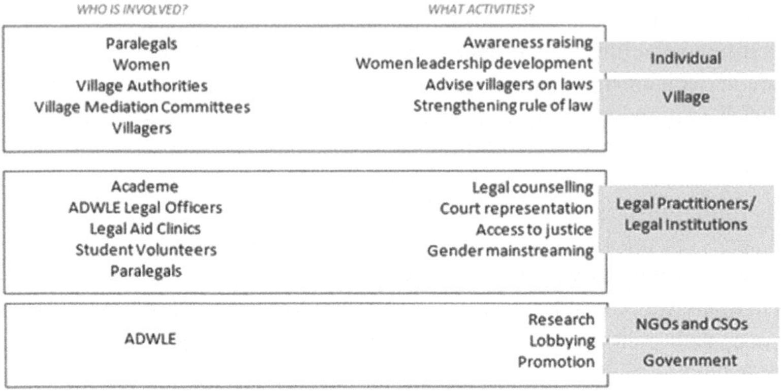

Fig. 16.1 Persons involved in the program streams and activities of ADWLE

n.d.). This is the very reason ADWLE aimed to implement three (3) key activities in recognition that the issue of VAW is both structural (i.e., access to justice, legal education) and cultural (i.e., traditional gender roles, discrimination based on gender): education and training, empowering women, and advocating change (see Fig. 16.1). ADWLE considers the material circumstances of the beneficiaries of their program streams, such as women's socioeconomic status (that affect their access to justice and education) and ethnicity.

ADWLE started with two staff and later in 2015 expanded to have six full time administrators, three lawyers, 12 paralegals, 24 village gender trainers, two international and national volunteers, and 72 teachers and student gender volunteers (Australian Embassy n.d.)

The UP CIDS Program on Alternative Development (AltDev) in 2019 visited ADWLE in Vientiane capital, Laos and interviewed Ms. Bouphasavanh who shared the history of the organization, gave an overview of the situation of women and women's rights in Lao PDR, and discussed the program streams in detail. At the time of the interview, ADWLE operated one (1) legal clinic to service 15 target villages in Xaythany District, Vientiane capital. The pilot areas were primarily peri-urban areas, including ethnic minority villages, whose daily source of income is based on agriculture and handicrafts. These legal aid clinics have been one of the major contributions of ADWLE in partnership with members of the legal community, including law schools and institutions, in order to strengthen the response to VAW, family conflicts and divorce, especially for the marginalized women. Ms. Bouphasavanh once said that ADWLE's initiatives lead to becoming a leading training center on Lao women's rights and a model that provides legal services to vulnerable women (Bielecki and Ertelt 2017).

Legal aid is the "provision of assistance to people who are unable to afford legal representation and access to the court system [and] is regarded as central in providing access to justice by ensuring equality before the law, the right to counsel, and the right to a fair trial" (BABSEACLE). The UN Principles and Guidelines on Access to Legal Aid recognizes legal aid as an "essential element of a fair, humane and efficient criminal justice system that is based on the rule of law and that it is a foundation for the enjoyment of other rights, including the right to a fair trial, as a precondition to exercising such rights and an important safeguard that ensures fundamental fairness and public trust in the criminal justice process" (United Nations General Assembly 2013, 2). There is also the recognition of the use of "community-based sanctions and measures, including non-custodial measures," and the "[promotion] of greater community involvement in the criminal justice system" (United Nations General Assembly 2013, 4).

Principle 10 or "Equity in Access to Legal Aid" of the UN Principles and Guidelines on Access to Legal Aid explicitly provides that "special measures should be taken to ensure meaningful access to legal aid for women, children and groups with special needs" and that the "States should also ensure that legal aid is provided to persons living in rural, remote and economically and socially disadvantaged areas and to persons who are members of economically and socially disadvantaged groups" (United Nations General Assembly 2013, 7).

ADWLE decided to pilot their legal aid clinics in rural ethnic communities to "[gain a] deeper understanding of the challenges faced by women in rural ethnic communities and [...] support them [in their] access [to] their rights" (ADWLE n.d.). Gender disparities and vulnerabilities are particularly glaring in remote rural areas, with rural women and ethnic minorities facing more significant constraints in inclusion and access to services (Asian Development Bank and World Bank 2012). Lao rural women are also more likely to face "time poverty" constraints due to the rigid traditional gender roles which expect them to allocate the majority of their time in arduous but unpaid reproductive care and domestic work (UN Food and Agriculture Organization, International Fund for Agriculture Development, & International Labor Organization 2010).

Figure 16.1 illustrates the persons involved in ADWLE's program streams. In order to train duty bearers and educate the community, Village Authorities[4] are given a "training of trainers" which focuses on gender sensitivity and perspectives, CEDAW, laws relating to women's rights, and human trafficking to raise awareness at the village level. Upon accomplishing the training of trainers, the goal is for Village Authorities to raise awareness at the village level and "leadership development" for village women.

Village Mediation Committees[5] and paralegals are also capacitated on laws relating to women and how to apply formal laws (versus customary laws) and

[4] In the local level, Lao PDR is divided into three administrative tiers and corresponding units: first-level (province and prefecture), second-level (district), and third-level (village). Villages are administered by village chiefs or village authorities (Chapter VIII, Article 75, Local Administration, Lao People's Democratic Republic 1991 [rev. 2003]). https://www.constituteproject.org/constitution/Laos_2003?lang=en.

[5] In 2016, Lao PDR "had adopted the *Prime Minister Decision on Village Dispute Resolution Committees* to formalize village mediation services and integrate local dispute resolution" and in 2018, the "Prime Minister approved the Legal Aid Decree to expand legal aid offices across the country" (Wassel 2019). Village mediation remains as "an effective and... preferred way to resolve

implement gender-sensitive mediation; paralegals, on the other hand, serve as legal counselors for villages and assist in investigations. Capacitating Village Mediation Committees and paralegals on laws to protect women's rights and gender-sensitive mediation ensures that villages will also be advised on such laws, that laws are applied correctly, and the rule of law is upheld. Village Mediation Units (VMUs) are categorized under the semi-formal justice system (UN Women 2016). The state has given jurisdiction to local authorities (i.e., village chief, village administration) through VMUs "to collect information and evidence, educate complaining parties and perform mediations of local low level disputes" and "some judicial control to administer justice based on both the formal and informal justice systems" (UN Women 2016, 24).

Efforts to capacitate village authorities on laws on women's rights and gender sensitivity is essential to ADWLE's program streams, especially taking into consideration that village-level mediation is a traditional dispute resolution method used by Lao people and "conducted by elders or the head of the ethnic group based on the relationships, traditions and social norms of the group" (UN Women 2016, 24). "Mediation" is used to reach a compromise in order to maintain good relations and social harmony within the community—but gender-blind mediation and the non-recognition of unequal power relations at play may inadvertently cause harm to women who are victim-survivors of violence. Women may be forced to suffer in silence to "maintain" the harmony within the community. Moreover, women's political leadership and representation remains low in Lao PDR with only one (1) percent of village chiefs being women (UN Women Factsheet as cited in UN Women 2016, 32).

Moreover, Ms. Bouphasavanh narrated the challenge to persuade Village Mediation Committees to "apply formal laws, instead of customary laws that have been used for many centuries," which is critical for women to be "properly heard and given access to fair justice" (World Bank 2022).

ADWLE operated the legal aid clinic as a pilot project to respond to women's need for access to justice based on its research findings on Attitudes of Lawyers Towards Divorce in Lao PDR in 2014. At that time, ADWLE only had one (1) lawyer assistance to serve the target villages.[6] ADWLE, however, was unable to retain the lawyers who were charging commercial rates. While ADWLE was able to partner with law firm Law and Development Partnership (LDP) which offered to support the project and have in its roster a lawyer who specializes in human rights, "the part-time services of one or two lawyers were never going to be nearly enough to meet the demand for legal services in the areas that ADWLE had in mind" (Governance and Justice Group 2019). ADWLE eventually had to change their strategies due to the minimal

many local disputes" but formal mechanisms are "sometimes necessary to protect the individuals' full rights under the law."

[6] It is important to note that in 2019, the Government of Laos established government-run Legal Aid Offices under the Legal Aid Decree. ADWLE's Legal Clinics continue to specialize in legal aid for women and complements the established general Legal Aid offices of the Ministry of Justice at the district and provincial level.

and limited funding available to them and the equally limited number of practicing lawyers.

This is why paralegals play a crucial role in ADWLE's program streams. Paralegals "served as a bridge between vulnerable people and the justice system" (Governance and Justice Group 2019). Paralegals are trained on laws relating to women and serve as legal counsellors for villages, serve as the bridge with Legal Aid Clinics, and assist in investigations. Paralegals are volunteers and "ordinary approachable people in society […] trained to be available to listen to people's concerns and grievances with confidence" (Governance and Justice Group 2019). ADWLE ensures that the work of volunteer paralegals do not hinder them from their main livelihood (Governance and Justice Group 2019).

Tapping "ordinary" citizens as paralegals addresses the fears and intimidation rural women might feel when approaching legal institutions and law enforcement. Some concerns identified is whether they will be able to articulate their concerns eloquently due to language barriers[7] and experiencing bigotry and discrimination (Governance and Justice Group 2019). The role of paralegals, however, are not only limited to bridging the gap between women and the formal justice system. Paralegals in the community also monitor informal dispute resolutions and the processes by which these mediation forums operate within their communities; if something is remiss in how the dispute is handled, paralegals turn to the lawyers at legal aid clinics for legal assistance.

Villagers are also included in the program streams of ADWLE to exact "behavior and attitude change"[8] towards women and for a deeper appreciation of women's human rights. Activities include awareness-raising activities on gender, laws relating to women, human trafficking, and safe migration. "Gender groups" are also formed to share lessons with other villagers and events like Women's Day and End "Violence Against Women" Day.

On the provision of direct service, ADWLE legal officers and pro-bono lawyers are tapped to provide "legal counseling to women and GBV victims," support with legal documentation and court representation. Legal academic institutions are also engaged to conduct trainings on gender mainstreaming into curriculum and train student volunteers on gender. The goals of the activities under this program stream are to capacitate student volunteers to raise awareness in the class and community and the "gender sensitization" of future workforce and leaders. ADWLE's research stream focuses on gender inequality and women's issues, specifically divorce, CEDAW implementation, and sexual harassment, to raise the issues to the government.

---

[7] Language barriers stemming from unfamiliarity with the language used in formal settings like courts.

[8] This is taken from the authors' work and from information/data taken from field interviews and review of related literature.

16.6 Legal Education and Empowerment in Lao PDR as a Socio-Political Alternative Practice

In documenting *alternative practices*, UP CIDS AltDev has generally categorized such practices according to four types: political, economic, social, and cultural (Tadem et al. 2020). *Political alternative practices* are "politically, informal and formal networks of civil society organizations and social movements [...] operating [...] on issues related to environmental issues, women's rights, workers' and peasants' rights, human rights, human security, and many other concerns" while *social alternative practices* take in the form of "self-help groups [that] have long existed, and local networks [that] have coordinated their social protection activities" and "as in other practices, peoples' empowerment is a guiding principle and primary goal" (Tadem et al. 2020, 17–18). Using the above definitions, this section will look at how the program streams of ADWLE may be categorized as a socio-political alternative practice.

Figure 16.2 illustrates who is involved and the nature of activities, clustered by the unit it targets (i.e., individual, community, social institutions, organizations, government). During the interview with UP CIDS AltDev, Ms. Bouphasavanh explained that program streams are focused on working together with district and village authorities to strengthen the services available for local communities.

A noticeable strategy employed by ADWLE is involving both individual women and various social institutions and administrative units in their program streams. There is also an explicit recognition that addressing the root causes of violence against women is also the responsibility of the community and the State and not for women to bear alone.

As reported in the *Lao National Survey on Women's Health and Life Experiences*, the majority of women turned to informal and immediate networks instead of formal networks or agencies and persons in authority. With its focus on the village, as a community, rather than the individual, ADWLE's approach takes advantage of

Fig. 16.2 Nature of ADWLE activities

existing informal networks of support for intervention. However, formal reporting mechanisms must still be strengthened, and low reporting rates be addressed.

ADWLE also implements a two-pronged approach of addressing immediate needs through direct service provision and, in the long term, realizing strategic gender interests.[9] Acknowledging the difficulty in accessing legal services, ADWLE organizes legal aid clinics and mobile clinics, provides legal assistance, and offers court representation, which would otherwise not be a viable option for women because of their material circumstances (i.e., the dearth of financial resources, knowledge of the law protecting women's rights, confidence and self-esteem, the social stigma, and low educational status).

In the long term, what ADWLE appears to work on, as seen in its focus on capacity building activities, gender groups, training of trainers, and leadership development, is to transform the gendered cultural and societal attitudes towards women that lead to the tolerance of VAW, address the entrenchment of patriarchal beliefs and values within families and structures in local villages, and strengthen the protection of women's human rights, from the rule of law to social institutions like the community and family. For ADWLE, strengthening the rule of law means applying it fairly and with due consideration to women's subordinate position because of traditional, heterosexist gender norms.

References

Asian Development Bank and The World Bank. 2012. *Country gender assessment for Lao PDR: Reducing vulnerability and increasing opportunity.* https://www.adb.org/sites/default/files/institutional-document/33755/files/cag-lao-pdr.pdf.

Association for Development of Women and Legal Education (ADWLE). 2016. *Capacity assessment report: CEDAW implementation in Lao PDR.* http://adwlelaos.org/wp-content/uploads/2017/09/SRP001_Cedaw-Assessment_EF-14-07-2016-ENG.pdf.

ADWLE. n.d. *Strategic plan 2017–2021.*

Australian Embassy Lao People's Democratic Republic. n.d. Women's rights in Laos—a role model. https://laos.embassy.gov.au/vtan/Inthana-Bouphasavanh.html.

BABSEACLE n.d. Legal aid. https://www.babseacle.org/legal-aid/.

Bielecki, Darin and Meryl Ertelt. 2017. *Reaching out to Lao villages about womens' rights,* 12 March. International Republican Institute. https://www.iri.org/news/reaching-out-to-lao-villagers-about-womens-rights/.

[9] Moser (1993) borrows from Molyneux's (1975, 38) definition of gender interests as "those that women (or men for that matter) may develop by virtue of their social positioning through gender attributes. Gender interests can be either strategic or practical, each being derived in a different way and each involving differing implications for women's subjectivity." Gender interests move beyond the assumption of homogeneity in women's experiences and take into consideration women's intersectional identities, socio-economic contexts, class, ethnicity, religious structures, and power relations. Strategic gender interests or needs, when met, "transform" existing unequal power relations between men and women relating to "gender divisions of labour, power, and control and may include issues such as legal rights, domestic violence, equal wages, and women's control over their bodies" (Moser 1993, 39). "Meeting strategic gender needs" also "challenges women's subordinate position" in society by changing existing roles (39).

The Governance and Justice Group. 2019. Paralegals in Laos—Redefining legal aid and bringing justice closer. October 27. http://www.governancejustice.org/blog/2019/10/27/paralegals-in-laos-redefining-legal-aid-and-bringing-justice-closer.

Faming, Manynooch. 2018. Big (wo)man politics: Gender equality in Laos? *Asian Anthropology* 1–19. https://doi.org/10.1080/1683478X.2018.1463595.

Moser, Caroline. 1993. *Gender planning and development: Theory practice and training*. Routledge.

National Commission for the Advancement of Women. 2015. *Lao national survey on women's health and life experiences 2014: A study on violence against women*. Retrieved from https://laopdr.un.org/sites/default/files/2019-08/2014_Violence%20against%20Women%20in%20Lao%20PDR_English.pdf.

National Commission for the Advancement of Women, Mothers and Children. 2018. *Ending violence against children in Lao PDR: A multi-sectoral response to the national violence against children survey priority actions*. Retrieved from https://www.togetherforgirls.org/wp-content/uploads/UNICEF_Multisectoral-response_ENG_18jun18_NoBleed.pdf.

Tadem, Eduardo C., Karl Arvin. F. Hapal, Venarica B. Papa, Ananeza P. Aban, Nathaniel P. Candelaria, Honey B. Tabiola, Jose Monfred C. Sy, and Angeli Fleur G. Nuque. 2020. Deepening solidarities beyond borders among Southeast Asian peoples: A vision for a peoples' alternative regional integration. University of the Philippines Center for Integrative and Development Studies—Program on Alternative Development.

UNDP. 2011. Access to justice assessments in the Asia Pacific: A review of experiences and tools from the region.

United Nations, General Assembly (UN GA). 2013, March 28. United Nations Principles and Guidelines on Access to Legal Aid in Criminal Justice Systems, A/RES/67/187.

United Nations Entity for Gender Equality and the Empowerment of Women. 2016. *Access to justice for women in Lao PDR: Using customary rules and laws in ethnic communities in three provinces*.

UN Food and Agriculture Organization, International Fund for Agriculture Development, and International Labor Organization. 2010. *Gender dimensions of agricultural and rural employment: Differentiated pathways out of poverty*. Rome: FAO.

UN Committee on the Elimination of Discrimination against Women. 2018. *Concluding observations on the combined 8th and 9th periodic reports of the Lao People's Democratic Republic: Committee on the elimination of discrimination against women*.

Wassel, Todd. 2019. *A long road: Access to justice in Laos*. The Asia Foundation. https://asiafoundation.org/2019/06/19/a-long-road-access-to-justice-in-laos/.

World Bank. 2022. Creating a sustainable tomorrow in Lao PDR: Inthana Bouphasavanh. 8 March. https://www.worldbank.org/en/news/feature/2022/03/08/inthana-bouphasavanh.

Angeli Fleur G. Nuque was Junior Research Analyst at the University of the Philippines Center for Integrative and Development Studies (UP CIDS) Program on Alternative Development (AltDev). She is an alumna of the Department of Women and Development Studies of the UP College of Social Work and Community Development. Her graduate these focused on feminist gerontology, aging, and sexuality.

Ananeza P. Aban is currently a Ph.D. researcher of the Department of Conflict and Development Studies and a recipient of the Special Research Fund doctoral scholarship of Ghent University in Belgium. She also serves as a senior research analyst of UP Center for Integrative and Development Studies Program on Alternative Development. Among her major tasks are coordinating the field research across Southeast Asia, co-editing the Alternative Practices monograph series (2020, 2023), and ensuring the international linkages of the program. She finished her Master of Community Development degree from the University of the Philippines in Diliman.

Part VI
Towards a Peoples' Solidarity Economy

Chapter 17
Building Sustainable Social and Solidarity Economy in Southeast Asia: Toward More and Better Jobs for Women in the Informal Economy

Rosalinda Pineda Ofreneo and Josephine Parilla

Abstract The case study discusses the "social solidarity economy" (SSE) model as practiced by PATAMABA-WISE (Workers in the Informal Sector Enterprise), a community-based women-led social enterprise in Angono, Rizal, Philippines. This is situated in the broader advocacy of the regional network, Homenet, in promoting fair trade and a solidarity economy. PATAMABA-WISE members are involved throughout the stages of production, marketing, and financing. In order to maximize access to state resources, it also participates in local governance and taps into public funding for support services for its projects. The SSE practice of PATAMABA-WISE is anchored on the three principles of "people, planet, and profitability." (The Pambansang Kalipunan ng mga Manggagawang Impormal sa Pilipinas (National Network of Informal Workers in the Philippines) or PATAMABA is a national network of home-based and other informal workers defending and advancing the collective interests of such workers. WISE (Women in the Informal Sector Enterprise) is a community-based, women-led social enterprise in Angono, Rizal, Philippines. PATAMABA WISE is an affiliate of HomeNet Philippines, a broad coalition composed of 27 membership-based grassroots organizations. Its advocacy agenda includes recognition and representation of home-based and other informal workers,

Sections of this paper are derived from Rosalinda Pineda-Ofreneo, "Engendering Social and Solidarity Economy in the Context of the 2030 Sustainable Development Agenda," *Philippine Journal of Social Development* 11 (2019), https://base.socioeco.org/docs/ofreneo-sse_and_gender_equality_april_-_2019.pdf; and Rosalinda Pineda-Ofreneo, "Building Sustainable and Social Solidarity Economy in Southeast Asia: Towards More and Better Jobs for Women in the Informal Economy," in *Building Peoples' Movements in Southeast Asia: Towards an Alternative Model of Regionalism,* 2nd Southeast Asian Conference on Alternatives (Quezon City: UP Center for Integrative and Development Studies, 2021), 11–13, https://cids.up.edu.ph/wp-content/uploads/2022/02/2nd-sea-conference-alternatives-proceedings-online-ver.pdf

R. Pineda Ofreneo (✉)
University of the Philippines, Diliman, Quezon City, Philippines
e-mail: rpofreneo@gmail.com

J. Parilla
PATAMABA Workers in the Informal Sector Enterprise, Angono, Rizal, Philippines

J. M. C. Sy and A. P. Aban (eds.), *Reimagining Development in Southeast Asia*, Asia in Transition 21, https://doi.org/10.1007/978-981-97-9140-8_17

social protection (including maternity protection) for all, and access to productive resources.

Keywords Women, home based workers · Social solidarity economy · Sustainable livelihoods · Participatory governance

A solidarity economy is a quilt, a woven patchwork of many diverse economies that are centered on life-values instead of profit values. Solidarity economics is the process of identifying, connecting, strengthening and creating grassroots, life-centered alternatives to capitalist globalization, or the Economics of Empire (Ethan Miller n.d.)

Solidarity economy emerged in Latin American countries at a time when their economies were reeling from the impact of structural adjustment programs imposed by multilateral institutions. It grew elsewhere in the global South as progressive social movements which sought to respond to neoliberal globalization in diverse ways in the context of their own values and experiences.

The Social and Solidarity Economy (SSE) movements have become global, expanding throughout the world. It has reached more than 1.2 billion members, counting 2.94 million cooperatives employing almost 280 million individuals (CICOPA 2018). Cooperatives in Africa and Latin America have been revived, and social enterprises in Europe and Asia have been on the rise. In South Asia, SSE takes the form of self-help groups, and in India alone there are 2.5 million of them. The fair trade producers and markets have also been expanding because of SSE, reaching the $6 billion mark in 2013 (Utting 2013). Fair trade towns, schools and universities are playing an increasing role in the campaign.

In terms of global presence, the RIPESS Intercontinental Network for the Promotion of SSE ties together continental networks committed to the promotion of SSE, with the Asian Solidarity Economy Council (ASEC) serving as its base in Asia (Quinones 2014). Its goal is to promote SSE, foster intercontinental cooperation, and advocate SSE at different levels. In terms of mainstreaming within state structures, SSE ministries and national secretariats have been established in Brazil, France, Luxembourg, and Colombia.

The United Nations system has also created an ILO SSE Academy and the UN Inter-Agency Task Force on Social and Solidarity Economy (UNTFSSE) to raise awareness about the growth of SSE in the international platform and in policy making bodies.

UNTFSSE defines SSE as "the production of goods and services by a broad range of organizations and enterprises that have explicit social and often environmental objectives and are guided by principles of cooperation, solidarity, equity and democratic self-management." (UNTFSSE 2014). It is characterized as "an attempt to reassert social control over the economy by prioritizing social objectives above profit maximization, recognizing key role of collective action and active citizenship for both economic and political empowerment of disadvantaged groups in society, and reintroducing notions of ethics, sharing, equality and democracy in economic activity." (Utting 2015).

In simple terms, Social Solidarity Economy is a form of economic democracy where people actually learn how to manage the production of goods and services to serve their own social and environmental objectives autonomous from the state.

17.1 Overlapping Crises as Context: Women in Informal Economy as Most Disadvantaged

Overlapping crises have been generated by the current neoliberal development model. It has failed miserably in addressing the goals of sustainable human development as 45% of the world's population still live in poverty. The existence of extreme inequality is evident as 26 super-rich people own the wealth equivalent to that of the poorest half of humanity (Oxfam 2019). About 150 million children are stunted because of malnutrition and hunger (World Bank 2018). There is also an eminent existential threat caused by climate change that has now become a climate emergency (UNFCCC 2019) leaving humanity with only 11 years to go to avert a global catastrophe.

Because women are largely tied to the home, the value of their reproductive work is statistically invisible since it is generally not reflected in the National System of Accounts. Many of them (almost 50% in the Philippine case) are classified as "housewives" and considered not in the labor force, meaning that they are not considered economically active. Globally, 48 women out of 100 are in the labor force, compared to 75 out of 100 men (ILO 2018). For many of those not in the labor force, the time they spend on doing unpaid care work leaves them no opportunity to engage in work that earns a clear and visible income. This phenomenon, called "time poverty," is more apparent among grassroots women in rural and urban poor areas, who have no access to basic utilities and social services, and who cannot afford hired help or labor-saving devices. Such deprivation can have harmful effects on health and even on life itself. According to global data released by UN Women, 80% of household water collection, which can be very time-consuming and back-breaking, is done by women and girls. Inefficient stoves using combustible fuel within households and resulting in harmful indoor pollution claimed 4.3 million lives in 2012, 60% of whom were those of women and girls (UN Women 2018: 6–7).

Grassroots women who are able to engage in productive work, despite their reproductive burden, are found mostly among the working poor in the informal economy in both urban and rural areas. Among their ranks are home-based workers and store owners, vendors, small farmers and fisherfolk, unpaid or contributing family workers in household farms and micro-enterprises, waste recyclers, domestic and other service workers doing laundry, massage, hair and nail care, etc. They do work that are compatible with or akin to reproductive work, and since this work is considered secondary or supplemental to the productive work that men do, the pay is usually dismally low, working conditions are substandard, and social protection is lacking or completely absent.

Regional data on South and East Asia show that around 60% of those who work in the non-agricultural sector is in informal employment, and the range is from 42% in Thailand to 84% in India (ILO 2014: 12). In the Philippines, one credible source claims that more than 80% (33 million out of 40.7 million) of employed workers are informal (PIDS 2018). Many women who are employed are in the bottom rungs of the informal economy. Their vulnerable and unprotected situation magnifies the issues emanating from their dual status as women and as workers deprived of rights and benefits enjoyed by men and by those who are in formal employment.

The scale of informal employment and non-realization of rights resulting from this has given rise to the global discourse on transitioning from the informal to the formal economy embodied in the 2014 report of the International Labour Conference. In the transitioning document, solidarity economy is identified as an important component of informal employment and as a strategy for facilitating formalization (ILO 2014). In other words, it is seen as a pathway toward more and better work that is recognized and protected, that is productive yet sustainable, that pays more, that is more regular, that is more marketable, and can provide flexibility in meeting family responsibilities while maintaining work-life balance (Pineda Ofreneo 2014).

17.2 Advancing and Practicing Fair Trade and Solidarity Economy

In the face of all the aforementioned challenges, informal workers through HomeNet Philippines, HomeNet Southeast Asia, and other networks have been pushing for fair trade and solidarity economy in the region for more than 15 years. HomeNet has been active in various forms of fair trade advocacy in collaboration with trade unions, business groups, and civil society organizations. Promoting fair trade principles help create opportunities for disadvantaged producers and provide them fair prices for their products; ensure workers' rights to freedom of association, just remuneration, job security, social protection, occupational health and safety, and good working conditions; respect and protect the environment; eliminate child labor and all forms of discrimination while advancing gender equity and women's empowerment.

HomeNet is currently working in six countries in Southeast Asia. In Cambodia, HomeNet's focus is organizing people with disability and HIV/AIDS as actors of development in a fair trade context. In Thailand, HomeNet established the HomeNet Thailand Association. It is a membership-based organization with about 4000 members from small producers' groups. Some 80% of their members are women. In 2016, 25 producers' groups planned together to establish "HomeNet Thailand Brand" as a social enterprise using the SSE framework.

Meanwhile in the Philippines, SSE initiatives of varying scale and levels of success have been introduced by PATAMABA chapters in many localities, including the Angono municipality in the province of Rizal.

17.3 A Community-Based SSE Model with BUB Support: The Case of PATAMABA Angono

PATAMABA chapters in the municipality of Angono, Rizal province are found in two coastal barangays highly susceptible to flooding, namely, San Vicente and Kalayaan. The Kalayaan sub-chapter was established in 1992 and San Vicente followed in 1995. That time, majority of the women members were involved in smocking and embroidery for the export market. Eventually, when demand for smocking declined, other livelihood activities were explored either individually or as a group like making Christmas balls, doormat, round rugs, bags made of beads, etc.

The organizing stage of the sub-chapter in Kalayaan started in 1991 through a project about child labor. Data gathered from the project saw the need to organize the parents of working children in order to curb child labor. In the beginning, members underwent capacity building and leadership training to strengthen the newly formed organization. Later, livelihood training was introduced and included skills training on production of peanut-butter, candles, chocolate candies, etc.

In 1998 the Chapter was introduced to gender concerns when it participated in para-legal training concerning violence against women. The knowledge about gender was deepened through an orientation on Gender and Development (GAD) conducted in 2004, and later through awareness-raising sessions on various topics, including national and local instrumentalities that respond to gender based violence.

The PATAMABA Rizal governance and advocacy framework consists of the following:

1. Social Protection—which focuses on social security, health insurance, security in the workplace, and access to justice.
2. Human Development—through the conduct of capacity and capability building;
3. Asset Reforms—focusing on the issue of workplace security (land and housing, common area of production);
4. Participation and Recognition—developing voice and representation of informal workers;
5. Employment and/or enterprise building.

Disaster risk reduction and management (DRRM) was not originally part of the advocacy of the organization although the members live in coastal areas. Up until 2007, the cycle for big flood happened every 10 years but since Milenyo (2006), huge flooding occurs every year either because of a typhoon (like Ondoy in 2009) or Habagat. Recurrent and worsening experiences of flooding pushed PATAMABA members to address their vulnerability through disaster risk reduction and management (DRRM) training and practice.

PATAMABA members participated in the Zero-Squatter program of the municipality and were relocated in San Vicente. The plan was for them to avail of the community mortgage program. Unfortunately, in order to qualify for CMP, 80% of the houses should have been constructed in the area, a requirement which they were not able to meet. For now, the 216- strong PATAMABA Housing Association (HOA) members directly buy the lots from the owner. More and more members have already been awarded certificates for having paid their lots in full. The HOA now enjoys a stable and accessible supply of water and electricity from service providers. They have worked to achieve this vast improvement since the time when water was being rationed and electricity had to be bought through sub-meters. They have also promoted the development of urban gardens for food security in the area.

Backyard gardening is a normal practice in barangays Kalayaan and San Vicente but is not a primary source of income because these communities are deep in flood waters for 5–6 months in a year. But for those who sell their produce, they do not have to go to the market to bring their products because they are not even enough for the communities' demand. Those who maintain gardens for personal consumption argue that planting vegetables help reduce the budget for buying food and at the same time assure the quality and freshness of what the family eats.

17.4 Focus on PATAMABA WISE

Angono producers are organized under the PATAMABA-WISE (Workers in the Informal Sector Enterprise) which is the livelihood arm of PATAMABA Rizal, the provincial local chapter of the National PATAMABA. It was registered with the Department of Labor and Employment (DOLE) and the municipal government of Angono. It began with 29 members contributing P200 each in 2009, and since then expanded to 238 (overwhelmingly women, with less than ten men). The strategy of PATAMABA WISE for expansion is through engagement in the child labor program of the municipality and the Department of Labor and Employment.

Angono, being one of the 1233 poorest municipalities in the Philippines, was covered by the Bottom Up Budgeting (BUB) program in 2013. The Angono BUB was spearheaded by the Local Poverty Reduction Action Team (LPRAT) composed of 28 members (14 from government and 14 CSO) which was chaired by then Mayor and now Vice Mayor Gerry Calderon, and co-chaired by Josephine "Olive" Parilla of PATAMABA WISE. LPRAT is mandated to facilitate the drafting, finalization and approval of the Local Poverty Reduction Action Plan (LRAP) of the municipality primarily identified by the Civil Society Organizations (CSOs). A Local Project Management Team (LPMT) was also organized to oversee the day to day implementation of the LPRAP which includes livelihood projects.

The LPMT is a smaller group headed by Sir John Arys Manaol (Project Manager hired by the municipality.) Members of the LPRAT include the BUB Cluster heads: Jocelyn "Josie" Lipio (for the production of rags), Emilina "Emmie" Nacario (for homecare products), Isabel "Sabel" Torres (store management), Russel Pinon (for doormat), Mary Dela Cruz (for fashion accessories) and Efren Aquino (for boxes, frames and eco bags). Five out of the six cluster heads are PATAMABA WISE members.

17.5 Participation in Livelihood Project Operations

PATAMABA WISE members are primarily involved in production, marketing and/ or investment (pertaining to their initial contribution of Php 200).

Membership requires a contribution of Php200 which is allocated for capitalization. Financial sustainability strategies also include buying and using products of PATAMABA by its members and promoting and marketing of their products to non-members. Quality control is done by the Project Team. Home care products were also tested and certified by the Bureau of Product Standards (BPS) of the Department of Trade Industry (DTI).

During peak operation, the products are sold with at least 30% mark-up. With the gross income, the organization is able to pay the monthly rental of the productivity center, administrative expenses, giveaways during Christmas gift-giving, aside from the compensation of the workers.

PATAMABA WISE managed to produce rags, doormats, homecare products, picture frames (from water lilies), *higantitos*, accessories, and ready-to-wear clothes. They conceptualized a tag line—buying a piece helps a lot of women. They managed to access funds from DOLE which further improved their capabilities and access to resources. As soon as the BUB was adopted by the LGU of Angono in 2013, PATAMABA WISE became the leading organization in partnership with the LGU in the development of the municipality of Angono. They facilitated the formation of conduits for the local organization toward spearheading the economic activity of the municipality. PATAMABA is in-charge of the local economy and economic activity for the vulnerable and marginalized sectors in the municipality. PATAMABA leaders teamed up with the community training and livelihood office for the training and capacity building activities.

17.6 Marketing Networks

The main buyers are from the community (side street stores, employees of the municipality and some friends). Some of the products are also bought by members and supporters of the Homenet Producers Cooperative which has a store on Maginhawa Street, UP Village; tiyangge and mall patrons outside the community are also able to buy Angono fashion accessories marketed there. Two BUB—run kiosks provided by the LGU in the Municipality Ground Plaza, municipal entrance, and other strategic areas in the municipality, sell to retail buyers. The Angono Community Training and Livelihood Office (CTLO) and Employees Cooperative help market the products.

Members can also sell rags. They find buyers in Angono; e.g., from PESO (Public Employment Service Office), a "middle man" who orders 1000 pieces per week for a construction site in Batangas, and other wholesalers. For home care products, the market consists of: cooperatives—Homenet and a Coop in Angono; and government agencies—PESO, CITECO, municipal office, DOLE-Cainta, on cash on delivery (COD terms).Members also sell on consignment basis. The municipal office helps promote products during events and also through social media. Buyers sometimes provide capital for pre-orders. They also help market products through social media, through word of mouth, and cooperation with other coops.

17.7 Adhering to a Multi-stakeholder Approach

PATAMABA's participation in local governance facilitates access to state funding, resources, and support services. It helps that the Angono Municipality, which has won the Seal of Good Local Governance for four years as well as a Gawad Kalinga award, is committed to the BUB project process even after the exit of the Aquino government. The BUB was renamed as Assistance to Disadvantaged Municipalities (ADM) and later as Assistance to Municipalities (AM). Despite leadership turnover in the last local elections, continuity is evident in the operations of the municipality productivity center. Irrespective of the election outcome, PATAMABA WISE was confident their production clusters could survive with or without government support, since many of these clusters had been in existence even before the BUB project started (Parilla et al. 2019).[1]

For the Angono BUB, the Department of Labor and Employment (DOLE) provided Php 4 M plus in 2014, Php3.369 M in 2015, Php 6.2 M in 2016 for livelihood funding support which included training on occupational safety and health. Aside from DOLE, other agencies also provided support to the municipality. The Technical Education and Skills Development Authority (TESDA) conducted skills training in massage, hair cutting, automotive, solar energy, etc. The Department of Trade and Industry (DTI) provided training in water hyacinth production, entrepreneurship, and local economy development. The Department of Interior and Local Government (DILG) focused on gender and development (GAD) which included the drafting of the Angono GAD Code. The Department of Agriculture (DA) assisted fisherfolk (with fish cage and terminal), and farmers (through a small irrigation project). The Department of Health DOH supported community health teams/barangay health workers, the dengue program and epidiomology center; Department of Education (DepED) took charge of schools facilities; the Department of Social Welfare and Development (DSWD) ran the 4Ps Sustainable Livelihood Development; and the Department of Agrarian Reform (DAR) provided a cooperative training program.

[1] The authors conducted an interview with E. Nacario, and N. Amano on 31 March 2019, at the Barangay San Vicente Municipal Council Hall, Angono, Rizal.

PATAMABA members in Angono benefited from many of these programs, projects and activities.

For PATAMABA in particular, the Department of Labor and Employment (DOLE) granted WISE the amount of Php1.4 M for its enterprise building (training cum-production) in 2010, and livelihood activities, including a bakery, food carts, and production of home care products, rugs and boxes. All these activities continued until 2015, except for the bakery which was temporarily shelved due to unavailability of a production site.

In 2015, PATAMABA WISE was tapped by the Angono LGU and DOLE Rizal to run a livelihood project under the child labor program. It went into production of homecare products and rugs which engaged 175 individuals. This led to organizing the parents of child laborers in the municipality to become affiliates of PATAMABA WISE. With this intervention, Angono was recognized as the "First Child Labor—Free Municipality" in the country.

PATAMABA also has had a good experience relating with the private sector. For example, PATAMABA WISE forged an agreement with Lafarge cement factory to use the latter's waste materials to produce *higantitos'* souvenirs which the Angono municipality used as giveaways to guests. Production, however, has stopped pending payment of receivables from the Municipality (Parilla et al. 2019).

As to tie-ups with NGOs, PATAMABA WISE arranged for Women in Sustainable Economic Action (WISEACT) to conduct training on Occupational Safety and Health (OSH), Product and Plant Assessment (PPA), and product development to comply with the requirements of Bureau of Product Standards.

WISEACT helped PATAMABA WISE develop the quality of their soap. From 2006 to 2007, WISEACT staff took members on a Lakbay-Aral to Alay-Kapwa to learn about soap packaging; to the Department of Science and Technology (DOST_ for packaging retail and wholesale products; and to San Miguel for packaging and labeling.

Relations with academe have developed through the years. For example, the Department of Women and Development Studies of the UPCSWCD has fielded practicum students in the area since the early 2000s, with partnership forged in gender awareness training and GAD planning, organizational development implementation of the anti-VAWC law, sexuality education for youth and older persons, feminist leadership, and social marketing. The Ateneo de Manila University engagement with PATAMABA Angono began with DRRM training, and is now moving on to youth-focused skills training in livelihood development, starting with the production of decorative boxes.

17.8 The Triple "Bottom Line" of PATAMABA

The social and solidarity economy practice of PATAMABA WISE is anchored on the triple "bottom line" of "People, Planet, and Profitability."

17.8.1 People: Participatory Governance, Social Development, and Women's Empowerment

PATAMABA is a membership-based organization adhering to democratic principles. As explained in the case study, "it employs participatory governance in the management and operation of livelihood projects. The members own and govern their business because they are not mere members of the organization, but they are also the investors who are themselves involved in the production, marketing and management of their business" (Parilla and Royandoyan 2017: 5). Opportunities for members to develop themselves are implied in the statement of its President, Josephine Parilla: "In participatory leadership, I want everyone to shine, not just the leader." (Parilla and Royandoyan 2017: 6).

What holds the organization together is not the experience of increased incomes, which at best remains supplementary but a sense of solidarity and belongingness which are the values the members hold dear. Members know that they can rely on each other for mutual aid during critical times. They also put to life the traditional Filipino value of *tangkilikan*, which is important for SSE success. Parilla explains *tangkilikan* in this manner: "We produce, we sell and buy our own products. In return we provide our members the opportunity to earn. The more the members sell and buy, the more they earn" (Parilla and Royandoyan 2017: 6).

PATAMABA has contributed a lot to social development by ensuring that their products first of all address community needs. It provides fair compensation, encouraging incentives for product patronage, and flexible working conditions to women who also have to contend with child care and other family responsibilities. When they were working together in a common production center, producers made sure that space was reserved for children to play in. Surplus is plowed back to the members in the form of mutual aid and educational funds. Members benefit from awareness-raising on gender sensitivity and GAD, the Reproductive Health law, and the Kasambahay law, Exclusive and Continued Breastfeeding, among others. Some have also undergone certificate courses on dressmaking, welding, and solar energy provision through cooperation with the Technical and Skills Development Authority (TESDA). These courses opened up better employment opportunities for the certificate holders.

PATAMABA WISE facilitates income augmentation for women which helps increase family income. They provide economic as well as leadership and management opportunities for women which entails a level of empowerment for them. Women are also able to make decisions because they earn for their families. They now have decision making power on household expenses and personal effects.

Aside from economic development, women also developed greater awareness about their rights. PATAMABA facilitates Gender and Development training and awareness raising about violence against women and children for their members. PATAMABA recognizes the inclusion of poor and socially disadvantaged sectors, such as older women and youth.

17.8.2 Planet: Environmentally Friendly Products

Home care products of PATAMABA WISE are environment friendly, as they recognize the need to advocate for environmental protection, conservation and preservation. Home care products are biodegradable. Findings of the Bureau of Product Standards of the Department of Trade and Industry show that WISE powdered laundry detergent soap and liquid dish washing soap have no hazardous chemical content. These are high quality community-based products. BUB home care products are made of coconut and sodium sulfate, and are safe for watering plants. Rags and doormats are made using recycled waste materials from nearby factories. Water lily from the lake is made into picture frames and bags while waste from aggregates are produced into *higantitos* and key chain out of *banlik* (local name of waste from aggregates).

17.8.3 Profitability: Impact on Poverty Reduction

Through the efforts of PATAMABA WISE, women and men are able to generate income for their families. Members also learn to develop livelihood skills and work in teams. The livelihood projects available for their members offer an alternative productive work and income for the unemployed. These allow different vulnerable sectors to participate in livelihood activities.

The spending capacity of their members also increased. They can now send their children to school, secure daily meals and provide basic needs for their family. For the elderly, income generated from livelihood activities has discouraged dependency from other members of the family.

PATAMABA leaders, however, admit that they still have a long way to go before they achieve economic sustainability. They still rely on grants, and need to move toward self-reliance and better marketing through product development and wider networking. They have had productive relations with national agencies. local government, women's organizations, and academic institutions such as the Ateneo de Manila University and the College of Social Work and Community Development of the University of the Philippines. Over time, and with the support of allies, they hope to truly achieve full SSE status.

17.9 Key Elements of the PATAMABA WISE SSE Model

In summary, the following may be cited as the key elements of the PATAMABA WISE SSE model, which reflect those embodied in other SSE initiatives of the Homenet national and regional networks of home-based and other workers in the informal economy:

- Primacy of democratic and inclusive membership-based organizations (in this case, women in the informal economy) as the key actors ready to work with other stakeholders in the community.
- Sharing responsibilities and accessing resources from the public or state sector as well as from NGOs/civil society and private groups.
- Ensuring product quality and development, sustained networking for marketing, SSE promotion and advocacy.
- Participation in local governance, which facilitates access to state funding, resources, and support services (e.g., the BUB).
- Addressing the primary economic needs of the community such as food and home care products.
- Recycling, environmental safety, and conservation connected to DRRM and climate change adaptation.
- Integration of a gender perspective—recognizing and developing the productivity and potentials of women in poverty toward their empowerment and gender equality.
- Readiness to improve, expand, develop, upgrade, upscale, and converge, to ensure sustainability, resilience, and plowing back of greater benefits to people and planet.

References

CICOPA. 2018. Cooperatives and Employment Second Global Report 2017 [EN, PDF] https://www.cicopa.coop/publications/second-global-report-on-cooperatives-and-employment/

ILO. 2014. Transitioning from Informal to Formal Economy https://www.ilo.org/wcmsp5/groups/public/---ed_norm/---.relconf/documents/meetingdocument/wcms_218128.pdf

ILO. 2018. World Employment Social Outlook—Trends for Women Global Snapshot. https://www.ilo.org/wcmsp5/groups/public/---dgreports/---dcomm/---publ/documents/publication/wcms_619577.pdf

Oxfam. 2019. 5 shocking facts about global inequality and how to even it up. https://www.oxfam.org/en/even-it/5-shocking-facts-about-extreme-global-inequality-and-how-even-it-davos

Parilla, J., Nacario, E., and Amano, N. 2019. Group interview conducted by the authors, March 31, Barangay San Vicente Municipal Council Hall. Angono, Rizal.

Parilla, J., and Royandoyan. M. 2017. Aspiring Towards Social Solidarity Economy: The Case of PATAMABA–WISE. Presented by J. Parilla at the ASEAN SDG-SSE Dialogue, UP Hotel, Diliman, Quezon City, and later submitted as a paper by M. Royandoyan to Social Development 311, a course under the Doctor of Social Development Program of the College of Social Work and Community Development, University of the Philippines, Diliman.

Philippine Institute of Development Studies. 2018. Social Protection in the Informal Economy. https://www.pids.gov.ph/details/social-protection-in-the-informal-economy

Pineda Ofreneo, R. 2009. "Solidarity Initiatives from the Ground Up: Learning from the Homebased Workers of Southeast Asia," paper presented during the First Asia-Pacific Regional Conference on Community Development as a Discipline and as a Profession – Renewing Committed Service for Social Transformation, 7–8 May 2009, College of Social Work and Community Development (CSWCD), University of the Philippines, Diliman, Quezon City, and published as part of the proceedings of the conference.

Pineda Ofreneo. 2014. Solidarity Economy in the Philippine Setting: Towards More and Better Work for Women in the Informal Economy. Paper presented during an international workshop on "Promoting Sustainability through Social and Solidarity Economy" at the Yonsei University, Wonju Campus, South Korea.
Quinones, B. 2014. Rediscovering Social Solidarity Economy in Community-Based Supply Chain. *Philippine Journal of Social Development* 6 (1): 1–30.
UN Inter-Agency Task Force on Social and Solidarity Economy (UNTFSSE). 2014. Social and Solidarity Economy and the Challenge of Sustainable Development. http://unsse.org/wp-content/uploads/2014/08/Position-Paper_TFSSE_Eng1.pdf
UN Women. 2018. Turning Promises into Action. Gender Equality in the 2030 Agenda for Sustainable Development. http://www.unwomen.org/en/digital-library/sdg-report
UNFCCC. 2019. *Climate Impacts Top WEF's 2019 List of Risks.* https://unfccc.int/news/climate-impacts-top-wef-s-2019-list-of-risks
World Bank. 2018. *Nearly Half the World Lives on Less than $5.50 a Day.* https://www.worldbank.org/en/news/press-release/2018/10/17/nearly-half-the-world-lives-on-less-than-550-a-day
Utting, P. 2015. *Realizing the 2030 Development Agenda through Social and Solidarity Economy.* https://ccednet-rcdec.ca/sites/ccednet-rcdec.ca/files/realizing_the_2030_development_agenda_through_social_and_solidarity_economy.pdf
Utting, P. 2013. *Power point presentation during the ASEAN Leaders' Forum.* In RIPESS, 2013. Compilation of Presentations and Proceedings. 5th RIPESS International Meeting on Social Solidarity Economy, University of the Philippines.

Rosalinda Pineda Ofreneo is professor emerita, University of the Philippines. She served as dean of the UP College of Social Work and Community Development for one term. She has been active in the women's and other social movements for half a century, working mainly with grassroots women in the informal economy. She served as volunteer regional coordinator of Homenet Southeast Asia, a network of homebased workers in five countries, for nine years. She is now in the Advisory Committee of Homenet Philippines and is one of three co-convenors of the Asian Solidarity Economy Council (ASEC) Philippine Network.

Josephine Parilla is the treasurer of Homenet Philippines and is the President of PATAMABA Workers in the Informal Sector Enterprise (WISE). She has been a leader of PATAMABA since the early 1990s. She served as Council Member of the Workers in the Informal Sector Council (WISC) of the National Anti-Poverty Commission since 2002 and was appointed as Sectoral Represetantive 2005–2010. She has also been part of the Regional Development Council of Region 4-A since 2002.

Chapter 18
Collective Enterprise as a Workplace: The Case of Payoga-Kapatagan Multipurpose Cooperative (PK-MPC)

Benjamin R. Quiñones Jr.

Abstract The objective of this study is to highlight and draw lessons from the economic and finance model of a hitherto charitable organization responding to the distressed conditions of poor peasants in a province constantly threatened by risks arising from natural calamities, exploitative practices of landlords and traders, and not so responsible local governance. Realizing the un-sustainability of its benevolent activities, the charitable organization re-organized into a cooperative and adopted the triple bottom line goals of people (community engagement), planet (ecological conservation), and profit (economic and financial sustainability).

Keywords Cooperatives · Social solidarity economy · Collective enterprise · Organic agriculture

The objective of this study is to highlight and draw lessons from the economic and finance model of a hitherto charitable organization responding to the distressed conditions of poor peasants in a province constantly threatened by risks arising from natural calamities, exploitative practices of landlords and traders, and not so responsible local governance. Realizing the un-sustainability of its benevolent activities, the charitable organization re-organized into a cooperative and adopted the triple bottom line goals of people (community engagement), planet (ecological conservation), and profit (economic and financial sustainability). Since becoming a cooperative, the Payoga-Kapatagan Multipurpose Cooperative (PK-MPC) evolved as a workplace that consciously reconnects with communities, improves and expands the livelihood base of its members, helps them transition from an informal to a formal economy, creates and promotes decent work, and transforms landless peasants into entrepreneurs. The people's collective enterprise consists of the production and distribution of organic fertilizer aimed at shifting local farming practice from inorganic to organic agriculture, with the accompanying support of financial services, technical assistance, marketing support, and capacity-building organized jointly by the people

B. R. Quiñones Jr. (✉)
Asian Solidarity Economy Council, Quezon City, Philippines
e-mail: benquinonesjr@gmail.com

J. M. C. Sy and A. P. Aban (eds.), *Reimagining Development in Southeast Asia*, Asia in Transition 21, https://doi.org/10.1007/978-981-97-9140-8_18

through the cooperative. Being economically and financially sustainable, PK-MPC has developed the capacity to become a major advocate of organic farming, thereby contributing to lesser carbon emissions in the province, the greening of Isabela farms, the creation of new decent job opportunities, and the advancement of sustainable farming.

In this case study, the term ‘collective enterprise’ (i.e., an enterprise co-owned and co-managed by many individuals who jointly establish, work together, and mutually operate the enterprise), is used to portray an economic unit of social solidarity economy (SSE). SSE is a process of economic activation and mobilization of the labor of the impoverished and marginalized sectors to foster alternative solutions to the crisis of employment and the welfare state that is systemic in the neoliberal economy. Lack of financial means on the part of the poor, marginalized, and socially excluded makes labor the main contributory factor to resource mobilization and investment in SSE. In addition to fleshing out the strategic dimensions of SSE through the experience of the PK-MPC, the study also makes an initial attempt to enumerate the contributions of SSE to the achievement of the Sustainable Development Goals (SDGs).

18.1 The Origins

The Payoga-Kapatagan Multipurpose Cooperative (PK-MPC) was originally a development non- governmental organization (NGO) registered with the Securities and Exchange Commission (SEC) in 1985 as the “Payoga-NGO.” The initiators of the organization were Belgian missionaries who, in partnership with the Parish Youth of St. Rose de Lima Parish Church in Gamu, Isabela, saw that share tenants in the province of Isabela had no voice, as they were under the economic and political control of politicians and big landlords engaged in the business of trading and moneylending. The initiators determined that the mission of the Payoga-NGO was to “strengthen the voice of farmers” and “empower them by providing livelihood opportunities.”

The main intervention of the Payoga-NGO consisted of providing farm households in the municipality of Gamu with stock grants of male and female goats. The farm households were expected to return to the NGO three offspring, regardless of gender, within 12 months. The stock grant, coupled with technical and financial services provided pro bono by the NGO were warranted by the poor conditions of the farmer-beneficiaries and the fact that Isabela is a disaster-prone province. In 2013, for example, Typhoon Lando struck the province causing the loss of Php 33 million worth of crops and adversely affecting 180,000 farmers. A year later, the province was again in a state of calamity, this time a drought, due to El Niño.

However, the stock dispersal program was not sustainable. The farmer-beneficiaries treated the stock grant not as a livelihood opportunity, but as a source of *pulutan* (side dish or appetizer, such as goat meat, that is eaten with alcohol) for special occasions such as Christmas, weddings, birthdays, baptisms, festivals, homecoming of family members from overseas, and other celebrations.

In 1992, the founders, together with some farmers, established the Kapatagan Multi-purpose Cooperative (K-MPC) as an alternative means of combating poverty, and registered it with the Cooperative Development Authority (CDA) with its own Board of Directors. Initially, only 100 farmers joined the cooperative while the rest continued to rely on the dwindling resources of Payoga-NGO. In 1999, the Board of Payoga-NGO voted to merge with K-MPC, and K-MPC was re-registered with the CDA (as amended in 2011) as the Payoga-Kapatagan Multi-purpose Cooperative or PK-MPC. Today, PK-MPC is a growing organization with a strong membership of 3,408 farmer members spread all over Isabela Province. More than 50% of its members are women.

18.2 Features of the Collective Enterprise that Depict the SSE Dimensions

Certain dimensions of SSE set it apart from the market-oriented neoliberal economy. Among its strategic dimensions are the following.[1]

18.2.1 Dimension 1: Socially Responsible Governance

The collective enterprise is a social-mission oriented organization that recognizes the limits of environmental exploitation for the purpose of sustaining society's ability to meet people's needs. In the case of PK-MPC, the marginalized and socially excluded people (landless farmers) are the owner-managers of the cooperative. The cooperative has a bottom-up organizational structure (see Fig. 18.1) where the Member Clusters comprise the dominant power blocs in the General Assembly. PK-MPC has evolved into a strong and healthy organization with a responsible Board of Directors (BOD) that is accountable to the General Assembly. It has a highly capable management team and commits members who work together with the cooperative's officers.

PK-MPC members are grouped into 225 clusters with 15 members per cluster. Each cluster elects its own cluster leader who, in turn, elects the BOD and the Committee members. If found incapacitated, any BOD or Committee member can be terminated by the General Assembly. The BOD and Committees consult members

[1] For an elaborate discussion of the dimensions of the social solidarity economy, see Benjamin R. Quinones, Jr., "Social Solidarity Economy (SSE): An Alternative to Neoliberal Economy" (Speech delivered at the 40th Anniversary and Symposium on Solidarity Economy of the Pacific Asia Resource Center (PARC), Tokyo, Japan, March 23, 2014), http://www.alliance-respons.net/bdf_fiche-document-123_en.html.

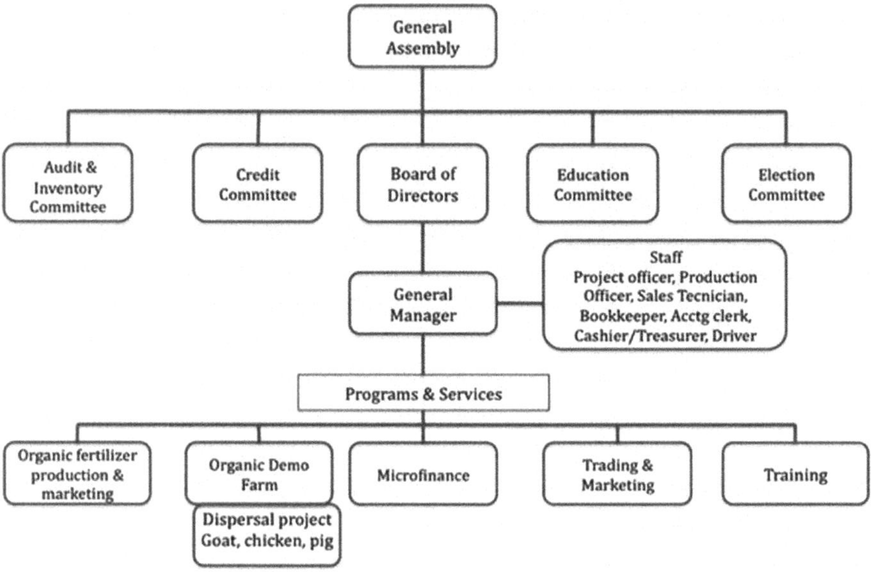

Fig. 18.1 Organizational structure of PK-MPC

through the clusters when deciding on policy matters concerning deposits, loans, marketing assistance, and other cooperative services. As a matter of fact, the adoption of organic farming was decided upon by the General Assembly, and only after did the BOD take steps to implement it.

Gender-wise, the General Assembly is composed predominantly of women. Of the total membership, 1755 (51.5%) are female, and 1653 (48.5%) are male. The Board of Directors (BOD) is also predominantly female (4 of 5 board members are female). The BOD Chairman is female. Of the eight staff members, six are female, two are male. The General Manager is female.

The cooperative's basic approach to ensure gender equality is PK-MPC's membership policy of non-discrimination when accepting members. Gender equality (i.e., women representation) is actively considered in recruiting members, decision-making, the management/operation of the cooperative, hiring/promoting staff/ workers, and training.

Notably, the PK-MPC was born out of a long-drawn process of conflict resolution between the Payoga NGO and the Kapatagan MPC which exposed the membership to sharp contradictions in interests and opinions. Credit is due to the responsible leaders, officers, and committed members for resolving conflicts and rising above the contradictions, thus ushering in a new era for the cooperative. Based on the dynamics of their current governance, it does appear that PK-MPC has learned a great lesson about the profound importance of bottom-up, participatory governance, and participatory dialogue in decision-making and in conducting the cooperative's day-to-day activities.

18.2.2 Dimension 2: Edifying Ethical Values

The collective enterprise upholds and defends the basic human rights of all community members against injustice, including exploitation and psychological and physical harm. The socially responsible practices of the collective enterprise reflect the edifying ethical values of its leadership and management. The edifying ethical values inculcated by the cooperative among its members include: *bayanihan* (solidarity, cooperation, working together); respect for the rights of others; rejection of all forms of violence and discrimination; people and planet over profit; self- management of the collective enterprise, people's participation in decision-making, resource allocation and wealth redistribution; responsibility, accountability, reliability; and efficiency, viability, and sustainability.

There were initial objections from local farmers, among them hitherto non-members of the cooperative, to the adoption of the organic farming method. Many of them thought they could not achieve good harvest if they discontinued the use of chemical fertilizers. Instead of ignoring the objections, the coop's Board and management worked hard educating the farmers about the necessity and benefits of protecting the environment through organic farming.

In addition, the cooperative diligently provided technical and marketing assistance and financial (deposit and loan) services to the farmers in support of organic farming. With more local consumers patronizing organic farm produce, PK-MPC eventually proved that organic farming is capable of raising farm productivity. It is more profitable and acceptable to local consumers.

Thus, the education campaign eventually paid off. To date, 2650 rice and corn farmers from the total 3408 farmer members of PK MPC have shifted to organic farming and are able to increase their farm yields and incomes. Subsequently, the production of Greenfriend organic fertilizer increased from 45,000 bags (50 kg (kg) per bag) in 2006 to 350,000 bags in 2016.

18.2.3 Dimension 3: People-Centered Development

The collective enterprise builds its own capacity to produce and/or access products and services that address human needs and contribute to the general well-being and social protection of community members.

Promotion of organic farming. The main business enterprise of PK-MPC is the production and distribution of organic fertilizer with the view to shifting agriculture in the Gamu municipality from the rampant use of chemical fertilizers and pesticides to the more sustainable and socially responsible practice of organic farming. The use of organic fertilizer also better prepares farm households for the effects of climate change (Fig. 18.2).

PK-MPC produces its own brand, Greenfriend, made up of local biodegradable raw materials such as chicken, bat, and carabao manure, and rice straw mixed with

Fig. 18.2 The collective enterprise at work (Photo courtesy of the Foundation for a Sustainable Society)

carbonized rice hull, agricultural lime, legume, and enzymes in 80% water. It is processed and stored in a concrete warehouse with the capacity to store 500,000 bags inside a seven-hectare farm owned by the cooperative along the highway of District 3, Gamu, Isabela.

Pooling of community savings to provide sustainable financial services. Instead of relying on donor funds, PK-MPC embarked on mobilizing community savings by offering deposit services not only to cooperative members, but also to the public at large. While access to donor funds encourages dependence of households on outside help, the mobilization of local savings enabled PK-MPC to reconnect with the local community and provide an alternative use for local resources. Many depositors opted to become members of PK-MPC because of the opportunity to access productive loans which, according to the Cooperative Law, can be made available only to members.

Any PK-MPC member, male or female, can avail of production loans from Php 10,000 (USD 200) to Php 30,000 (USD 600) payable within a minimum of six months to a maximum of two years. Loans are primarily for the purpose of purchasing organic fertilizer, thus supporting the cooperative's organic fertilizer business. Instead of cash, farmers get organic fertilizer at the standard quantity of 15 bags per hectare. PK-MPC sells its organic fertilizer for Php 230 (USD 4.60) per 50-kg bag, much lower than the prevailing market price of Php 300 (USD 6.00) to Php 350 (USD 7.00) per 50-kg bag.

Members who prefer to use chemical fertilizers do not actually lose their borrowing privileges. However, with the cooperative's interest rate policy geared toward supporting organic farming, loans for the purchase of organic fertilizer are on a preferential basis of 1% interest per month, while loans for the purchase of chemical fertilizers are charged a higher rate of 2.5% per month. In effect, those who use chemical fertilizers have to pay more for the cost of damage to the environment.

Integrating farm households into the collective enterprise's value chain. PK-MPC integrates farm households into the value chain of its collective enterprise by providing further incentives to both members and non-members who can supply raw materials for its organic fertilizer business. Rice straw costs Php 500 (USD 10) for 250 kg, and animal manure (chicken, bat, or carabao) fetches Php 30 to 45 (USD 0.60 to 0.90) for 50 kg.

Individuals (members or non-members) who gather the raw materials and mix them in the fertilizer factory can earn Php 4 (USD 0.08) per 50-kg bag. Those who volunteer as sales agents for the cooperative can earn Php 20 (USD 0.40) for every 50-kg bag of fertilizer they sell.

PK-MPC also hosts the "Summer Jobs for Kids" Program for young people aged 10 to 11 years old during summer vacation in partnership with the Department of Education (DepEd). The students are assigned to work in the PK-MPC Organic Vegetable section of the coop's demo farm. "This helps our youth appreciate and love organic farming," said PK-MPC General Manager, Ms. Madrid.

PK-MPC distributes native pig and free-range chicken to deserving farmer members (i.e., members with excellent credit and savings records and have shifted to organic farming). The livestock dispersal program is not on pro bono basis. The livestock grant must be repaid in kind.

PK-MPC's support to groups at risk: Anecdotal evidence. Among the 70 seasonal workers of the organic fertilizer production, the oldest are female—Clara Beronia, 85, and Villamor Lintao, 74. The youngest, who is 20 years old, does the same work as the senior citizens. Explaining why PK-MPC hires senior citizens, Ms. Madrid said, "It is one way of recognizing their valuable contribution to agriculture, and keeping them in our workforce enables them to maintain their vigor. There are many Claras and Villamors among our members, but hiring them is not a viable option because they reside far from our plant and transportation will be a big problem for them. Senior citizen workers also receive an additional incentive of Php3 for every bag of fertilizer they make."

18.2.4 Dimension 4: Environmental Conservation

The collective enterprise deliberately chooses the products and services it offers to serve the needs of community members, while it is constrained by its social responsibility to protect and enhance the local ecosystem and biological diversity.

The environmental conservation measures undertaken by PK-MPC include the following:

1. An active campaign against chemical fertilizers and pesticides through the promotion of and support for organic agriculture.
2. Provision of a loan facility geared toward supporting organic farming. A lower interest rate is offered to borrowers who shift to organic farming and use the loan for the purchase of organic fertilizers/pesticides. But borrowers who use the loan for the purchase of chemical fertilizers/pesticides are charged a higher interest rate.
3. Use of locally available waste materials (animal manure, farm waste, etc.) as raw materials for organic fertilizer production.
4. Conduct of seminars on environmental conservation in partnership with government agencies and environmental NGOs.

As a result of PK-MPC's dedicated advocacy toward the greening of Isabela province, it is now the official representative of all farmers' organizations in Region II (Cagayan Valley Region) to three Regional Councils: Organic Farming, National Disaster, and Small and Medium Enterprises. Being economically and financially sustainable, PK-MPC has become a major advocate of organic farming, thereby contributing to the lessening of carbon emissions in the province, the greening of Isabela farms, the creation of new decent job opportunities, and the advancement of sustainable farming.

18.2.5 Dimension 5: Socio-Economic Sustainability

The collective enterprise achieves economic and financial sustainability by producing products and services that address people's needs and are therefore valued and patronized by the community members.

The collective enterprise gives new meaning to work in society. Being both owner and worker of the cooperative, members view their work as a creative and productive endeavor to meet/serve their own needs and ends. To the extent that tenant farmers wisely use the financial and technical resources provided by the cooperative, the crucial option is wide open for them to rise above being tenant-workers and become entrepreneurs (Fig. 18.3).

The concept of "cooperative work" is attractive and liberating for tenant farmers who do not own the land they till and who henceforth must work hard in order to make others live. The cooperative enables the asset-less tenant to accumulate wealth and achieve financial independence by helping him establish off-farm enterprises that add revenues to the family income.

Compared to a for-profit enterprise, the collective enterprise integrates and reconciles in one person the functions of worker, owner and manager, which, under capitalism, is assigned to three persons. The for-profit enterprise pays a higher salary to top management and lower wages to workers, and distributes the profits to shareholders. However, the collective enterprise pools capital resources from members

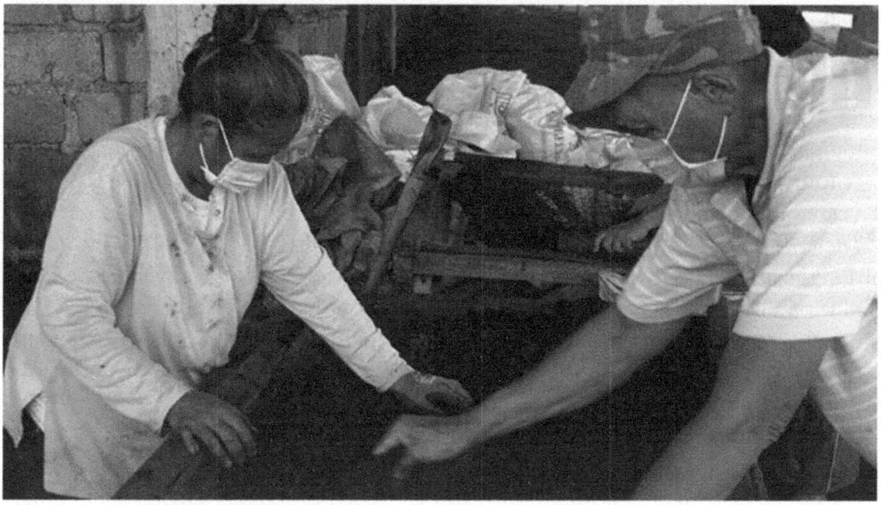

Fig. 18.3 The products made in the collective enterprise seek to address the needs of community members (Photo courtesy of the Foundation for a Sustainable Society)

and transforms these into loans to support a common undertaking—organic farming. Profits of the organic farm are fully retained by the members while PK-MPC retains only the cost of capital. Furthermore, the members enjoy the benefits of patronage refund plus a share in the profits of the cooperative.

In the field of contract growing (considering livestock dispersal as an old form), for-profit enterprises charge contract growers for the cost of raw materials (e.g., pigs, goats, chickens), while PK-MPC gives them out as stock grant to its members. For-profit companies buy the matured livestock generally at low farmgate prices. In contrast, PK-MPC gets paid in kind in the form of weaned pigs, goats, and spring chicken—the market value of which are comparatively lower than that of the matured livestock that was earlier dispersed.

The collective enterprise reconnects workplaces with communities. One of PK-MPC's strategic actions to reconnect with communities is to offer deposit services to the public at large, and not only to members. Depositors eventually opt to become members of the cooperative to take advantage of its loan services.

The cooperative's financial services enable tenants to shift from monocrop production to integrated farming (i.e., combining crop production with livestock raising). Furthermore, access to cooperative loans helps tenants upgrade their livestock enterprise.

Another strategic action is livestock dispersal, which is reserved for "deserving members". To a neighboring asset-less tenant family who happens to be a non-member, becoming a livestock owner- operator is a come-on to join the coop as members because the program provides a means by which the family can create and grow wealth.

A third strategic action is job creation in support of the organic fertilizer business. Not only cooperative members but also non-members are welcome to earn extra money by selling raw materials required for organic fertilizer production to the cooperative and selling "Greenfriend" products to farmers in neighboring municipalities.

A fourth strategic action is the exposure to organic farming of youth, ages 10–11 years old, during school summer vacation.

The collective enterprise promotes decent work. All PK-MPC staff work a standard eight-hour day, from 8:00 a.m. to 5:00 p.m., Monday to Friday. They receive regular pay at salary rates required by law. They enjoy a social protection package that include: Social Security System (SSS) benefits, access to low cost-housing finance through Pag-ibig, medical and hospitalization benefits through Philhealth, and 13th-month pay.

In addition, the staff/workers enjoy profit sharing, performance incentives at the end of the year, rice allowance in cash, and a 10% sales commission.

Engagement with social impact investors. PK-MPC actively engages and enters into partnership with social impact investors such as the Foundation for a Sustainable Society Inc (FSSI), ICCO Cooperation, HEKS (an ecowaste coalition), the provincial government of West Vllanderen (Belgium), the Philippine Center for Agricultural Research and Rural Development (PCARRD), and the Department of Agriculture (DA).

Engagement with the government and private sectors. PK-MPC helps members market their organic produce in local markets, as well as to private companies and government agencies at marked-up prices. Payment for the organic produce is coursed through the cooperative, which enables the latter to deduct the amount of loans outstanding from the sales proceeds. Compared to the traditional system where farmers sell their produce at a very low price, the cooperative ensures that farmers are paid reasonable prices for their quality organic crops. On top of marked-up prices, members are given patronage refunds and dividend shares equivalent to around 70% of the total earnings of the cooperative at the end of the year.

Regional branches of government agencies such as the DepEd, the DA and the Department of Environment and Natural Resources (DENR) have entered into partnership with PK-MPC to buy seeds and organic fertilizer on a regular basis. At the municipal level, PK-MPC is actively involved in the implementation of the municipality's environmental conservation policy.

Total assets, liabilities, and net worth. In terms of financial capacity, the total assets generated by PK-MPC increased from the Php 1 million start-up capital in 1999 to Php 68.87 million in 2016, of which Php 47.6 million are liabilities (see Table 18.1). PK-MPC's net worth as of end of 2016 registered at Php 21.27 million, or more than 21 times its original capitalization.

Financial sustainability. PK-MPC's revenues increased from Php 40.94 million (USD 818,900) in 2014 to Php 70.54 million (USD 1.4 million) in 2015, against total expenses of Php 34.86 million (USD 697,246) in 2014 and Php 52.81 million (USD 1.05 million) in 2015. Over 97% of the expenses were administrative and selling costs. Net revenues amounted to Php 6.08 million (USD 121,620) and Php 17.73

Table 18.1 Total assets, liabilities, and net worth of PK-MPC (in Php), 2014–2016 (*Source* Payoga-Kapatagan Multipurpose Cooperative)

	2014	2015	2016
Total assets	36,810,213.70	60,555,219.85	68,868,299.63
(% increase)		(64.51)	(13.73)
Total liabilities	18,989,902.98	36,256,315.04	47,598,706.51
(% increase)		(90.92)	(31.28)
Total net worth	17,820,310.72	24,298,904.81	21,269,593.12
(% increase)		(36.36)	(–12.47)

Sources of funds. PK-MPC finances its operations mainly from internal funds and revenues of its organic fertilizer business, which accounted for 99.9% of its total revenues in 2014, 2015, and 2016. PK- MPC typically sells organic fertilizer on credit to members. The other sources of income for PK-MPC (accounting for 0.1% of its total revenues) are service charges from training, membership fees, income from bank deposit, and tractor services

million (USD 352,617), implying a return on investment (ROI) of 14.85% in 2014 and 25.12% in 2015.

18.3 Contributions to Sustainable Development Goals (SDGs)

The United Nations regularly monitors the contributions to SDGs of the State and private companies. While SSE is initiated not by the government or by private corporate entities but by local citizens, and since SSE sets itself as an alternative development model, it is imperative to identify and measure the cooperative's contributions to the SDGs. It cannot be overemphasized that SSE achievements are, in fact, contributions to sustainable development of organized groups of ordinary citizens, thus representing a third pillar for SDG realization.

The SDG indicators are selected from among the various indicators developed by the United Nations (UN). Only qualitative information is provided in Table 18.2 since quantification of SDG contributions is beyond the scope of the study. It can be deduced from the tabulated information below that collective enterprises of ordinary citizens do contribute to sustainable development and their contributions have great potential for achieving the SDGs.

Table 18.2 SDGs addressed through the SSE features of PK-MPC

Feature of the collective enterprise	SDG being addressed	SDG indicators
Socially responsible governance	SDG 5 (Gender Equality)	Higher proportion of female than male in: (a) total membership of the cooperative; (b) Board of Directors; and (c) management and staff
	SDG 16 (Peace, justice, and strong institutions)	Members are satisfied with the leadership and management of the cooperative; members actively participate in decision-making
Edifying ethical values and People-centered development	SDG 16 (Peace, justice, and strong institutions)	Women are empowered, resulting in reduced proportion of women subjected to physical, psychological, or sexual violence
	SDG 1 (No poverty)	Decrease in the proportion of community members with income below the poverty line
	SDG 2 (Zero hunger)	Decrease in the proportion of community members who are undernourished
	SDG 10 (Reduced inequalities)	Decrease in the number of community members who have no access to finance
People-centered development	SDG 8 (Decent work and economic growth)	Decrease in the number of unemployed among community members
Environmental conservation	SDG 13 (Climate action)	Increase in the proportion of community members engaged in organic agriculture, waste recycling, and use of renewable energy

(continued)

Table 18.2 (continued)

Feature of the collective enterprise	SDG being addressed	SDG indicators
Socio-economic sustainability	SDG 12 (Responsible production and consumption)	Higher proportion of community members with access to basic services (e.g., housing, water, electricity, public transport)
		Higher proportion of community members who practice waste sorting, recycling/use of waste materials for composting, organic fertilizer production and use

18.4 Challenges

At the present stage of their development, SSE initiatives such as the collective enterprise of PK-MPC, are small in scope and largely localized. A number of challenges have to be overcome if SSE is to be mainstreamed as an alternative development model. A few of these are enumerated below:

- ***Inculcation of long-term perspective and development.*** PK-MPC members need to have a longer-term perspective of developing their community and not focus only on short-term production goals. They need to have access to long-term finance to enable them to make substantive investments in infrastructure and capital assets. In particular, PK-MPC needs to develop a plan for the effective transfer of ownership of land to the tillers.
- ***Awareness of the kind of economy co-created by cooperative members.*** There is a need to foster greater ideological understanding and awareness among cooperative members and the local citizens in general about the relevance of building SSE as an alternative to the neoliberal market economy.
- ***Mapping of SSE initiatives.*** It is important to identify SSE stakeholders at the provincial and regional levels—who they are, what their business models are, what they do/what products and services they produce. Such information is needed for networking among SSE stakeholders and for scaling up SSE initiatives.
- ***Building synergies among SSE stakeholders.*** Collaborative partnerships need to be promoted among stakeholders doing similar activities in the same strategic part of the value chain (e.g., production, financing, logistics, marketing/trading, capacity-building, research/advocacy, etc.).
- ***Support continuing studies on linkage-building for scaling up.*** Linkage-building among SSE initiatives need to be supported for the development of value chains at the national and international levels, that are owned and operated by the marginalized and socially excluded sectors of society.

Benjamin R. Quiñones Jr. is a Board Member of Live green International Inc., a social enterprise engaged in the production and distribution of organic farm produce in the Philippines. He founded the Asian Solidarity Economy Council (ASEC) (2007-present), a member of the Intercontinental Network for the Promotion of Social Solidarity Economy (RIPESS) which aims to promote the social and solidarity economy (SSE) ecosystem in Asia and the Pacific. ASEC evolved from the Coalition of Socially Responsible Small and Medium Enterprises in Asia (CSRSME Asia) which the author established early on to foster the development of community-based social enterprises while serving as Programme Coordinator of the Asian and Pacific Development Centre (APDC) in Kuala Lumpur, Malaysia (1996-2003), an inter-governmental organization established by the UN Economic and Social Commission for Asia and the Pacific (UN-ESCAP). He developed and pilot tested in India, Indonesia, Nepal, the Philippines and Thailand the Self-Help Group Linkage Banking program while serving as Secretary General of the Asia Pacific Rural and Agricultural Credit Association (APRACA) in Bangkok, Thailand (1985-1993), a regional body established by the UN Food and Agriculture Organization (FAO). He holds a Ph.D. in Organization Development from the SAIDI School of Organization Development, an MS in Agricultural Economics from the University of the Philippines in Los Baños, and an AB Economics from the School of Economics, University of the Philippines in Diliman, Quezon City.

Chapter 19
Workers' Information Center in Cambodia: Transforming the Narrative of Garment Workers

Bianca Ysabelle E. Franco

Abstract The women-led Workers' Information Center (WIC) in Cambodia strives to create for women workers the spaces for empowerment, development, grievances, and for collective resistance in the garment sector. The Center operates as a "non-union" to avoid government intimidation and union busting while still generating collective power for their demands. It advocates not only for the rights of garment workers but also for equal access to public services for all Cambodian women. In order to do this, it works with other member organizations of the United Sisterhood Alliance in Cambodia.

Keywords Women · Gender · Labor · Garment industry · Corporations

19.1 Introduction

"Wake up! Do not live like a frog in a well" it is important to learn your rights!" (Workers' Information Center 2015a) was the Som Pov's message to her fellow garment workers. As a key organizer of the Workers' Information Center (WIC) in Cambodia in 2015, her journey is one that exemplifies the goals of WIC, a non-government organization of women workers in the garment industry. As a self-organized association, it caters to the collective needs and issues relevant to workplace rights violations. More importantly, it provides space for grievances, training, leadership development, and empowerment—necessary ingredients in making a difference in women's lives. The garments industry in most parts of the world thrives on the back of female labor. Though it provides much-needed jobs and livelihood especially for the rural poor, challenges associated with the current global capitalist production regime and its practices, along with gendered assumptions about women remain. Cambodia's women garment workers' journey exemplifies this. For this reason,

B. Y. E. Franco (✉)
Independent Scholar, Quezon City, Philippines
e-mail: biancaysabellefranco@gmail.com

J. M. C. Sy and A. P. Aban (eds.), *Reimagining Development in Southeast Asia*, Asia in Transition 21, https://doi.org/10.1007/978-981-97-9140-8_19

WIC's story is important in unraveling collective resistance and transforming the narrative of garment workers.

19.2 Cambodia's Garments Industry

The garment manufacturing industry has been a key driver of Cambodia's economy for nearly 30 years. It accounts for 16 percent of the country's gross domestic product (GDP) and 80 percent of its revenues from export (International Labour Organisation 2017). Based on the Cambodia Investment Review last year, from 2018 to 2020, there is a dramatic decrease in the garment export product from 76.16% to 68.77% (Badzmierowski 2022). Back in 2018, the European Union is its biggest market, followed by the United States, Canada, and Japan (Rastogi 2018); data from the International Labour Organization report in 2022 states that "the decline in Asia's market share since 2015 was absorbed primarily by Europe as the share in exports of other regions (Netherlands, Germany, Poland, Spain, and Italy) increased" making the garments sector a lynchpin of its participation in the globalized economy. In 2018 alone, the garments and textiles industry comprised over 60 percent of total exports (AQF Team 2019) and will increase in 2021 as 73% of the global revenue is from the "consumer spending on clothing and apparel" (Grand View Research 2021).

These figures result from Cambodia's massive efforts to rebuild its economy after the civil war in the 1990s. After independence in the 1950s, the economy relied on the agricultural sector. With much of the country reeling from high poverty levels, especially in the rural areas in the 1980s, economic transformation became urgent. The 1990s were also the period in which much of Southeast Asia pursued rapid industrialization to catch up with their East Asian neighbors, such as South Korea, Singapore, and Taiwan. Under the World Bank and the International Monetary Fund auspices, Cambodia embraced the neoliberal growth model by reforming its laws to attract foreign direct investments (FDI) amid the promise of cheap labor and other incentives. Via its Law on Investment in 1994, the Cambodian government promised a competitive business environment for FDIs despite challenges inherent in a developing country (Bargawi 2005) from this, the preferential access provided by the United States catalyzed the growth of the industry (Ear 2013). Despite liberalization and the abolition of the Multi-Fiber Agreement in 2005, the industry remained competitive throughout the years (Yamagata 2006) and immensely contributed to Cambodia's economy.

Garment factories in Cambodia are foreign-owned and controlled. Exporters mainly come from China, Hongkong, Singapore, Malaysia, and South Korea, a situation made possible by Cambodia's strong bid for overseas capital and investments. Under Cambodian laws, foreign exporters based in the country can expand their businesses. Such a legal framework also allows foreign investors to skirt their own countries' quota thereby capturing a bigger share in the global market (Hall 2000) with Cambodia at over 43.84% shares globally (WITS 2023).

Aside from this conducive business environment, Cambodia has one of the lowest wages in Asia, according to a recent report (AQF Team, 2019). In 2019, there is a 45:49 ratio of regular wage and salaried workers to irregular, temporary or unknown working arrangement (International Labour Organization 2022). While this is attractive to investors, this has been a source of tension between workers and employers, resulting in strikes and deaths. In 2014 (BBC News 2014), three workers were killed when police gunned down protesters demanding a pay hike. There are also studies and reports indicating the negative impact on workers' health (Yamagata 2006). There are also reports of job stress, isolation, and depression (McKay and McKenzie 2020). The Human Rights Watch has also reported various types of labor rights violations such as forced overtime, sexual harassment, and anti-union discrimination (Human Rights Watch 2015). Yamagata finds that working in garment factories has been beneficial in terms of poverty reduction (2006). However, the job is vulnerable to fluctuations in demand and crises in importing countries. Using statistical analysis, Yamagata (2006) finds that the wage gap between men and women rises when demand decreases. According to the Independent Property Services' report, in Cambodia, "men earn about 50% more than women" (Castro 2022). Therefore, these findings indicate the vulnerabilities and uncertainties that garments workers face even in the midst of precarious work.

19.3 Women's Work: Global and Local Gender Norms

The brief overview above will be incomplete without discussing the role of women workers in Cambodia's garment industry. Garment and textile industries employ large numbers of women in the developing world. Cambodia is a case in point in which workers are usually women from the rural areas. While the argument is that the country's bid to globalize its economy is to provide them jobs in the cities, their situation is problematic from the standpoint of feminist literature on globalization and women's work in the manufacturing sector.

According to feminist political economy scholarship, it is not that low wages result in companies employing women workers (Elias 2008). The lure of women workers in the global assembly line is a result of constructing garment labor as feminized, a role usually ascribed to women (Elias 2008, 406). In their analysis of women's work in the global South, Elson and Pearson (1981) argue that women are also attractive to exporters because of their "nimble fingers", docility, and the assumption that their income is secondary to their husbands—socially constructed traits that are beneficial to international capital. In other words, not only is it more efficient to hire women because of their perceived "natural" trait in garment work, they are also not keen to unionize and therefore less likely to disrupt the production cycle.

Aside from the links between global factories and women's work, local meanings behind women's work in Cambodia are also instructive. Studies indicate that the figure of "good daughters" within the family is central to the deployment of young women to garment work (Hall 2000, 128). When women's work became in great demand in the 1990s, daughters' exodus from the rural areas to the cities was largely equated to their role as "family's saviors" from poverty. However, this becomes problematic as this might also prevent women from asserting their rights and grievances at the shopfloor, given the expectations from their families.

For these reasons and based on the discussion and overview above, the work of the Workers' Information Center compels us to look at the ways in which women workers collectively transform their lives given the constraints placed upon them.

19.4 The Workers' Information Center: Collective Resistance and Empowerment

The Workers' Information Center (WIC) is an association of women garment workers (GWs) in Cambodia that promotes women's workplace rights. It advocates women's leadership, particularly in addressing issues that pertain to the garment sector, and space for GWs to express their grievances. The WIC also provides capacity-building opportunities for women to gain knowledge that can be used to realize their workplace rights. Legal assistance and healthcare are also available for urgent concerns. The issues addressed by the WIC include "low wages, hazardous working conditions, sexual harassment, discrimination, short term contracts, and limited social security provisions" (Workers' Information Center 2015b).

The WIC is part of the United Sisterhood Alliance in Cambodia. The United Sisterhood Alliance evolved from an organization established in 1999, the Women's Agenda for Change (WAC) to address the needs of women and other marginalized groups. The WIC started as a project on Women Garment Workers' Empowerment of the WAC. The issues that emerged from this project led to the establishment of the WIC as an association in 2009 (ESCR-Net n.d.).

The WIC is a worker-led organization. It is an organization that is by the workers and for the workers. It has what are called "drop-in centers" in factories. A drop-in center is a place where workers can meet with one another and discuss their grievances. Workers discuss issues on both working and living conditions. They often come to the drop-in centers after work (Aliyah, interview, 2019).[1]

The WIC also facilitates discussions between workers and their employers. In an interview with a garment worker named Aliyah who works at the Saulen Factory, she shared that these discussions are used to lobby the GW's concerns to their employers. For instance, initially, factory workers were only given 50% of the wage indicated in their contract. The workers started a thumbprint petition and demanded they be

[1] Researchers of the Program on Alternative Development (AltDev) at the University of the Philippines Center for Integrative and Development Studies (UP CIDS) conducted an interview with Aliyah, a garment worker at the Saulen Factory in Cambodia. The interview took place after the workshop on "Alternative Practices of Peoples in Southeast Asia Towards Alternative Regionalism" at the ASEAN Civil Society Conference/ASEAN Peoples' Forum 2019, September 11, 2019, Bangkok, Thailand. AltDev researchers include Honey Tabiola, Nathaniel Punongbayan, and Jose Monfred Sy. Focus on the Global South—Cambodia provided translation.

given the salary they are due. That petition is a success story amid the horrifying conditions faced by garment workers.

According to Aliyah, while the typical working day is eight hours, garment workers cannot find the time to rest or go to the restroom due to the high targets set by factory managers. They must work nonstop to meet the targets or risk verbal abuse from the managers. To make matters worse, if the already high target is met, that target is increased the following day. "They don't care about the worker, they just care about finishing the work," said Aliyah.

Further, factory owners employ illicit strategies to maximize profit at the cost of the garment workers. Factories in Cambodia can run for nine years tax-free. To avoid paying taxes, factories are closed down and started back up again under a new name, and all factory workers lose their jobs. Since the minimum wage increases yearly, factories end up hiring fewer and fewer workers to minimize the cost. This means that each worker gets more work assigned to them for the same salary. Meanwhile, workers cannot demand the recognition of their rights because they are not regular employees of the corporation. Contracts only last for three to six months, and after such time, workers must find new jobs. The WIC urges corporations to abide by Cambodia's labor laws and grant each worker the benefits they deserve.

The WIC does not only lobby for garment workers' rights but all Cambodian women's access to public services. These include farmers and sex workers, too. One of the main issues faced by these workers relates to privatization. The WIC is advocating for policy change because public services become inaccessible when controlled by private corporations (WIC 2017, 5).

The WIC works in solidarity with the other member organizations of the United Sisterhood Alliance. There are four organizations included in the United Sisterhood Alliance, including the WIC. One of the other alliance members is the Messenger Band, a musical group that performs songs that portray marginalized and impoverished groups in Cambodia. The Women's Network for Unity (WNU) is also a part of the United Sisterhood Alliance. The WNU caters to the needs of sex workers and promotes their access to social services. Lastly, the Social Action for Change (SAC) advocates for smallholder farmers and other marginalized groups to social services such as affordable medicines. These four organizations make up the United Sisterhood Alliance.

The WIC also monitors the implementation of labor laws in the garment and textile industry. To build awareness about the issues in the garment industry, the WIC published a report in March 2017 entitled "The Reality of the National Economic Backbone," which details the plight of garment workers in Cambodia. The report discusses the garment workers' access to health and hygiene, security and safety, and the standards of their accommodation.

The report is based on a series of consultations conducted by the WIC with the GWs (Workers' Information Center [WIC] 2017). They conducted interviews with 75 GWs and focus group discussions with 90 GWs from Chak Angre Krom, Posenchey, and Teuk Thla. The issues found by the WIC study pertain to health, security, and the accommodation available to GWs.

Among the health issues found by the WIC include the fatigue experienced by GWs due to long working hours, unhygienic workplaces, and food that contain harmful chemicals. GWs are forced to work eight to twelve hours a day at the factories to maintain the city's high cost. "I feel exhausted…" said a garment worker (WIC 2017, 11). Within these eight to twelve hours, workers must work extra hard to meet the productivity quota of their respective factories. Given these conditions, it is no surprise that workers report developing poor health since they started working in the garment industry.

Women flock to the cities to become GWs to provide better lives for their families. The salary they earn from working as GWs is barely enough for the rooms for rent in low-cost tenements in the vicinity of the factory. The rooms do not have access to sunlight, and the water available to them contains limescale deposits. GWs also have to pay an extra USD 1 per month to their landlords for the water supply, resulting from a rusty tank. Some GWs have reported finding fingernails and strands of hair in the food provided to them by their employers (WIC 2017, 11). They also do not have access to running water, and the public toilets at the low-cost housing buildings are often clogged, and soap is not available. Among the WIC's respondents for this study, 43% of GWs use public toilets, and 57% have private bathrooms in their rental rooms. The public restrooms are located in the open yard and unsanitary. In cases where there is stored water available, it is often "dark and muddy and [smells] really bad" (WIC 2017, 13). There is also no proper waste disposal mechanism. Lastly, while there are available state hospitals for medical concerns, GWs choose to visit private clinics close to their workplace because of the long working hours.

There are also several security issues faced by GWs. The walls and doors of the low-cost housing buildings are made from substandard wood which makes break-ins common. Theft is also usual, especially with clothing hung to dry. The locks on the doors are poorly maintained, but the GWs cannot afford to buy their locks. The GWs in the WIC study also reported that they fear going out at night due to violent gangs. More than half (57.7%) of the WIC's respondents reported violent gangs being present in the vicinity of their neighborhood. Many have also reported experiences of sexual harassment—38.9% have been sexually harassed in the form of catcalling but received no help from local authorities. "If I could, I would not want to go out at night," a worker commented, referring to the dangerous neighborhood they live in (WIC 2017, 19).

Gambling is also common, with 78.9% of respondents reported having observed gambling activities in their area. The violence and gambling are not controlled because there are no police around low-cost housing buildings. Majority (83.8%) of the WIC's respondents also reported having no fire extinguishers in their buildings. This is alarming especially because the buildings are poorly ventilated and made with substandard materials—making it an obvious fire hazard.

GWs in Cambodia live in workers' blocks or compounds with more than 50 rooms in each building. Some buildings have more than 300 rooms. They live in rooms that range from 2.5 to 16 sqm with walls made from substandard wooden panels. While there is a policy that mandates employers to provide accommodation to their GWs, this is poorly implemented. Those employers who provide support give less than the minimum rental fee, so workers are forced to rent the cheapest rooms in the poorest conditions. In most of these workers' blocks, GWs only have access to public toilets. These toilets are poorly maintained, often clogged, and provide no proper waste disposal to users. Tenants of more than 300 rooms must share 16 public restrooms. GWs also report that rent is high while salary remains low, and the rent is fixed and usually non-negotiable. Only a small fraction (4%) of GWs reported having contracts with their landlords regarding rental contracts. This means that the landlords have the discretion to kick them out of their homes anytime. Delayed rental payment result to verbal abuse from landlords. Complaints about landlords were also common among the respondents of the WIC. "The landlord can get the room back any time and I do not know what to do," said one worker when asked about rental contracts. Another worker reported their landlord not returning the deposits tenants made for electricity and water, even though they agreed to do so.

Based on the findings of the report published by the WIC, they came up with recommendations addressed to the Royal Government of Cambodia in consideration to the landlords, workers, and employers. The WIC recommends the following: (1) the Cambodian government and the landlords continue to adhere to the country's Special Lease Law which guarantees that "all parties understand contracts, rent increases, accommodation standards, and rents" (WIC 2017, 29); (2) the government engage with the GWs to listen to their grievances; (3) encourage employers to provide adequate living conditions to GWs; (4) provide security from the landlords' property and premises, and cooperate with the local government with the reporting of thefts and robberies. Workers are in turn encouraged to continue reporting crimes of theft, robbery, and sexual harassment. Lastly, workers must actively cooperate in terms of keeping their living conditions satisfactory.

The Reality of the National Economic Backbone is an example of the WIC's worker-led campaign (ESCR-Net 2020). It is a bottom-up approach that exemplifies the garment workers' expertise over their own experiences—an expertise that cannot be embodied by researchers and academics. A team of thirty workers implemented the study from conceptualization to designing the questionnaire and conducting the interviews with their fellow garment workers. The report aims to build awareness about the realities of garment workers. Upon publication, the study results were presented to stakeholders to negotiate the conditions of the garment workers. Workers themselves used the data gathered through the study to lobby their concerns to local government authorities and community forums.

19.5 Recommendations

The story of WIC is the journey of Cambodian women's admirable goal to uplift their lives. While Cambodia's economic transformation has brought livelihood to its young women, it also comes with its contradictions. Even though women have jobs through foreign-owned factories, the precarity of this work on account of gendered assumptions of women's labor remains daunting. With much of the world linked to global capitalist production and its discontents, the WIC has remarkably managed to make a difference.

Nonetheless, it would also be beneficial in the longer term for WIC to look beyond rights. While the rights framework has certainly gained traction among multinationals owing to advocacy and shaming campaigns from transnational civil society and the media, there are other ways to move forward. Robinson's (2006) global care ethics might be one step in the right direction. It espouses to veer from looking at the "workers" and instead look at them as people with sets of care responsibilities outside of work. This framing takes a hard look at the unequal division of care work for men and women and then compels companies to consider this in designing programs that combine both productive and reproductive roles of the two sexes in the household. According to Robinson (2006), this is not necessarily out of range of multinational companies as some have already incorporated these themes in their corporate social responsibility. For instance, work-life balance has already entered the lexicon of corporate jobs. Concepts such as "fair trade," "clean clothes," etc. have already been around recently. The idea is to combine productive and reproductive work in company ethics and the involvement of both men and women in both of these spheres.

19.6 Conclusion

The garment sector has been one of the motors of Cambodia's economy—luring foreign capital with low wages and weak protection of workers' rights. Navigating the tensions between foreign capital and labor protection and braving the challenges that come with garment work are the workers who are mostly women. This feminization of garment work is shaped not only by the garment sector itself which is often associated with women and women's perceived docility but also cultivated by local gender norms that often groom women, especially daughters, as "saviors" of their families from poverty.

Amidst the backdrop of the gendered political economy of garment work in Cambodia, workers at the WIC, most notably the women workers, carved out an alternative space where the workers can exercise more agency and address the issues and concerns of their sector. Contesting the usual expectation that women usually do not collectively organize to claim what is due them, members of WIC showed that they can collectively act to confront the issues that plague the garment sector. As an association of women garment workers, WIC is a formidable force in the promotion of women's workplace rights and women leadership. They champion on workers' welfare through providing accessibility to legal assistance, healthcare, and public services, while also addressing issues related to low wages, hazardous working conditions, sexual harassment, and other grievances. With its "drop-in centers," WIC cultivates a safe space for workers to acknowledge their concerns, discuss the issues, and work toward solutions. WIC as a "non-union" form of organization is unique that it attempts to skirt the possible intimidation that governments and businesses usually throw at unions—while still generating collective power that can exert pressure to advance their demands.

Moreover, this work's success is made possible through the solidarity of various organizations under the United Sisterhood Alliance. This alliance collectively fought for and overcame challenges faced by women in various sectors—impoverished groups, smallholder farmers, sex work, and the garment sector; this only shows that the struggles of many working women across various sectors are interconnected and can be overcome collectively.

References

AQF Team. 2019. *Is Cambodia Poised to Become a Major Manufacturing Hub?* Asia Quality Focus. 23 October. https://www.intouch-quality.com/blog/is-cambodia-poised-to-become-a-major-manufacturing-hub.

Bargawi, Omar. 2005. Cambodia's Garment Industry—Origins and Future Prospects. *Working Papers 13, Economics and Statistics Analysis Unit (ESAU), Overseas Development Institute.* 1 October. https://odi.org/en/publications/cambodias-garment-industry-origins-and-future-prospects/.

BBC News. 2014. Cambodian Textile Workers Killed during Pay Protest. *BBC News*, January 3. https://www.bbc.com/news/av/world-asia-25593553.

Badzmierowski, Brian. 2022. Cambodia's garment sector 'on the brink of change'. Cambodia Investment. 30 March. Review. https://cambodiainvestmentreview.com/2022/03/30/cambodias-garment-sector-on-the-brink-of-change/.

Castro, Mizpah Grace. 2022. 2023 Update: Average Salaries in Cambodia. *Independent Property Services.* 14 October. https://ips-cambodia.com/cambodia-average-salaries-2022/.

Ear, Sophal. 2013. Cambodia's Garment Industry: A Case Study in Governance. *Journal of Southeast Asian Economies* 30 (1): 91–105.

Elias, Juanita. 2008. Hegemonic Masculinities, the Multinational Corporation, and the Developmental State: Constructing Gender in 'Progressive' Firms. *Men and Masculinities* 10 (4): 405–421. https://doi.org/10.1177/1097184X07306747.

Elson, Diane, and Ruth Pearson. 1981. 'Nimble Fingers Make Cheap Workers': An Analysis of Women's Employment in Third World Export Manufacturing. *Feminist Review* 7 (1): 87–107. https://doi.org/10.1057/fr.1981.6.

ESCR-Net. 2020. Interview with Channsita Mark, Workers' Information Center (Cambodia). https://www.escr-net.org/news/2020/interview-channsita-mark-workers-information-center-cambodia.

ESCR-Net. n.d. United Sisterhood Alliance. https://www.escr-net.org/member/united-sisterhood-alliance.

Grand View Research. (2021). *Textile Market Size, Share & Trends Analysis Report By Raw Material (Cotton, Wool, Silk, Chemical), By Product (Natural Fibers, Nylon), By Application (Technical, Fashion), By Region, And Segment Forecasts, 2022–2030.* https://www.grandviewresearch.com/industry-analysis/textile-market.

Hall, John A. 2000. Human Rights and the Garment Industry in Contemporary Cambodia. *Stanford Journal of International Law* 36 (1): 119–174.

Human Rights Watch. 2015. "Work Faster or Get Out": Labor Rights Abuses in Cambodia's Garment Industry. Human Rights Watch. https://www.hrw.org/report/2015/03/11/work-faster-or-get-out/labor-rights-abuses-cambodias-garment-industry.

International Labour Organisation. 2017. *Cambodian Garment and Footwear Sector Bulletin 5.* https://www.ilo.org/asia/publications/WCMS_541288/lang--en/index.htm.

International Labour Organization. 2022. *Employment, wages and productivity trends in the Asian garment sector.* https://www.ilo.org/wcmsp5/groups/public/---asia/---ro-bangkok/documents/publication/wcms_848624.pdf.
McKay, F.H., and H.J. McKenzie. 2020. Life Outside the Garment Factories: The Lived Experiences of Cambodian Women Garment Factory Workers. *International Journal of Migration, Health and Social Care* 16 (4): 415–427. https://doi.org/10.1108/IJMHSC-03-2019-0041/full/html.
Rastogi, Vasundhara. 2018. Cambodia's Garment Manufacturing Industry. *ASEAN Briefing*. https://www.aseanbriefing.com/news/cambodias-garment-manufacturing-industry/.
Robinson, Fiona. 2006. Beyond Labour Rights. *International Feminist Journal of Politics* 8 (3): 231–342. https://doi.org/10.1080/14616740600792871.
Workers' Information Center. 2015a. "Story of Change." Workers' Information Center. https://www.wic.unitedsisterhood.org/?page=detail&ctype=article&id=794&lg=en.
Workers' Information Center. 2015b. "Worker's Information Center (WIC)." https://wic.unitedsisterhood.org/?page=detail&ctype=article&id=790&lg=en.
Workers' Information Center. 2017. *The Reality of the National Economic Backbone.* Workers' Information Center. https://focusweb.org/publications/the-reality-of-the-national-economic-backbone/.
World Integrated Trade Solution (WITS). 2023. *Cambodia Trade Summary 2020 Data.* https://wits.worldbank.org/CountryProfile/en/Country/KHM/Year/2020/Summary.
Yamagata, Tatsufumi. 2006. The Garment Industry in Cambodia: Its Role in Poverty Reduction through Export-Oriented Development. *Institute of Developing Economies.* https://www.researchgate.net/publication/5141229_The_Garment_Industry_in_Cambodia_Its_Role_in_Poverty_Reduction_through_Export-Oriented_Development.

Bianca Ysabelle E. Franco holds a master's degree in sociology from the University of the Philippines Diliman and formerly served as a research associate at Ateneo de Manila University's Development Studies Program, focusing on human rights, deliberative democracy, and conflict studies. Her written contributions have been featured on prominent academic blogs and news platforms worldwide. In May 2024, her book chapter titled "Experiencing Grief: Duterte's Lethal Drug War and Its Widows" in the edited volume *Games, Changes, and Fears: The Philippines from Duterte to Marcos Jr.* was published by ISEAS.

Chapter 20
Collective Action to Attain a Better Life: Maigting na Samahan ng Panlipunang Negosyante ng Towerville

Rowena E. Osal, Jennifer A. Macabudbud, Evelyn D. Galangera, and Regie Anne D. Placido

Abstract On the one hand, this chapter describes the processes of relocation and the extreme poverty it causes. On the other hand, it revisits the processes of creating associations and businesses. Yet most of all, this paper reflects the voices of people—the members of Igting. Ma-Igting na Samahan ng Panlipunang Negosyante ng Towerville (Igting, United Association of Community Entrepreneurs of Towerville) is a group of mothers who, beginning in 2011, have put efforts for the growth of their association and businesses. While Igting has become a stable organization with a growing enterprise, the process of achieving has not been easy; the path to it is not always straightforward.

Keywords Social enterprise · Informal settlers · Relocation · Community organizing

Giving voices to the members of Igting as they tell their stories comes from many contexts. One of these is the desire to show their experience to help and inspire other associations that are being formed. This also includes sharing important lessons not only in entrepreneurship but also in forming associations. However, the most important point to draw from this sharing is the desire of the members of Igting based on their understanding and insights. The members of Igting see this as important because, for a long time, their stories were written and shown to them, but these stories did not come from themselves.

Therefore, this paper is a product of collective research and Igting's writings. The contents of this paper come from the members' testimonies or diaries or products of

R. E. Osal (✉) · J. A. Macabudbud · E. D. Galangera
Ma-Igting na Samahan ng Panlipunang Negosyante ng Towerville, San Jose del Monte, Bulacan, Philippines
e-mail: igting.org@gmail.com

R. A. D. Placido
Center for Asian Mission for the Poor (CAMP)—Asia, San Jose del Monte, Bulacan, Philippines

J. M. C. Sy and A. P. Aban (eds.), *Reimagining Development in Southeast Asia*, Asia in Transition 21, https://doi.org/10.1007/978-981-97-9140-8_20

focus group discussions (FGDs). The process of writing this paper was accomplished through participatory research workshops conducted by the Program on Alternative Development (AltDev) of the Center for Integrative and Development Studies (CIDS) of the University of the Philippines (UP) with some members of the Center for Asian Mission for the Poor–Asia (CAMP).

This paper is divided into three parts. The first part describes the life of the members of Igting before and after they were relocated to Towerville. Describing this situation is important because it serves as the context that guides the majority of the discussions. The second part describes the processes that Igting underwent in forming the association and in entrepreneurship. This sharing includes their organizing initiatives, the challenges that they faced, and the way they overcame these challenges. The final part presents the important lessons that Igting has learned for seven years spent organizing. These lessons cover establishing relationships with people, the importance of strengthening their association, and the collective growth of the organization.

20.1 "We have homes, but we don't have businesses": Living Conditions in Towerville

Towerville is a housing project of the National Housing Authority (NHA) in San Jose del Monte, Bulacan. This project was started in 1996 under the government's Manila Urban Development Project. Eventually, Towerville served as a relocation site for poor families. Other families were victims of calamities such as Typhoon Reming in 2006 and Ondoy in 2009. Others were evicted from their homes because they were considered obstructions to development projects such as the North and South Luzon Expressway and the Northrail Project. Relocated families came from many areas of Metro Manila such as Malabon, Navotas, Manila, Pasay, and Quezon City. Towerville has six phases. Phases 1 to 6 were built in Barangay Minuyan Proper, while Phase 6 is part of two barangays: Barangay Gaya-Gaya and Barangay Graceville. Since Towerville was built, relocations have been continuous until its population grew to 70,000 families.

To recall, many of those relocated were poor families who have lived in lands that they do not own. They are popularly referred to as *informal settlers* or *squatters*. They often oppose relocation, saying that "You are already poor in the place where you came from, but you will be brought to a place that will make you poorer. Here, while the conditions are poor, it is near work, school, hospital—everything." Others are really stubborn, and they disregard the dangerous conditions of their environment. This situation shares similarities with the experiences of the residents of Bagong Silangan, Quezon City who were relocated to Towerville after Typhoon Ondoy.

The lives of the residents of Bagong Silangan were indeed more comfortable back then: it is near work, school, the hospital, and other basic needs. However, when Typhoon Ondoy struck the area, residents bore the burden of calamity. Many houses of residents were carried away by the flood and were left afloat in the river; other houses were carried away along with those who lived in them. Many died due to floods, while others climbed trees or the roofs of their houses to survive; others made themselves float using large water containers. Along with those that were drifted away by the flood were animals like chickens, pigs, goats, and cows, which later died. Residents were put in the worst situation—they had no food, water, and milk for their children. Those who survived dwelled for three months in the covered court that served as a temporary evacuation center before they were relocated to Towerville.

Some reasons for people to accept their transfer to a relocation site are a family's dream to have a livable house, own land, and a good environment and keeping themselves safe from the dangers associated with natural disasters (such as floods in the case of Bagong Silangan). Thus, through the continuous efforts of the government to persuade residents, some residents decided to accept living in Towerville. They did not realize that the relocation will lead them to experience deeper poverty and will decrease their hopes for a better future.

The process of relocation was never easy for the members of Igting, who were part of the thousands who transferred to Towerville. Hundreds of families boarded buses to this relocation site located in San Jose del Monte, Bulacan. Many of them had to demolish their old homes and rebuild their new homes in government-provided land. Others were given grocery items of PHP 2,000. Residents of Bagong Silangan were only given one food pack. Others were given financial assistance, but according to the experiences of relocatees, it would not last long and it would not be enough for them.

The relocatees experienced challenges as they began settling in Towerville. They found that there was no water and electricity in Towerville, and the houses were constructed using substandard materials. Worse, they have no source of livelihood there. Many had to return to the city. Most of them were men who travel to work and return home weekly to save money for fare. According to one of the relocatees, the expectations for their new homes were high, but as they arrived in Towerville, some basic services were missing. Therefore, relocatees would not avoid thinking that they were thrown to Towerville like garbage. The dismal situation has led residents to an internalized feeling of hopelessness, and gasps for air would only be heard.

Rowena Osal, a member of Igting from Marikina, tells her story in Filipino:

> Napakahirap ng kalagayan namin. Kubo lang ang naitayo namin. Wala kaming tubig, walang kuryente, at bababa pa kami sa ilog para maglaba at mag-igib. Minsan, nakikiigib din kami sa mga kapitbahay na may tubig. Kanya-kanya sa amin ang pagpapakakabit ng tubig mula sa NAWASA (National Water and Sewerage Authority). Depende sa layo ng lugar, umaabot ng Php 5,000 ang kailangang bayaran. [. . .] Kailangan din mag-stay-in kami ng aking asawa sa Marikina para sa aming trabaho. At bilang ina, napakahirap sa akin na lagi kong inaalala ang kaligtasan ng mga anak ko. Paano ang pagluluto ng pagkain nila? Ang pag-aasikaso sa kanila sa tuwing tutungo sa paaralan? Parang laging dinudurog ang puso ko sa pag-aalala sa kanila.
>
> [We had the most difficult situation there. We only built a hut. We don't have water and electricity, and we would only go down to the river to wash clothes and fetch pails of water. Sometimes, we would only fetch water from our neighbors who have it. We apply for water service from the NAWASA (National Water and Sewerage Authority) individually.

Depending on the distance of the area, people have to pay up to PHP 5,000. [. . .] My husband had to stay in Marikina to work. And as a mother, it is too difficult for me to think about the safety of our children. How would I cook food for them? How would I help them get ready for school? It seems like my heart breaks as I think of them. (our translation)]

Jocelyn Monsende also said:

Nang kami ay malipat dito, walang kuryente, walang tubig, at puro talahib. Ang giniba naming bahay dati ang siyang itinayo namin dito. Stay-in ang asawa ko sa trabaho at linguhan lang siya kung umuwi, kaya kami lang ng anak ko ang magkakasama palagi. [. . .] Sobra kaming natakot minsan isang gabi nang may nagtangkang butasin ang dingding namin. Simula noon, hindi na ako makatulog. Makaraan ang ilang linggo, nawalan ng trabaho ang asawa ko. Kailangan kong gumawa ng paraan upang matustusan ang pangangailangan ng pamilya ko. Namasukan ako ng katulong sa pinanggalingan naming lugar.

[When we moved here, there was no electricity, there was no water, and the area was full of wild grass. In this area, we rebuilt the house that we demolished earlier. My husband stays in his workplace, and he only goes home once a week. So, I only live with my child. [. . .] We were afraid because one night, someone tried to make a hole in our walls. After that night, I was unable to sleep. After a few weeks, my husband lost his job. I had to find a way to satisfy my family's needs. I had to work as a house helper in the place where we came from. (our translation)]

According to Jennifer Macabudbud, a fire victim:

Ang trabaho ng asawa ko ay maglalako ng mga istante na ginagamit sa tindahan. Kapag walang bumibili, wala ring kita. 'Pag walang kita, hindi siya makakauwi kaya madalas nangungutang ako sa tindahan. Kapag hindi kami makabayad, pagagalitan kami ng may-ari. Kapag wala na talagang masaing, manghihingi ako ng kaunting kanin na masasaing sa kapitbahay. Kahit hindi na ako kumain, importante na makakain ang mga bata. Kapag wala namang baon ang mga bata, hindi ko na lang sila pinapapasok. Minsan umabot na sa puntong gusto nang sumuko ng asawa ko dahil sa kalagayan namin dito. Ang lahat ng karanasang ito ay sobrang naging hamon sa mga pamilya, lalo na sa usaping hanapbuhay. Umasa na lang kami sa paggabay ng Panginoon kung paano mairaraos ang buhay sa araw-araw.

[My husband sells shelves used in stores as his job. When no one buys them, we won't earn money. Without money, he could not go home, so I would often owe money from the store. If we couldn't pay, the storeowner would get mad at us. If we couldn't cook rice, we would as for rice to cook from our neighbors. It would be okay for me not to eat; it's important for the children to eat. If the children didn't have their allowance, I would not let them go to school. At one point, my husband wanted to give up because of the living conditions there. Everything we have experienced here was challenging for families, especially in terms of finding a livelihood. We instead put our hopes in the Lord's guidance for how we would survive every day.]

The effects of the relocation did not only cover livelihood. Like the stories of Rowena, Jocelyn, and Jennifer, the effects of the relocation branched to further make the lives of the families, especially children, difficult. This also deepened issues regarding security and caused the loss of dignity among people. The prevalence of illnesses also became an issue. Residents suspected that this concern was caused by the way their houses were adjoined and the hot temperatures in these houses, which make them feel like they were baking ovens. As such, residents often experience illnesses such as asthma, fever, and hypertension, and with older residents being the most affected. Respiratory illnesses are often the cause of the deaths of the residents. Health centers and hospitals are far for their immediate medical needs. And the need for some mothers or fathers to stay in Manila further contributed to the destruction of families. Some families lived separately, while others had engaged in other relationships regardless of sex.

Some families were unable to bear the brunt of living in Towerville under dire conditions, and they decided to sell the houses that were granted to them to go back to Manila and rent a house there. Many families did not have a choice but to stay in Towerville and hope that the government will notice and respond to what is lacking there. Undeniably, Towerville is surrounded by a beautiful environment—with scenic mountains and green plants. But residents feel the loss of hope more. The residents could not help but think about why Towerville seemed to be left behind while other areas are being developed and technology continues to evolve. Perhaps this relates to issues regarding weak and ineffective governance and politics.

Since the beginning, Towerville is a government project—the government initiates and implements it. Moreover, regarding governance, saying that that the government fulfilled its role in implementing this project is a grandiose lie. From the processes of transfer and housing distribution, the government has had many flaws. For example, some relocated families were granted free housing by the government, while others only received land. These things happened despite the government's promise that the relocatees would be granted housing. Financial assistance for those relocated to Towerville was distributed unequally. According to the residents of Towerville, the amount of aid would depend on the city or municipality of origin and the reason for relocation. For example, people who came from wealthier cities receive a higher amount of financial assistance. People who had to relocate because of infrastructure projects (such as road widening projects) often receive more money than those who were evicted for being informal settlers even when they were victims of fires and floods.

Like what was mentioned earlier, the way people received basic services was not at its best. Rowena Osal experienced having to connect illegally to gain electricity. After the residents talked to Meralco in 2006, the residents were then provided legal electric services. Thus, the people themselves found a way to satisfy this basic service. The residents also shared a similar experience regarding other basic services such as water.

It is possible to say that the residents only feel the presence of politicians during elections. Politicians actively promise many things like development, work, businesses, infrastructure, and more. During those times, politicians in Manila remember the people who were thrown away to relocation sites. Others bring vehicles to fetch people. They are being fed, given a small amount of money, and will be promised many things. The residents of Towerville are only important to them during elections. Afterward, these politicians seem to become blind, deaf, and mute with regard to the demands of common people.

To be fair, the government provided assistance such as the 4Ps (Pantawid Pamilyang Pilipino Program (Assistance for Filipino Families Program) and other livelihood training programs such as tocino-making, candle-making, soap-making, and other "makings." But these were not enough. For example, in the case of

4Ps, people did not do much because they were merely waiting for their benefits. Regarding training programs, there was not enough assistance such as starting capital, production equipment, and other needs for starting a business. Thus, they were not fully taken advantage of. For example, Rowena Osal attended a tocino-making program by the city government. However, nothing was given after the session for attendees to start a business. In Rowena's case, it was difficult, if not impossible, to start a business because they did not have electricity and a refrigerator, she did not have enough capital to purchase the required equipment, and she did not know a supplier of ingredients. Abby had a different encounter. In her experience, people who have connections with the local government receive proper training and assistance in finding jobs. She expressed, "*ika nga nila, kapag malapit sa kaldero, madaling maulingan*" ["As the saying goes, if it is near a cauldron, it easily catches black soot,"] our translation).

While some services come from time to time, they are still not enough, or from what Abby experienced, only a few receive them. As such, poverty continues to affect Towerville despite the promises of the government and of politicians. In essence, if the relevant authorities only considered the situation and the demands of relocatees regarding their livelihood, the situation in Towerville would have been different. If the government had an appropriate response, people would have been more productive as they sustain a better life for their families and serve as models for other residents and communities.

At this point, the discussion shall proceed with Igting's experiences with organizing. It is important to remember that the context of poverty, inadequate services, and the loss of hope is the prevailing situation that has affected the initiatives of organizing the members of Igting since 2010.

20.2 Initial Organizing Work and Awareness of the Situation

The history of Igting's organizing work started in 2010 when the Center for Asian Mission for the Poor–Asia (CAMP) entered Towerville. CAMP started as a faith-based organization that proclaimed the Gospel of the Christian faith. Though CAMP started in Tondo in Manila, its exposure in Towerville pushed the organization's founder, Reverend Chulyong Lee, to focus on the area to help the poor relocatees. CAMP began its work with a survey. People were asked about the suitable forms of livelihood for mothers in Towerville. According to the survey's results, tailoring was seen as the most suitable form of livelihood. Through the survey, 40 individuals expressed their interest to join CAMP's livelihood project. To initiate the project, the first group needed officers, and when its leaders were selected, they decided on the eventual name of the organization, and the group was named the United Towerville People's Organization. The project began on July 15, 2011, with a training session in tailoring. The group had to study for four months. CAMP hired a trainer from the

Technical Education and Skills Development Authority (TESDA) to teach mothers in Towerville the skills in tailoring. In the first month of studies, all in the group were surprised to find that each of them would receive some grocery items as aid. This gave enthusiasm for the group members to continue their training. Although as the training continued, some were eager to learn tailoring, yet some would only focus on receiving aid. The training program ended in November 2011, and in December 2011, a graduation program was conducted, and each trainee sewed dresses that were showcased through a fashion show. On that same day, TESDA gave each trainee their National Certificate II (NC II).

After graduation, everyone began asking questions. "*Ano na ngayon ang mangyayari sa atin ngayong tapos na ang training natin?*" ("What would happen to us now that training has ended?") Daisy, one of the graduates of the training program, asked. Ate Weng responded, "*Tiwala lang, siguro naman may mabuting kahihinatnan ang training na ito. Ito nga may NC II certificate na nga tayo, ngayon pa ba tayo panghihinaan ng loob?*" ("Have trust. Perhaps this training will result in better things. We already have our NC II certificates, now, will we still lose hope?") Ate Terry responded, "*Oo nga naman*" ("Yes indeed."). Ate Weng added, "*Siguro naman ang training na ito ay hindi matutulad sa mga unang training sa atin na nasalihan, tulad ng tocino-making, soap-making, longganisa-making, at iba pa. Paano tayo lalago kung wala namang sapat na ayuda?*" ("Perhaps this training will not end up being like the previous training we have joined, such as the tocino-making, soap-making, longganisa-making, etc. How will we grow if we do not have enough aid?")

The following week, the group and CAMP held a meeting. And in this meeting, CAMP said that the group will run a business. "*Anong negosyo po?*" ("What business?") an officer asked. "*E di siyempre tahian, kaya nga kayo dinaan sa masusing pag-training upang matutong manahi*" ("Of course, tailoring. This is the reason you underwent rigorous training so you learn how to sew"), CAMP said. When the group heard these words, their eyes had a spark of hope—a hope for Towerville residents to have a livelihood. They first assigned a production supervisor, a cutter, and a purchaser in the sewing center in Towerville. During the first two months of production, the group did not keep a wage, and because of this, some of their members lost the determination to continue. Of the 40 individuals who were accepted in training, only 20 were eager to stay because some only wanted the grocery items for training. They did not persevere to keep working. Others certainly wanted to have a business. And because only 20 persons remained in the group, it had to elect a new set of officers, and they decided to change the name of the organization to Ma-Igting na Samahan ng mga Panlipunang Negosyante ng Towerville, Inc. (United Association of Community Entrepreneurs of Towerville, Inc.)

During its second month, the honorarium for members was arranged. The officers, supervisors, and CAMP decided that each member would receive PHP 150 every day for eight hours of sewing. As production continued, they did not become complacent. One time, a client was recommended to the sewing center and ordered customized T-shirts. Therefore, the sewing center needed a label to be placed at the back of the T-shirt. Because the group partnered with a Korean non-government organization (NGO), and at that time, the group could not decide for themselves, Igting relied heavily on the NGO. An event would also be held in Korea around that time; therefore, the label had to be made there. CAMP, along with some members of the association, went there, and they discussed the name in the label of Ma-Igting na Samahan ng mga Panlipunang Negosyante ng Towerville, Inc., and finally decided a shorter name IGTING. In addition, the association was to be registered with the Securities and Exchange Commission (SEC).

While the SEC registration and the naming of the association were successful, the association was still trying to find its way into managing the business. Four months of training were not enough to gain expertise in tailoring and running a business. Starting the enterprise was rough, and unavoidable crises happened every day, and it seemed that the members of Igting were only going along with the flow of water. In addition, maintaining the unity of the members turned out to be difficult. Perhaps this was because of the diverse behaviors and reasons of people who now had the time to converge in one area and belong to the association only. However, the desire to succeed in this business was a paramount desire among members.

Aside from CAMP, one of Igting's partners was the College of Social Work and Development (CSWCD) of the University of the Philippines (UP). Through its Field Instruction Program (FIP), the college sent students to immerse and collaborate in Igting's development. One of the aims of the FIP is cultivating knowledge and skills regarding the association and social enterprise. This process of enrichment encompasses many training programs and participative activities.

At the start of the collaboration with FIP, Igting members had a hard time understanding words or terms related to forming associations. Perhaps it was caused by the members' lack of experience of membership in an association. In response to this challenge, meetings on this topic were frequently held. These meetings started with simple chats about their families, their challenges in life, their experiences as members of Igting, and their plans for developing the association. However, not everyone has understood the process that the FIP has proposed. Terry once asked, "*Bakit may mga estudyante . . . ano ang pakay nila dito?*" ("Why are there students . . . what do they want here?") Daisy answered, "*Baka mga espiya!*" ("They might be spies!") Olive said, "*Bahala na kung ano ang pakay nila . . . mahalaga may trabaho [tayo].*" ("Their aims would not matter. The important thing is we have a work.")

To alleviate these points of concern and to deepen the understanding of FIP's aims, students sought to live communally with the members of Igting. This was done by having students dwell in the members' houses to further become acquainted with the mothers of Igting and their families. The students went to almost all members. The mothers let students do domestic chores such as cooking, taking care of children, and helping with their assignments. This immersion aimed to better understand the situation of the members of Igting, whether in its personal or organizational aspects. This is not only to foster strong relationships with the members of Igting but also to ensure that the participative process that they propose is appropriate and relevant.

Along with immersing in the lives of the members of Igting are frequent meetings regarding organizational matters (such as the way other members behave), family matters, and challenges that they face in life. There were also orientations and workshops on creating and managing an association and a business. The week would not end without a meeting where they continue talking about their roles and responsibilities. From these organizations, Igting was gradually shaped as an organization.

Perhaps one of the key contributions of the FIP in Igting is the creation of committees. At first, the mothers thought that it was just a game. This is because the process that members underwent in forming committees involved creative workshops. In one meeting, students drew circles that symbolize committees. After the students explained the roles of the committee, the students asked the members to jump into the circle where they wanted to stay. Because the mothers thought that it was just a game, some of them jumped into circles without knowing that they were already becoming part of the committee. At that point, it was possible to say that members did not completely understand what they were doing. This fault manifested in Igting's problems with supplies of materials. Because members did not fully understand committees and their roles, such as marketing, the organization had significant problems with finding material supplies and clients. In the end, this had a significant effect on their income as an association. Because the association did not earn as much, many people lost courage. However, CAMP provided aid to each member to receive a daily pay of PHP 150.

CAMP's involvement was not limited to providing assistance for the wages of the members of Igting. Overall, CAMP oversaw the management of the sewing center. Because of this, even when members had tasks, they were afraid of deciding for themselves or for the organization. Therefore, the members of Igting yielded to CAMP because it holds the finances, and they treat themselves as newcomers in business. At that time, the collective leadership of the organization was considered weak, and many members were not active in the organization. CAMP's financial assistance also contributed to the dependence of some members and complacency regarding the sewing center's operation because they would still earn whatever may happen.

Looking back at the past, this system was a big mistake because it led members of Igting to be motivated in the wrong ways. The motivation of some members turned out to be income and not the development of the organization and the business. As such, the group was in chaos, and a day without fighting rarely happened. Rowena, who serves as the association's production supervisor, was a witness to this issue. Fights were often about the quality of products. As production supervisor, Rowena critiques the members' works. Oftentimes, some members would not wholeheartedly accept these critiques. Therefore, Rowena faced difficulties that led her to the point of wanting to give up. Thus, for some, the observation that the success of businesses run by community organizations was rare. However, other members only needed to strengthen themselves to make their livelihoods flourish.

This was Igting's situation from 2012 to 2013: a shallow understanding of the association's aims, the unequal accomplishment of tasks, the lack of participation among members (e.g. working on their own), and frequent misunderstandings between members. Despite these issues, some members did not give up. Alongside this, training and workshops continued to develop not only the business but also the organization. In the succeeding processes, the vision, mission, and goals of Igting were established. Members saw these as things that could help in fostering and strengthening relationships among each other by setting forth their aims that they all decided on.

20.3 The Development of Skills and Awareness

Certainly, Igting has many elements for establishing a developed and vibrant social enterprise. However, it still has significant flaws. One of the observations of the organization is the need for rigorous capacity building not only for leaders but also for the members of Igting. Capacity building aims not only to hone individual skills but also to encourage the association to assert its decisions. In the end, Igting wants to stand on their own feet and become responsible for the individual roles of members without disregarding the organization's aims.

Igting's capacity building activities focused on two aspects: developing a business and an organization. Regarding business, there was training in social enterprise, business management, creating excellent business models, managing human resources, marketing and product development, financial management, and quality control. There were also special topics such as waste management. These training sessions were conducted through collaborations with many institutions such as the Department of Trade and Industry (DTI), UP Institute of Small-Scale Industry (ISSI), and other experts in entrepreneurship. These partnerships with institutions and individuals were inaugurated by CAMP. Aside from the aforementioned training sessions concerning business and production, the Training Committee of Igting also conducted training sessions on basic skills in tailoring such as cutting, advanced pattern making, weaving, and basic and advanced machine maintenance.

Regarding organizational development, FIP became Igting's major partner. Through FIP, Igting conducted many training sessions and workshops about values formation, community organizing, community development, social enterprise, gender sensitivity, violence against women and children, and team building. Igting, through CAMP, coordinated with the local government of San Jose del Monte in conducting an orientation on cooperatives. Alongside this, field trips to many social enterprises and cooperatives such as NOVADECI and Pinagbuklod Multi-Purpose Cooperative. Through these field trips, there was sharing of experiences with the challenges of the organization and potential solutions to them. There were also field trips to many sewing centers for members to learn effective production systems and processes, especially in terms of quality control, fabric cutting, and fabric waste management.

These partnerships with institutions and individuals were not only for capacity building. It was also conducted for networking. They were done to widen the organization's access to opportunities that would help Igting in its production, organization, and products. For example, partnering with DTI in 2017 served as a way for the organization to participate in the said agency's One Town, One Product (OTOP) Program. In OTOP, DTI designated designers to help Igting design their own products. Igting also participated in Kapatid Mentor ME (KMME), which is also a program of DTI), in 2018. This program provides free business training seminars for small enterprises.

These aforementioned skills and activities greatly helped every member of Igting. But change did not come fast. Perhaps this is because of the diversity of trainers and speakers. Some topics were complex at first, thus requiring continuous training and orientation to facilitate a better understanding of ideas such as what social enterprises are. The absence of some members in these activities resulted in misunderstandings. These activities help illuminate and understand the principles in forming an organization. Collaborations with institutions and individuals helped in broadening access to other forms of training. Internal training sessions also helped not only with organizational development but also in members' self-confidence and relationships with each other.

Overall, all lessons that members learned became the foundation and basis of Igting in improving its system and processes. For Igting, the most important lesson that members learned is making themselves open to understanding social issues such as the lack of basic services and livelihood for the poor in the community. The members also became aware of the importance of human rights and of being independent and standing on their own. By deepening their understanding of these issues, members have learned to respect others and cooperate for change. They also learned that they can be involved in addressing issues that affect them to correct mistakes and to become humane. Finally, members better understood the true relevance of an organization that runs a community enterprise that is concerned with the community's development.

20.4 Rising and Facing Challenges

In 2014, Igting faced its most significant challenge. In a meeting, CAMP broke the news that the funding that supports Igting has come to an end. Despite this bad news, CAMP promised that it will continue supporting the organization, but in different aspects. In other words, while CAMP will continue guiding Igting, there was no money to fund it. The association now had to stand on its own feet, and Igting is now responsible for many of its decisions, especially in finding ways to continue running the business. But before financial support ended, CAMP provided Igting an initial fund of PHP 450,000. In addition, CAMP also let Igting use the sewing center.

This event was not easy to accept. For some members, "*Ayon sa ilang mga miyembro, "Kapag nagkakaron kami ng pagpupulong kasama ang CAMP, ang sabi ay 'we are all family' . . . may tatay, nanay, anak, apo, at iba pa . . . saan banda ang Igting sa sinasabi nitong family?*" ("Whenever we have meetings with CAMP, they say that 'we are all family' . . . there is a father, a mother, a child, a grandchild, and others . . . where is Igting in what they consider family?") Therefore, members felt failure. Perhaps this was how people understood that CAMP and Igting are partners in projects. Others believed that Igting did not fail CAMP, especially when it "designated" projects. There was also an issue concerning the sewing center. The sewing center was founded using the funds of donors and was constructed inside CAMP's compound in Towerville. While the sewing center was built for the mothers of Igting, the issue of ownership became unclear because it was situated on land owned by CAMP. However, at this point in Igting's existence, this was not the most pressing issue. Instead, it was the loss of funds to support Igting. Yet in the end, this issue would emerge again.

One of the primary concerns that emerged in this situation is the wages of members. Therefore, this refers to how the initial funds shall be used. However, these meetings resulted in arguments because members have not fully understood what a suitable system of salaries is. Eventually, the organization decided that each member would receive a daily wage of PHP 180 provided that they reach at least 80% of the quota. However, the initial funds only lasted for three months. According to Myrna, the decision to maintain a daily wage and provide salaries for reaching 80% of targets was not well thought of. Thus, the enterprise went into a critical situation; it was in danger at that point. This situation resulted in a series of meetings where they discussed what would happen to Igting. One of the biggest questions were: Do we let our livelihood be gone? Many were saddened by this situation. Others cried. While the situation was critical, the members gained strength and decided to change Igting's wage system. From a daily rate, they adopted a per-piece rate.

Because of this decision, many members left the organization. Here was their reason, "If I would only earn PHP 100 every day, it would be better for me to stay at home." According to Racel, some members left because they thought that production would stop because there was no support from CAMP. Glo told this story, "*Kaunti na lamang ang pumapasok, nangangapa ang lahat kung paano kami makakarami sa paggawa.*" ("Only a few went to work; everyone was trying to catch up in producing a lot.") Overall, many members became skeptical, and asked—would Igting last long? But because of this challenge, Igting—or its remaining members—realized the truth. They had to do everything to survive and help the organization rise up; they needed strategies and plans. Members saw that the need to emphasize the importance of the organization and of everyone's participation.

From 2014 to 2015, Igting conducted a series of empowerment workshops. They were inaugurated by the Training Committee of Igting with FIP. In two months, many empowerment workshops were done. However, concerns regarding the lack of participation among some members remained. They would say, "You should attend these workshops; we'll work here instead." But many were not swayed by members who did not want to participate. Workshops were scheduled every Monday from 8 a.m. to noon. Before workshops began, participants had to introduce themselves. The techniques used in the workshop were considered creative and were considered games. But for Igting, the lessons learned from these activities were deep, especially their relevance for understanding the status of the organization. After these games, members would talk about how these activities relate to their roles in Igting. These lessons were written in metacards, which include the causes of problems, the solutions to them, and the lessons learned from each activity. Every workshop ends with a communal lunch.

After two months, the group had an evaluation. The members assessed the impact of the workshops. Overall, they significantly helped especially in making everyone's understanding open. One of the most prominent lessons is the importance of unity. According to Racel, "*Kailangan ng mas malalim na pagkakaisa . . . sapagkat napakalaking tulong ng hanapbuhay na ito sa pamilya.*" ("It is important to have deeper unity . . . because this business helps the family.") Others noticed developments within themselves. Tess said, "*Natuto akong makipag-usap sa ibang tao nang maayos.*" ("I learned how to communicate well with other people.") Gloria Gerardo saw herself change. She said, "*Dati mapagmataas ako, ngunit natuto namang magbago. Hindi mabubuo ang samahan kung may maninira. Kailangan pangalagaan ang trabaho kahit kaunti lang ang sahod.*" ("I was full of pride, but I learned to change. The organization wouldn't be firm if someone destroys it. It is important to treasure work despite receiving a low wage." Others like Nelia considered the importance of learning from of experiences of others and of herself. She said, "*Huwag gayahin ang maling gawain ng iba. Patuloy dapat ang paalala ng layunin ng Igting sa mga miyembro.*" ("Do not imitate others' wrongful acts. Always remind every member of the aims of Igting.") Others emphasized sacrifice and dedication. Gloria Artigo said, "*Dapat maging masipag, matiyaga, may dedikasyon sa trabaho, makisama sa iba, at habaan ang pasensya. Kailangan ayusin nang sama-sama ang Igting. 'Pag may problema, kailangang bigyan ng solusyon.*" ("Everyone needs to be industrious, to persevere, to show dedication in working, to get along with others, and to lengthen their patience." Rose Gabriel added, saying, "*[Kailangang] matuto tayo magsakripisyo. Dahil sa Igting, naalagaan ko pa rin ang aking mga anak at malapit pa rin sa pamilya.*" ("We have to learn how to sacrifice. Because of Igting, I was able to take care of my children and stay close with my family.") With sacrifice came the importance of expanding their understanding. Myrna said, "*Kailangan ng pag-unawa sa mga miyembro, pagsasaayos kaagad ng conflict, pakikiisa at pakikisalamuha, pagkakaroon ng special meeting kung saan nalalarawan ng lahat ang mga updates, at pagkakaroon ng team building.*" ("It is important to better understand the members, to immediately resolve conflicts, extend solidarity and socialize with others, conduct special meetings in which everyone can describe the updates, and conduct team building.") Eventually, everyone agreed that Igting would not succeed if its members did not act collectively. Abby said, "*Sama-sama dapat ang pagkilos, sama-samang aangat hindi lang ang sarili, dapat lahat kasama sa Towerville ayon sa VMG ng Igting. Kailangan hatakin ang lahat pataas.*" ("We should work and rise up together. Not only should we uplift ourselves but everyone in Towerville according to Igting's VMG [vision, mission, and goals]. We have to lift everyone up.")

Along with workshops comes paying attention to the organization's structure. The officers, the board of directors, and the business development group held an extensive meeting. They talked in detail about their roles, especially in the way officers participate in processes pertaining to entrepreneurship. Before, officers seemed to be display figures. To help the organization, officers sought to strengthen their roles in entrepreneurship.

In improving organizational structure, Igting paid attention to the committees and clarified their roles. For instance, in the new system, the membership committee attends to the implementation of policies in the sewing center. These include monitoring absences and mediating in times of misunderstanding. The community affairs committee attends to actions that cater to the community such as gift-giving activities and ensuring that visitors (for example, Koreans) were understood well). The training committee researches various opportunities that are relevant to developing technical skills such as entrepreneurship training or business development. The marketing committee continues to persevere in networking and finding clients. Meanwhile, the product development committee attends to the development of the association's own products that can be sold outside.

The overall system in the sewing center was revamped. An inventory recorder was designated to effectively monitor the exit and entry of products such as materials through subcontracting. This was important to ensure that the number of materials purchased is correct and to avoid shortages. Stockroom personnel were also assigned to monitor the exit and entry of Igting's own products such as uniforms, T-shirts, bags, and others. This also included monitoring materials such as cloth, needles, threads, buttons, garters, and others. The organization's strengthened monitoring was done not only to make the system work but also to avoid excessive expenses because of wasting materials, especially fabric.

Regarding tailoring, industry and perseverance were emphasized in production. This was done with the help of line leaders and production supervisors. In revamping the system, Igting ensured that the process of production will generate high-quality products at the right time. It ensured that workers would not need to ask for work. The rigorous monitoring of line leaders and production supervisors was required to ensure that work was continuous and of high quality.

Through all these changes, the sewing center's system of production was better and more effective. To maintain the organization's income, it set a target of PHP 100,000. The marketing committee worked to ensure that orders and clients continue to come at the right price. Meanwhile, those in production had the task to reach these targets. Regarding finances, the coordination of the auditor, the treasurer, and the finance administrator of Igting was strengthened. Before 2013, only one person belonged to the finance committee. And through the continuing coordination of the organization's committees and officers, the operations of the enterprise were improved.

The efforts of the members of Igting gradually generated results, and others noticed them. For instance, the City Cooperative Development Office (CCDO) noticed Igting's work in 2018. Therefore, Igting had the chance to present their work at Starmall Pasalubong Center. Afterward, the products of Igting were showcased in various consignment stores such as DTI Go Lokal Store in Makati, NAIA Terminal 3, and Islas Pinas in Double Dragon Plaza. It also resulted in continued invitations to training sessions on business management, financial management, and basic costing and accounting for non-accountants from the DTI.

At present, Igting's situation was better compared to when it started. Members were more confident in their tasks, unity, and most importantly, decision-making. Officers strengthened their sense of responsibility. With the membership committee, Igting paid attention to instilling discipline among members. Concerning entrepreneurship, the production team continues to hold accountability in the business, with the help of the marketing and finance committees. Training sessions also helped deepen the knowledge of members.

However, Igting still does not say that everything is perfect. Every day, the organization continues to face challenges. But they would not fall from these challenges. We can say most of Igting's members are simply mothers, and they are not professionals in terms of running a business, the continuing experience and perseverance further molds each one of them. Now, Igting aims to help citizens by providing livelihood opportunities. Therefore, it launched what they call a "home-based sewing project," still with the help of CAMP. It seeks to provide stay-at-home mothers access to livelihood. Igting also plans to continue its growth by forming a cooperative. This way, each member will generate income every year. The members of Igting took the initiative to establish a cooperative seriously because it was uncertain if CAMP would eventually give the sewing center to Igting.

20.5 The Organization's Strengths and Its Social Enterprise

At this point, Igting's story of organizing and entrepreneurship was told. What lessons can be learned and can help organizations and social enterprises in a similar situation? For the members of Igting, significant lessons can be learned from three aspects. These relate to interacting with people, establishing the importance of a strong organization, and running a business.

Interacting with people—Igting did not undergo an easy organizing process. Perhaps this was linked to the nature of the community that was formed in Towerville, where residents came from various areas with different behaviors and beliefs. In addition, poverty in Towerville caused the division of families and of the community. As such, it was difficult to foster unity among people in the community, especially in participation because of skepticism and the need to find ways to survive. This led others to depend on receiving immediate aid or benefits from others. This manifested in Igting's experiences. For instance, in dressmaking training, many participated at first, but they did not continue because they only desired aid in exchange for training. This behavior was detrimental to running a business because it causes a lack of perseverance, a need to lengthen patience in honing skills, and finally, giving up. Therefore, others wanted gains more easily. This did not only manifest in Igting's experience; this was also shown in joining the activities of politicians where they are only used for their campaigns. Certainly, they would be paid a huge amount, but it would be temporary and would not last long.

Aside from the various characteristics and behaviors of people, the lack of experience in joining an association was also a challenge. In Igting's experience, some members would be highly sensitive, hard to talk to, have a high sense of pride, or lack discipline. Perhaps this is because they have not understood the situation and have not been aware of the organization's system. The lack of experience in forming and managing an organization because the source of misunderstanding within the group. In addition, some members create issues that undermine order within the organization. Some members would not accept critiques of their mistakes. Their lack of awareness or their narrow-mindedness causes the proliferation of negative behaviors. These become obstacles to organizational development because it affects the flow and quality of production. Finally, apt behaviors in the organization and in running a business are important—everyone has to act professionally at all times.

Despite innate difficulties in forming an organization and in running a business, this great experience led people to act. Therefore, despite having a hard life and despite some surrendering, many continued to struggle and find solutions. But this was made possible not only through Igting's efforts but also with the help of other people and organizations such as CAMP and FIP. Yet the efforts and the perseverance of the members of Igting were still more decisive in leading the way from poverty to a better life.

This effort manifested in the active participation of members in workshops in which they became more aware of the situation. Through the discussions with FIP, members gradually understood terms that became topics for reflection such as a better life, poverty in Towerville, forming an organization, and social enterprise. These terms were difficult to understand. However, through years of continuous studies, they gradually understood how every person has a significant role to change the way they live. The truth that there is something to do became the first stop for the members to learn to unite with an organization that desires change. Rowena said, "Sa sarili ko . . . bilang isang miyembro . . . iminulat ko ang sarili ko na hindi porke't mahirap ako at walang pinag-aralan wala na akong magagawa." ("For me, as a member, I made myself aware that just because I am poor and lack education does not mean that I could not do anything.") This effort and awareness resulted in the unification and joining of talents and skills of members with one dream, mission, and goal to improve their lives and alleviate poverty. This was needed by people who can lead, act, and stand.

Forming an organization—Like what was said earlier, Igting instilled the importance of establishing an organization. However, when it was in its early years, Igting admittedly did not give it enough attention because members were more focused on running a business. The lack of a deep understanding of the importance of an organization resulted in different issues. Some members did not want to participate because they thought they would not gain anything from the organization's activities. This resulted in significant problems in the decision-making process. The lack of interest among others to join the association's activities such as meetings led to a disproportionate understanding of its activities and business. Eventually, it resulted in conflicts and issues. One of these involves the process of assigning tasks. For example, other members only understood that the concept of equality only relates to ensuring that they receive wages equally. They failed to understand that the word *equality* refers to the democratic processes within the organization, participation, and deciding on the organization's development.

However, many still understand the aims of the organization, especially leaders. They would continue guiding and encouraging members to show unity in the organization's activities. Leaders also took the initiative in ensuring that members treat the organization's business as something important. Through the continued sharing of learnings from training and acquired knowledge from workshops and seminars, members would begin to realize the importance of Igting as an organization. Through Igting's partners such as CAMP and FIP, empowerment workshops were conducted for members, and these workshops helped in broadening their knowledge until the chance for them to participate in such activities came. Alongside this, to strengthen the organization further, Igting created a more effective and democratic structure where all members play roles in the organization and are not dominated by only a few people. Through these established committees that respond to different tasks in the organization, collective governance in Igting was strengthened further.

Running a business. Similar to interacting with people and forming an organization, running a business was difficult. For Igting, this difficulty stems from the members' negative mindset. For example, when Igting was still helped, members were dependent and became complacent, thinking that even without output, they would still receive wages. Thus, some members did not put efforts into improving their skills. There was also the lack of initiative to strengthen and enrich themselves and the business. Therefore, ensuring the optimal quality of products became a huge problem. The low quality of products affected the business's income. Thus, when external funding ended, many left because of the fear that they would not earn and that Igting might fall.

However, it was only the beginning for Igting to struggle in the field of entrepreneurship. Tight business competition made the procurement of products difficult. For example, in making T-shirts, it was difficult to negotiate prices with clients because they quote low prices. Many other sewing centers would offer wholesale and low prices for their products. It was also difficult to gain a market for its own products because the marketing committee lacked knowledge in strategizing. Because Igting was not yet known in the market as a maker of bags and clothes, it was diffcicult to compete with known and big companies.

Despite this difficulty, Igting's partnership with different agencies and organizations such as the DIT and the local government of San Jose del Monte helped with the business's operations. Through these partnerships, Igting gained opportunities to further enhance and improve its products. In addition, through various training sessions, leaders developed a broad knowledge of entrepreneurship, and the quality of Igting's products became better. There were also field trips to other cooperatives and associations that are similar to Igting, and from there, the organization got ideas for them to improve processes such as the correct way to cut cloth, the correct way to distribute tasks, and others. Also, because of conversations with other organizations, Igting gained ideas concerning organizational management.

These opportunities, exposure, and training experiences also helped improve the processes of production, marketing, and finance. The association also put efforts into strengthening its communication and coordination with the business development group to make the flow of production better and ensure that it generates the right output at the right target date. All the training, system improvements, sacrifices, perseverance, and efforts of each member in developing the organization and its business did not lose relevance because they caused changes in every member to trust each other's abilities. Therefore, a business that an organization keeps or operates should dedicate more time to strengthening good and amicable relationships among each member because the business and the organization would not develop without a strong foundation in the strength and the power of people.

Through the continuous efforts of Igting in organizational development, Igting has also taken the lead in providing projects and services for the community. Based on Igting's experience, for a project for the community to be effective, guidance at the beginning is important, especially in terms of running a business or working. Support in training and production equipment such as materials and facilities are important. We all know that in a community, people come from different states of life: some were able to study, some were unable to; some have jobs; some have none; those who are jobless face the worst. However, there are people who, despite poverty, do not lose hope in their lives. At one point, it came into the life of Igting, especially when its members were newcomers to Towerville. Thus, CAMP's assistance, with the UP CSWCD, is certainly laudable; they did not leave Igting when it was still unable to stand on its own. Before, the concept of social enterprise was not yet fully understood, but because of the unceasing guidance of people to the members of Igting, little by little, they have begun to understand the importance of an organization that runs a business. The most important things are having one dream, contributing skills and knowledge, conviction, unity, compassion, understanding, and getting along with different personalities in fulfilling any goal.

In the organization Igting, members are considered to be part of the margins of society who do not have voices regarding development. One member of Igting said, "Who are we? We are ants compared to giant entrepreneurs." Igting is composed of simple mothers who faced great challenges in life, especially in terms of finding work. From this experience, mothers took with them hope and the ability to make livelihood flourish in Towerville. It never became easy; Igting was able to do so with the help of people and institutions that guided them. However, with assistance comes the importance of everyone's conviction to become active in their roles and responsibilities. It is not acceptable to only receive help; it is also important to act collectively. Without action, all this help would be wasted. It is not acceptable to become dependent only because some help them. Any opportunity that they have received should be enriched. For Igting, a social enterprise is for people who had to be made aware of their situation in life—that even when they were sent to a place where work was scarce, a sacrifice has to be done to develop their skills and knowledge. Igting proudly shows how for eight years, its members fulfilled their roles in running its business. Igting believes that poverty can be addressed if action, cooperation, and most importantly, the effort to act collectively for development come from each individual. Compared to big businesses with huge capital, in a social enterprise such as Igting, members have equal rights, all voices are heard, and policies and protocols—which also come from members and are decided on by the organization—are observed. Igting wants to tell the group's story to everyone—that they are living proof that simple people, even when they did not receive high education, can still develop as they desire if they persevere and have faith in God. Through the amalgamation of skills and knowledge of everyone and through having an aim for development, these efforts for a better life would be made easy.

Rowena E. Osal is originally from Masbate. At the age of 9, she moved to Manila. She only completed Grade 5 because her parents could not afford to send all of their children to school. She worked as a maid when she was still single. At 17, she got married and settled by the river in Marikina City. Her husband worked as a tricycle driver, and they had five children.

Living close to the river, their house was frequently damaged by floods during storms and heavy rains. This situation led them to relocate to Towerville. They moved there in 2004, and life was challenging at first, especially in terms of finding work while all their children were in school. Despite the difficulties, Rowena did not give up; she continued to hold on to her faith in God and remained hopeful.

In 2011, Rowena became a member of Igting, which marked the beginning of positive changes in her life. It gave her the opportunity to be part of the organization and she eventually took on the role of Production Supervisor for Igting. Over her 13 years as a member, she has fulfilled her dream of providing for her children's education, thanks to the livelihood the job has provided.

Jennifer A. Macabudbud is 46 years old and from Binangonan, Rizal. She grew up without parents and was adopted by her cousin, who supported her education and allowed her to finish high school. Jennifer worked in Manila for several years, got married, and has four children. The couple faced many challenges, as her husband struggled to keep a steady job and often switched employment. Their life was difficult, especially when the children fell ill. They were victims of a fire in Tatalon, Quezon City, which led them to move to Towerville. Their experience in Towerville was not easy; they endured a lot of hardship during that time. While her husband stayed in Manila, Jennifer and the children were in Towerville, where there were times when they had no food to cook or money for the children's needs. This caused Jennifer a lot of stress as she tried to support her husband.

Eventually, she learned about a training program by Camp Asia/Igting in Towerville and decided to participate. She underwent training for 3 months and, upon completing it, sought to be absorbed into Igting. This marked the beginning of her learning a new trade. As a member of Igting, she had the opportunity to be elected as an officer, becoming the Public Relations Officer (P.R.O). Currently, she is responsible for inventory and serves as a line leader for Igting. With Igting's help, Jennifer and her husband were able to work together and improve their livelihood.

Evelyn D. Galangera is originally from Capiz and completed high school. She moved to Manila to seek better opportunities, as in her province, their livelihood was primarily farming.

Her mother passed away when she was just a baby, and her father remarried, so she was raised by her grandmother. In Manila, she had the chance to enroll in a vocational course but was unable to complete it due to illness. Later, she had the opportunity to work abroad in Brunei Darussalam because her uncle was employed there. She worked there for 7 years.

After her contract ended, she returned to Romblon, where she got married and had children. Her husband was a security guard, so they moved back to Manila because his job was based there. They settled in Bagong Silang and became victims of Typhoon Ondoy, which forced them to relocate to Towerville. In Towerville, the family faced significant hardship, especially since her husband was ill, which required Belen to find work. She worked in Manila but had to stop to care for her husband.

It was during this time that she discovered Igting. She became a seamstress, which greatly helped her family. She also became the President of Igting. Even now, despite being a senior, she continues to work to support her family.

Regie Anne D. Placido is a Conservation Specialist at Haribon Foundation for the Conservation of Natural Resources Inc. She is a community organizer and an environmental advocate with over 10 years in the development and environment sector. Prior to Haribon, she has worked at CAMP Asia, Inc. as a livelihood program manager facilitating organization development and capacity strengthening of the Igting organization from 2014 to 2019.

Chapter 21
Worker-Driven Social Responsibility as an Alternative: Interrogating Export Zone Workers' Struggles in the Philippines

Benjamin B. Velasco

Abstract The problem of decent work among workers in global supply chains is well-documented. The debate revolves around what effective mechanisms can resolve decent work deficits. Using interviews of unionists and organizers, desk review, participant observation, and case studies of struggles of workers in several export zones in the Philippines before and during the pandemic, the study sought to explore the state of rights at work, to understand the dynamics of workers' responses, and interrogate these as resistance to the dominant global supply chain model premised on wage suppression and labor repression. The experiences reveal the limits of self-regulation and corporate social responsibility by global brands and local suppliers, as well as State regulation and reforms to ensure labor compliance. Nonetheless, the struggles illustrate the potential of worker-driven social responsibility as an alternative—of unions exercising leverage to compel brands to forge binding agreements. The study concludes with a set of recommendations to enable labor rights and human development for export zone workers in the Philippines.

Keywords Export zone workers · Corporate social responsibility · Worker-driven social responsibility · COVID-19 pandemic · Global supply chains · Decent work

Among the various characteristics of neoliberal globalization has been the transformation of transnational corporations from a very vertical to a more horizontal organization with high-value-added production in the core industrial countries and low-value-added assembly in the peripheral Global South. As such, a concentration of global supply chains of transnational corporations is located in Southeast Asia to take advantage of cheap, docile, and female labor. The Philippines is no exception

B. B. Velasco (✉)
School of Labor and Industrial Relations, University of the Philippines, Diliman, Quezon City, Philippines
e-mail: bbvelasco@up.edu.ph

J. M. C. Sy and A. P. Aban (eds.), *Reimagining Development in Southeast Asia*, Asia in Transition 21, https://doi.org/10.1007/978-981-97-9140-8_21

as it hosts special processing zones with electronics and apparel as among its top exports.

The problem of decent work among export zone workers is well-documented. Global brands and their regional suppliers are able to leverage competition among Global South nations for foreign investments in a race to bottom in wages and working conditions. Workers respond to the decent work deficits through protests, strikes, unionization, and collective bargaining which, however, is suppressed by the various actors in the global supply chain ecosystem—brands, suppliers, and governments.

Using key informant interviews of unionists and organizers, desk review, participant observation and case studies of the struggles of workers from different factories in several export zones in the Philippines before and during the pandemic, the study sought to explore the state of rights at work, to understand the dynamics of workers' responses, and interrogate these as resistance to the dominant global supply chain model premised on wage suppression and labor repression. The study reflects on the applicability of worker-driven social responsibility as an alternative model and an effective mechanism to resolve decent work deficits.

21.1 Decent Work Deficits in the Ecozones

Economic zone or ecozone is the term used in the Philippines for an export processing zone (EPZ) or special economic zone (SEZ). Similar to EPZs or SEZs elsewhere, ecozones are enclaves where foreign investors are provided incentives such as tax holidays and other fiscal subsidies for the ostensible rationale of generating employment. Further, lax enforcement of labor and environmental standards is also an inducement for the entry of locators in ecozones. The Philippine Economic Zone Authority (PEZA) regulates the operation of several publicly owned ecozones and hundreds more of privately managed SEZs. The labor department retains jurisdiction of labor relations and enforcement of labor standards within ecozones.

The literature finds decent work deficits to be a sustained concern for ecozones in the Philippines. Among these labor rights issues are wages well below the cost of living, excessive production quotas, gender-based workplace violence, non-standard forms of employment, and unsafe workplaces (Certeza et al. 2012; Edralin 2001; Ofreneo et al. 2007; Partido Manggagawa 2010; US Department of State 2010; Velasco 2021).

Numerous studies also find these challenges to decent work to be prevailing in Global South regions where assembly and production sites of global supply chains have nestled, such as Central America, South Asia, and Southeast Asia. Amidst the pandemic, global brands suddenly cut off orders from suppliers, or otherwise arbitrarily decreased them, which meant the burden of the COVID-19 lockdowns and disruptions in the supply chain were passed on to subcontractor factories located in ecozones, and down the line to home-based work. The first and lower-tiered contractors then externalized the costs to their workers through mass retrenchments, wage theft, decreased wages, and job speedup. The lessons of the pandemic show in stark relief the power asymmetries that characterize global supply chains. Brands or lead firms rake in the substantial profit generated in production while suppliers extract as much of the remaining surplus by squeezing workers, especially those in the bottom rungs of the supply chain (Anner 2020; Asia Floor Wage Alliance 2020, 2021a, b; Business and Human Rights Resource Centre 2020, 2021; Clean Clothes Campaign 2020; Kyritsis et al. 2020).

The experiences of workers in EPZs and SEZs in other countries are also reflected in the plight of ecozone workers in the Philippines before COVID-19 and during the pandemic. Narratives of ecozone workers engaged in struggles reveal the labor response to the deterioration of decent work deficits. An epidemic of workers' rights issues erupted as capitalists weaponized the economic crisis and later the pandemic lockdowns to cheapen wages, mastermind wage theft, layoff workers, move to contract work, and bust labor unions.

The cases cited here are certainly a minute sample of the total population of factories in the Philippines that are tied to global supply chains. Nonetheless, the subdued fighting mood of the working class in the country means that workplace grievances are more widespread but do not break out into open disputes as workers would rather not complain or fight. Many workers prefer to suffer in silence despite being victims of discriminatory policies by the government and misbehavior by factory management.

21.2 Stories of Disputes Before the Pandemic

The following are illustrative case studies of disputes that transpired in a major government-run ecozone located in an industrial province just outside of Metro Manila from the period 2014 to 2018.

21.2.1 Factory A

Company A is a transnational company manufacturing electronics parts. There are around 500 workers, almost all of them men. Half of the workforce are union members, and the entire union leadership is male.

Occupational safety and health concerns prodded workers to start organizing a union. The hazardous nature of machine cutting and lack of safety devices led to workers losing fingers. When management learned of unionization, it moved to cancel the union registration. Despite workers voting overwhelmingly for the union, management filed a protest and used that as an alibi to refuse to bargain. It took the threat of a strike for management to finally open negotiations for a collective bargaining agreement (CBA).

Over the course of more than seven months of negotiations, Company A management did not bargain in good faith and instead harassed the union by firing active unionists and suspending top union officers. It took a two-day strike in 2015 to break the deadlock and conclude a CBA.

Supply chain organizing was key in the victorious strike. The biggest customer of Company A is a subsidiary of a European company whose workers are progressive unionists. The union organizers in the Philippines engaged the union leaders of the European company to exert leverage upon Company A to settle the dispute in favor of the ecozone workers.

This was a major victory considering the hostility of companies inside the ecozone against unionism and strikes. There are very few workers with a CBA in the ecozone, and the union at Company A was arguably the first to hold a successful strike in the ecozone. The victory of the Company A union inspired workers from other factories to organize unions.

21.2.2 Factory B

Company B was a factory owned by a regional supplier based in South Korea. It was the largest garments factory in the ecozone before it shut down and left jobless some 1,000 workers, mostly women. Despite years of seniority, Company B workers received only the daily minimum wage. Workers who had been there since the factory started—called pioneers—got only a measly additional peso over the rest. Low wages and the lack of retirement pay were the impetus for unionization and later were the main demands in the CBA negotiations.

Union organizing began a year previous to the shutdown, and it was clear that management was determined to discourage it. Workers were questioned one by one and told to drop from the union. They were also offered money to resign. Workers, including contractuals who were being regularized, were instructed against enlisting with the union. The intimidation stopped only when workers took to one of the Company B's international customers, which has a public commitment to ensure freedom of association in its supply chain.

During the certification elections, management told workers to vote against the union, closure was threatened should the union win, and an anti-union leaflet was circulated. Still the union won a landslide victory.

It took some six months for the CBA negotiations to finally be concluded. The union even filed a notice of strike in order to break the deadlock with the management. However, five months after the CBA, Company B announced that it was closing down because of lack of orders. No evidence was presented for this claim. The workers alleged that closure was a ruse to bust the union. Later, it was found out that one of its biggest customers did not cancel any work engagements with the company during the year Company B closed down.

Aside from union busting and illegal closure, workers also complained of blacklisting. The names and pictures of Company B union members were allegedly given to at least two garment factories in the export zone to warn the companies against hiring them. When Company B workers marched in protest inside the ecozone, authorities tried vainly to stop them.

The labor dispute was settled after three months. The deal included provisions for rehiring its retrenched employees if the Company B reopens, additional financial assistance aside from separation pay, and several sewing machines for a livelihood project. Company B, however, did not operate again.

21.2.3 Factory C

Company C was a small company with a workforce of less than 100 and with an almost equal number of female and female workers.

The workers decided to organize to seek redress of violations of labor standards on wages, overtime pay, illegal deductions, and occupational health and safety. The management immediately tried to bust the union with support from the industrial relations (IR) head of the ecozone. Unionists were threatened with factory shutdown and enticed to accept separation pay. Several union officers were charged with spurious cases about producing defective products. One female union officer was disciplined, after she defended herself from physical attack by a non-union worker. She was brought to the ecozone IR office and questioned not about the incident but regarding union activities. Soon after, 20 union officers and members were terminated one after the other.

The union launched a strike for five days until all the unionists were reinstated. During the strike, union organizers engaged one customer of Company C. This connection proved significant in exerting leverage on Company C to finally settle the strike. The union moved for certification as the bargaining agent. Yet the labor department did not certify the union despite having complied with requirements. Not long after, the company filed for shutdown on the pretext of lack of orders even as it continued operating with non- union workers.

In desperation, the locked-out union members barged into the factory and occupied the shop floor. With the PEZA IR head and a regional labor department conciliator threatening the workers with cases for barging into the factory, the workers finally accepted a separation offer and ended the occupation.

21.2.4 Factory D

The Company D was established in early 2000s. A large majority of the 500 or so workers are female.

The Company D workers started organizing a union after getting in touch with the union in Company B. When management got wind of labor organizing, it moved to separate the identified union leaders from the rest of the workers by transferring them all to one production line. The company also called for a meeting to disparage the union. Another way that the management discouraged the workers from joining the union was by starting a "whisper campaign" of rumors that a union victory on the unfair labor practice case it filed will result to the company shutting down. Anti-union leaflets were distributed. Not surprisingly, the union lost the elections.

To finally bust the union, the company shut down temporarily for one month and offered termination pay. Company D justified this with claims of lack of orders. However, the union countered that the orders were just shifted to sister factories of Company D, and that the shutdown was meant to harass union members into accepting separation.

The factory reopened after several months and rehired many former workers who were non-union members. The 20 union officers and members who refused to accept the offer of separation were locked out from the factory.

With the support of labor groups in the area, the union launched protests at the ecozone. The Company D workers marched inside the ecozone despite harassment by PEZA security guards who wanted to take down the placards they were holding. As these mass actions were being held, international labor NGOs were also communicating with a major customer of Company D. The union president and seven other union officers were eventually allowed to return to work. The union believed that this was the result of remediation by the customer.

21.2.5 Factory E

Company E is a transnational electronics factory. Almost all of its 200 workers are women.

Workers organized a union to address grievances such as excessive production quota, verbal harassment, and violations of occupational health and safety standards. Another major complaint was the removal of the door of the women's restroom so that managers and supervisors can see if employees were loitering or chatting. The unreasonable time limits and other conditions on the use of the restroom led to numerous cases of workers suffering from urinary tract infection. The leadership and membership of the union at Company E was entirely female.

In response, management interrogated workers about their participation in the union. Work assignments of union leaders were reshuffled. Overtime was denied to regular employees and instead given to contractual workers. Three line leaders who supported the union were forced to quit.

The union countered by filing a complaint for union busting. With the labor department mediating, Company E agreed to the demand of the union that management release a memo that it will not interfere with freedom of association. However, it also instigated the formation of a company union, which intervened in the worker-initiated union's petition for certification as bargaining agent. The latter won the elections to represent the workers in the contract negotiations, but not without overcoming challenges to their victory.

At the start of union organizing, there were attempts to contact the customers of Company E through the mediation of international NGOs. Communications did not lead to favorable results. Two customer companies denied their link to Company E, while the others did not even respond.

Management continued discriminating against unionists. They were served disciplinary charges on minor grounds such as loitering in the comfort room. Unionists were excluded from receiving Christmas packs. The union president together with other officers, and members were later terminated. Likewise, scores of other union officers and members were suspended.

The union filed a notice of strike for union busting. Mediation got deadlocked as management was adamant in firing and suspending unionists. At the start of the strike, a fire truck along with guards went to the picket line area. Barricades were set up and workers were refused entry. During the first night of the strike up to dawn of the next day, the picket line was repeatedly attacked by armed men that resulted in injuries to two women workers. These actions were in flagrant violation of the existing rules governing the conduct of security personnel during labor disputes.

A settlement was finally reached after a series of meetings convened by the labor department. Company E agreed to reinstate 64 strikers but subject to an investigation by management. Later, Company E suspended then terminated the union officers. A case for illegal dismissal was filed by the dismissed union leaders but the labor court decided against the workers.

21.3 Narratives of Ecozone Workers Amidst COVID-19

The following disputes transpired in 2020 and 2021 at the height of the pandemic lockdowns in the Philippines in different ecozones in five industrial areas in the country.

21.3.1 Factory 1

Factory 1 was a Korean-owned company that manufactured clothing for fashionable global brands. The workers organized a union over issues like regularization of workers and non-payment of mandated benefits. The union won in the certification election and became the bargaining agent.

The company carried out a temporary shutdown of operations in October 2019, a week after the union submitted its CBA proposal. The union resisted by picketing the company gates. It opened again for a month but it failed to give the mandatory 13th month pay and wages due to workers in December 2019. Again, the union launched a picket protest.

The protest action continued even when the company eventually ceased operation. Local police together with ecozone security guards and village militia dismantled their encampment in the dead of the night of April 10, 2020. The dispersal of the picketline was rationalized as enforcement of COVID-19 lockdown protocols.

21.3.2 Factory 2

Factory 2 is a Chinese-owned maker of hand gloves for global brands. It has offices and factories in two industrial provinces surrounding Metro Manila.

Workers of Factory 2 successfully organized a union in late 2019 and got recognized as the bargaining agent in January 2020, at the eve of the pandemic. In February 2020, the union submitted its CBA proposal to the company. After two weeks, instead of responding with a CBA counter-proposal, management filed a petition to cancel the status of the union as workers' representative.

The lockdown led to the temporary closure of the factory which only resumed partial operations on May 5, 2020. By May 25, all workers were called back to work. However, by the middle of July, job rotation was implemented allegedly due to a lack of raw materials. In a decision dated September 25, 2020, the regional labor department affirmed the status of the union as the bargaining agent. The union received a copy of the decision two weeks after, which it believed was due to delaying tactics. Days ahead of the decision, on September 21, 2020, the company manager personally informed workers of the closure of the factory within 30 days and personally handed out separation checks.

Union officers refused to accept the separation pay as they believed the closure was motivated by union busting and refusal to negotiate. The union filed formal complaints but management did not change its position during the online conferences presided by the labor officials. Thus, on December 18, 2020, the union filed a notice of strike and held a strike vote after four days.

The union asked the labor department to facilitate the CBA negotiations. It also demanded the absorption of the union members in any of the plants still in operation, including a newly-built factory in another ecozone. The dispute remains unresolved as management has not responded to the union demands.

21.3.3 Factory 3

Factory 3 used to manufacture clothing apparel and accessories for export. It employed about 400 workers, including those in third party service providers, until it suddenly closed operations in the middle of 2020.

Workers were not paid their 13th month pay, overtime premium, holiday pay, the mandated minimum wage, and separation pay. Workers filed complaints at the labor courts but processing of the case was very delayed amidst the pandemic.

21.3.4 Factory 4

Korean-owned Factory 4 produced a range of items from dog carriers to bags, sold locally as well as for export before it stopped operating in February 2021. Sensing that their employer would evade paying its obligations, the workers organized a union on February 5, 2021. Their fears came true when termination letters "effective immediately" were handed to them on February 22, 2021 because it was shutting operations due to alleged serious financial losses brought about by the pandemic. However, the provincial office of the labor department neither received an establishment report nor a notice of permanent closure from the company.

Individual cases of non-payment of 13th month pay, non-implementation of the service incentive leave, and other labor standards complaints were filed at the labor department and the labor court but management did not respond to notices of conference. The union filed cases of illegal closure, illegal dismissal, and union busting. About 150 workers were thrown out of work by the sudden closure.

21.3.5 Factory 5

Factory 5 was a factory in a privately owned ecozone. It was engaged in bottle manufacturing for several local companies.

Some 200 employees, about a dozen of whom are women, were terminated after the firm filed for permanent closure on November 16, 2020. However, workers accused the company of union busting as the factory continued to operate with a reduced workforce. Likewise, the union—the sole, and exclusive bargaining agent for the workers—believed that the company was merely feigning closure as a way to replace regular workers with contractual employees who will work for less wages and benefits.

On March 16, 2020, the company effected a one-month temporary cessation of operations due to the restrictions brought about by the lockdown. The company then filed for the continuance of the temporary shutdown to October 16, 2020 due to market slump. When the momentary shutdown was about to end, the company announced that it was discontinuing its operations effective November 16, 2020. This time it cited serious financial losses caused by the COVID-19 pandemic.

The union contested the company's closure as it did not serve a 30-day notice, contrary to the mandates of the law. The union doubted the motive behind the shutdown as just an excuse to avoid the renegotiation of the CBA and that it was only using the pandemic as an alibi for the alleged losses of the company. The union filed complaints of illegal closure, illegal dismissal, refusal to bargain, and union busting at the labor court.

The union set up a picketline at the factory gates to protest the illegal closure and union busting. The picketline was later dispersed by village guards amidst the pandemic.

21.3.6 Factory 6

Factory 6 manufactures ready-to-wear clothing as well as children and women's apparel for local brands which are sold in the biggest mall chain in the country. The company employs approximately 300 workers. The company engaged in a host of labor standards violations such as wages and overtime pay below the mandated minimum, and non-payment of benefits such as 13th month pay and rest day premium.

Workers were engaged indirectly by Factory 6 through a placement agency organized as a "labor" cooperative. Workers started as probationary employees then became associate members of the cooperative. To become regular cooperative members, workers needed to shell out P 12,000 (equivalent to about a minimum wage earner's monthly income) within 2 years.

The factory shut down in March 2020 at the start of the lockdown and then accepted a few workers when it reopened in June 2020. Further, workers were accepted back to work on a piece-rate system instead of payment of wages per hour as was the norm. The workers not asked to report repeatedly talked to management, to no avail, for their return to work and to clarify their employment status.

Finally, workers organized a union in order to seek redress of their numerous grievances. When the company learned of the unionizations, it fired the leaders and members of the union. The union filed an illegal dismissal case and union busting at the labor court.

21.3.7 Factory 7

Factory 7 was a big garment exporting firm in a major ecozone in the central part of the Philippines. On November 27, 2020, Factory 7 fired 300 workers, including the union president. At that time, the union had a pending petition for certification election.

A rally of terminated Factory 7 workers was held on November 30, 2020, a traditional day of labor mobilization. As the protest action started, members of the local police arrived in the area and ordered the protesters to stop their activity because it was allegedly in violation of the COVID-19 protocols. The organizers tried to negotiate but the police denied their request to be allowed to hold a short program. Police officers violently dispersed the protesters and arrested the union president and four labor organizers. The arrested union leaders were detained and slapped with a criminal case of disobedience to persons in authority. They were later released as their case for disobedience was dismissed.

The labor department held mediation meetings between the management of Factory 7 and the union in December 2020 but the parties failed to reach any agreement. The union proceeded to conduct a successful strike vote. Intending to avert the strike, a negotiating team composed of the labor department, PEZA and city officials were formed. A series of conciliation-mediation meetings were held and an agreement was reached to hold the certification election in the factory on February 15, 2021, and to resolve the labor dispute through filing of illegal dismissal and union busting charges before the labor court.

In the certification election, the union campaigned on the basis of its demands for annual wage increases, company-subsidized health insurance coverage, 15-day sick and vacation leaves, retirement pay, job security, and other benefits and bonuses. However, since the union leaders and active union members were terminated, they found it difficult to campaign inside the workplace and respond to the anti-union campaign of the management. As a result, the union lost the election.

The Factory 7 management used alleged losses due to canceled orders as its alibi for retrenchment. The union countered that this is a lie as its main customer, a US brand, already exited from bankruptcy in September 2020. Further, production for the brand was in full swing, and workers were asked to report for duty even on holidays and Sundays at the time of the mass layoff. Aside from the US brand, the Hongkong-based company that owns the Factory 7 supplies to other global brands. Factory 7 later shut down but the union suspects that it moved its production to a runaway shop in a nearby city.

The firings at Factory 7 came on the heels of mass layoffs at other garment firms in the same ecozone at the height of the pandemic. Earlier, a giant conglomerate of garment firms retrenched 4,000 workers, another big apparel factory fired 2,000 workers, a smaller garment plant laid off 100 workers, and Factory 8 dismissed 67 workers including the union officers.

21.3.8 Factory 8

Factory 8 is a French-owned manufacturer of high-end fashion accessories for distribution to the Philippines, France, Italy, Japan, Hongkong, and the USA. Before the pandemic, Factory 8 employed some 500 regular workers.

A few years before the pandemic, Factory 8 workers organized a union in a bid to improve working conditions. Soon after registration and filing a petition for certification elections, management started harassing union officers and active members. Union officers who were identified by management were transferred to different departments. The union president was assigned to the stock room where she worked alone while other union officers were approached secretly by management representatives to tell them to withdraw from unionization.

The evening prior to the election day, management took all non-union workers to a resort where they threatened company closure should the union win the election. Seeing the incident as blatant management interference, the union boycotted the election and then later filed a case of unfair labor practice.

The case dragged on until the onset of pandemic, in the middle of which, 67 union members including the president were laid off allegedly due to the economic crisis induced by COVID-19. Once more the union filed cases of unfair labor practice and union busting but it was dismissed by the Office of the Labor Secretary. A notice of strike was filed by the union but the dispute was referred to preventive mediation which prohibited workers from proceeding with a strike.

Two union officers and around 40 members remain employed with Factory 8 but continue to be harassed by management. The company has also begun to hire new workers on short contracts of up to five months only.

21.4 From Bad to Worse: Labor Rights in the Ecozones Before and During the Pandemic

A rights-based framework highlights an essential feature of workers struggles in the ecozones. While workers as claim holders are aware of and assert their rights to unionize and negotiate for better work conditions from private actors, it is often the State as a duty-bearer which is amiss in ensuring that these rights are protected and fulfilled. At the State level, little action has been done by relevant agencies for the persistent and rampant violations of labor standards and labor rights in ecozones, nor has anyone been made accountable for the occurrence of specific cases such as the cases described above. It is worth noting that the cases of labor rights violations narrated above covered the terms of two Philippine presidents. In other words, impunity cuts across the particularities of governance by different presidents. In all cases, the intervention of government agencies has been limited, if not dismal, and skewed toward the interests of multinational companies operating in the ecozones to the detriment of worker rights protection.

Export processing zones are gated complexes where capitalists' interests are dominant but also where workers' struggles erupt. It is an arena where workers as rights holders assert their claims in the face of deficits of development due to the dereliction of duty by bearers such as government and employers. They are enclaves of poverty and repression when the right to unionize and bargain collectively is actively suppressed by capitalists, with the connivance of government agencies. Repression of freedom of association is a means to maintain cheap wages, insecure jobs, and docile workers, which in turn, redound to profits of foreign investors. Unions are expressions of empowerment but are continually being suppressed by profit-driven capitalists. The rule of law and other human rights-affirming policies should provide protection to workers against various forms of violations and abuses in the workplace and by employers, as well as give redress when these occur. However, this is not often the case.

Unionists are discriminated by management through termination and suspensions, with almost absolute freedom and no accountability. Human dignity is transgressed with widespread incidents of verbal harassment, lack of privacy, and forced overtime. Empowerment is possible through workers' self-organization into unions but this is suppressed through various union-busting and union-avoidance maneuvers. Apart from work standards that affirm the dignity of workers, labor laws ensure that corresponding penalties for erring companies are carried out, in the event of violations. But these remain unrealized in many cases in the ecozones.

Things went from bad to worse when the COVID-19 crisis erupted. Protocols enforced by the government during the pandemic discriminated against workers in the exercise of their rights. Social distancing instead of social dialogue was the norm during COVID-19.

The dispute resolution mechanism was suspended by the labor department when the lockdown was imposed in the middle of March until August of 2020. Thus, workers who suffered from workplace grievances and company abuses—which were widespread in that period—did not have any options for remediation. In contrast, other government offices continued to deliver public services by turning to work from home and online modality. This showed that the labor department could have kept the dispute resolution open for workers by providing its services online. In fact, this is what happened in the latter part of 2020 when the labor department started accepting complaints from workers. But it took several long months of pressure and calls from the labor groups before the labor department and the labor court finally reopened their doors in August 2020.

Nonetheless, even when the dispute resolution mechanism was operating once more, the labor department and labor court were slow in resolving the workers' grievances. For example, summons from the labor department did not arrive on time to the parties when hearings physically resumed. Justice delayed is justice denied was the result of the sluggish pace of the labor justice system at the height of the lockdown.

Several regulations issued during the pandemic clearly discriminated against workers and favored employers. One allowed employers to cut wages and shift to so-called alternative work arrangements. This could be done as long as affected workers gave their consent, which in many cases, companies did not comply with. This regulation opened the window for unionized firms to push for a moratorium on wages and benefits during CBA negotiations or even to utterly refuse to conduct collective bargaining. Such employer misbehavior is subject to complaints. But the suspension of the dispute resolution left unions, more so individual workers, powerless to resolve their grievances through the labor justice system. Another new regulation gave companies the leeway to place workers on furlough beyond the mandated six-month maximum. This was enforced by the labor department in the face of fierce opposition by the labor movement.

It is as clear as day that the labor department fell short of its mandate for social dialogue and social justice in the face of rampant labor complaints like forced leave, mass layoffs and company closures. As related in the stories of ecozone disputes, union busting remained common. Workers were victims of labor rights violations that were not redressed.

Given the closure of the dispute resolution mechanism, workers exercised their agency and claimed their rights on their own through direct actions and mass protests. However, police and other security personnel used the COVID-19 social distancing protocols as the pretext to clamp down on the right to strike and the freedom to assemble. Using the pandemic emergency powers, local police and village militia routinely prevented workers from striking and picketing. Further, the labor department refrained from intervening and enforcing the guidelines guaranteeing the exercise of the freedom of workers to to strike and protest.

An extreme and blatant instance of the labor repression during the pandemic was the attempt to militarize the ecozones ostensibly to contain workers' militancy. The national police and PEZA formed the Joint Industrial Peace Concerns Office (JIPCO) which was a scheme to place security personnel within ecozones to guard against militant unionism. JIPCO was renamed to Alliance for Industrial Peace Program (AIPP) in reaction to criticism that the scheme was a flagrant violation of the rules on the role of security personnel in labor disputes. The labor department sent a letter of concern to PEZA and the national police but did nothing more. The PEZA did retreat and declared the AIPP was on hold due to concerns raised by the European Union. The scheme to militarize ecozones fit in with the overall anti-insurgency and anti-communist witch hunt of the administration of then-President Rodrigo Duterte that escalated during the pandemic.

These rampant violations of workers' rights occurred despite reforms initiated in the early 2010's as a result of an International Labour Organization (ILO) High-Level Mission in 2009 to investigate killings of unionists, and transgressions of the freedom to organize and bargain collectively. In fact, in January 2023, another ILO High-Level Mission arrived in the Philippines to once more probe numerous incidents of murders of labor activists, harassment of trade unions, and violations of the right to associate and negotiate (Velasco 2023).

The new mission arose from outrage at the ILO annual conference in 2019 at the daylight murder of Dennis Sequeña, a member of the group Partido Manggagawa and a veteran union organizer. Sequeña was facilitating a labor rights seminar for workers of two garment factories in the Cavite Economic Zone who were preparing for certification elections when he was shot dead just a few days from the start of the ILO conference. His killing remains unresolved (Velasco 2023).

The latest ILO mission made a set of recommendations to the Philippine government. The reform proposals remain the subject of contestation as it has not been implemented and the labor movement continues to lobby for its enforcement.

21.5 From CSR to WSR: Ensuring Labor Rights Through Binding Agreements not Self-regulation

The widespread gaps in labor justice and deficits in decent work in global value chains have necessarily engendered extensive discussion about remediating the problem through proposals for social upgrading (Anner 2021). Gereffi and Lee (2016) propose six interventions for redressing global value chain labor issues: market-driven path, cluster-driven path, corporate social responsibility (CSR) path, public governance path, labor governance path, and multi-stakeholder path. Taking off from Gereffi and

Lee (2016), Anner (2021) argues that the first three are all dependent on positive action by global brands and their suppliers which is difficult to expect under conditions of extreme power asymmetries in global value chains. The profit-maximizing behavior of corporations militates against self-regulation. Thus, Anner (2021) posits that social upgrading is possible with the last three mechanisms. First, the public governance path or reforms by the State to ensure labor compliance. Second, the labor governance path or extensive collective bargaining by unions to improve working conditions. And finally, the multi-stakeholder path or binding agreements between brands, suppliers, unions, and other actors to guarantee enforcement of labor rights and standards. He provides Vietnam, Honduras, and Bangladesh as archetypes, respectively, of the three paths.

Anner (2021) maintains that the effective mechanism in any setting will depend upon patterns of worker resistance and types of labor regulation regimes that actually exist. Using his approach in interrogating the case of ecozones in the Philippines, the public governance path is a dead-end as the State is weak and captive to corporate interests, similar to Bangladesh. The lack of enforcement of the reforms instituted after the ILO mission in 2009 and the lack of action on the new ILO recommendations in 2023 are illustrative of the inapplicability of the public governance path for the Philippines. Meanwhile, the labor governance path is a steep climb as labor market conditions of mass unemployment and labor control regimes in the ecozones result in weak union organizing and thin collective bargaining coverage, again akin to the Bangladesh situation. The persistent difficulties of union organizing within and without the ecozones in the Philippines bear this out.

What holds promise for the Philippines is the multi-stakeholder path and the possibility of forging enforceable agreements between brands, factories, unions, and labor NGOs. This has been called worker-driven social responsibility (WSR). In contrast to the voluntary basis of CSR, WSR is founded on co-governance and active participation by workers in order to safeguard labor rights through binding provisions (Gladstone 2020).

Arguably, the most well-known model for WSR is the Bangladesh Accord which grew out of efforts by a broad front of national unions, international labor federations, and workers' rights NGOs after the Rana Plaza building collapse which killed 1,100 garment workers (Worker-Driven Social Responsibility Network 2017). Aside from the apparel industry in South Asia, WSR best practices include the farm sector in the USA (Worker-Driven Social Responsibility Network 2022, 2023). It can even be argued that the Maritime Labour Convention 2006 is an example of WSR and not CSR (Milde 2011).

Nothing similar to a Bangladesh Accord or a binding agreement among global supply chain actors exists in the Philippines. But union organizers involved in the labor disputes narrated above are inspired by the WSR model and are actively pursuing the possibility of achieving one in the ecozones.

The narratives reveal that engagement with global brands on the basis of enforcement of their codes of conduct and solidarity from international labor NGOs are among the repertoire of actions utilized by unions, complementing the mass actions of workers such as strikes and protests. Engagement with global companies through the mediation of labor rights NGOs assisted the struggles of the workers in varying degrees. This was key in the victory of the strikes at Company A and Company C. It led to the stop of the harassment of unionists at Company B and to the reinstatement of union leaders at Company D. In contrast, there was no engagement with the customers of Company E and this affected the fight of the women workers for better working conditions.

In the absence of stronger State measures to uphold labor rights, workers combined localized direct action such as union organizing, securing CBAs, strikes, and pickets, with engagement of global brands and international solidarity. The cases showed the complementary impact of engaging with global companies about their supply chain codes of conduct. In many cases, it is the leverage of these companies on their suppliers operating in the ecozones that force factories to remediate their violations and uphold rights at work. These global companies have the most to lose in disputes involving their supplier factories that will taint their carefully crafted reputation as good corporate citizens.

However, the pandemic was a notable period in which global brands became utterly unconcerned even with their ostensible good corporate citizen image as the profit squeeze led them to abandon their responsibilities to suppliers and workers in their supply chains. Engagement with global brands was an utter failure in contrast to the period prior to COVID-19 when lead firms—at times but not always—played a role in resolving disputes in favor of workers who fought for their rights. Global brands either did not respond to calls for remediation or paid lip service to their so-called codes of conduct. Many of the eight factories embroiled in pandemic-era disputes had codes of conduct and the brands they supplied also had their own compliance commitments. A universal provision in corporate codes of conduct is respect for freedom of association and the right to collective bargaining, which was extremely disrespected amidst COVID-19.

Thus, narratives of workers embroiled in labor disputes in different ecozones in the Philippines during the lockdowns reveal an epidemic of labor rights violations. Discriminatory state policies amidst the pandemic enabled employers to turn to precarious work, to bust unions with impunity, and to resort to wage theft. The labor department effectively exercised a policy of social distancing to the labor movement as it was blind and deaf to demands by workers for remediation of grievances. A series of pandemic-era regulations favored employers to the disadvantage of workers. The government suppressed workers' militancy through red-tagging, the dispersal of protests and strikes, and arrests and killings of unionists. This impunity was facilitated by a draconian Anti-Terror Law that was railroaded in the parliament during the pandemic which widened the scope of what can be considered terrorist acts to include legitimate protests.

These experiences reveal the fragility of self-regulation and CSR by brands and suppliers, on the one hand. On the other hand, it does demonstrate the realistic prospect for unions exercising leverage so as to compel brands to forge binding agreements, which is precisely the nature of WSR.

As part of advancing WSR, the following set of recommendations seeks to actualize the freedom to unionize and bargain that forms the foundation for the exercise of other rights at work. Without the agency of workers through their self-organization and empowerment through unionism, the labor standards and rights will remain unprotected, not upheld, and unfulfilled by duty bearers such as the PEZA, the labor department, and factory management. Thus, the importance of enabling measures so that workers can exercise their capability to freely choose to unionize as a means to improve working conditions and achieve human development.

1. Implementation of tripartite councils—with the mandate to discuss grievances of ecozone workers and recommend solutions—within all ecozones as provided for in the law that established the PEZA.
2. Information dissemination at all ecozones about the provisions of the Labor Code that (a) it is state policy to guarantee labor rights, (b) the law encourages trade unionism and collective bargaining, and (c) management interference in the right to organize is illegal and constitutes unfair labor practice.
3. Holding of seminars for ecozone workers, including managerial and supervisory employees, about labor rights, standards and welfare as a tripartite activity by the PEZA, labor department, employers and labor groups.
4. Implementation of the labor education law that mandates the inclusion of a course on labor rights, standards and welfare in the curriculum for senior high school and college students to prepare them for entry in the world of work.
5. Access for representatives of labor organizations to the ecozones and to company premises during break time for the purpose of union activities.
6. Formation of gender committees at the level of each ecozone with representatives from workers, employers and the PEZA to take up gender-related concerns such as workplace discrimination, sexual harassment, and violence against women.
7. Strengthening of gender education and women's rights orientation within local unions and labor groups.
8. Advancement of worker power through unity among various labor groups organizing in the ecozones and the use of innovative forms of actions such as social movement unionism and supply chain organizing together with old modes such as marches and strikes.
9. Abolition not just suspension of the AIPP, formerly the JIPCO, as it is a flagrant violation of freedom of association and right to free expression.
10. Implementation of 2023 ILO High Level Mission recommendation for a presidential body, and an independent group to assist it, with the mandate to resolve all pending cases of killings and abductions of unionists, and prosecute those involved.

References

Anner, M. 2020. Abandoned? The impact of Covid-19 on workers and businesses at the bottom of global garment supply chains. *Worker Rights Consortium*. https://www.workersrights.org/research-report/abandoned-the-impact-of-covid-19-on-workers-and-businesses-at-the-bottom-of-global-garment-supply-chains/.

Anner, M. 2021. Three labour governance mechanisms for addressing decent work deficits in global value chains. *International Labour Review* 160 (4): 611–627.
Asia Floor Wage Alliance. 2020. The Emperor Has No Clothes: Garment supply chains in the time of pandemic, issue II. https://asia.floorwage.org/wp-content/uploads/2020/05/The-Emperor-Has-No-Clothes-Issue-II-May.pdf.
Asia Floor Wage Alliance. 2021a. Money heist: COVID-19 wage theft in global garment supply chains. https://asia.floorwage.org/wp-content/uploads/2021/07/Money-Heist_Book_Final-compressed.pdf.
Asia Floor Wage Alliance. 2021b. A stitch in time saved none: how fashion brands fueled violence in the factory and beyond. https://media.business-humanrights.org/media/documents/A_Stich_in_Time_Saves_None_FINALpdf.pdf.
Business and Human Rights Resource Centre. 2020. Impacts of COVID-19 on supply chain workers in the electronics sector, June 8. https://www.business-humanrights.org/en/latest-news/impacts-of-COVID-19-on-supply-chain-workers-in-the-electronics-sector/.
Business and Human Rights Resource Centre. 2021. Wage theft and pandemic profits: The right to a living wage for garment workers, March 11. https://media.business-humanrights.org/media/documents/Unpaid_wages_v9.pdf.
Certeza, R., A. Adviento, and A. Reyes. 2012. *How fair they play?: A research report on working conditions and freedom of association in Adidas supplier facilities in Mactan Export Processing Zone, Philippines*. International Textile Garment and Leather Workers' Federation-Philippine Council.
Clean Clothes Campaign. 2020. Un(der)paid in the Pandemic: An estimate of what the garment industry owes its workers. https://cleanclothes.org/file-repository/underpaid-in-the-pandemic.pdf/view.
Edralin, D. 2001. Assessing the situation of women working in the Calabarzon, Philippines. *Philippine Journal for Labor and Industrial Relations* 21: 141–162.
Gereffi, G., and J. Lee. 2016. Economic and social upgrading in global value chains and industrial clusters: Why governance matters. *Journal of Business Ethics* 133 (1): 25–38.
Gladstone, A. 2020. Worker driven social responsibility agreements: A new future in labor rights protections. *Fordham International Law Journal* 44 (2): 549–603.
Kyritsis, P., G. LeBaron, and S. Nova. 2020. Hunger in the apparel supply chain: Survey findings on workers' access to nutrition during COVID-19. *Worker Rights Consortium*. https://www.workersrights.org/wp-content/uploads/2020/11/Hunger-in-the-Apparel-Supply-Chain.pdf.
Milde, P. 2011. The maritime labour convention 2006: An instrument to improve social responsibility in the cruise industry. In *Cruise sector challenges: Making progress in an uncertain world*, ed. P. Gibson, A. Papathanassis, P. Milde, 207–224. Gabler Reseach. https://doi.org/10.1007/978-3-8349-6871-5_13.
Ofreneo, R., J. Bellin, and M. Hega. 2007. *Women garments workers in selected Philippine economic zones: Towards an organizing and advocacy agenda in the context of trade liberalisation*. Committee for Asian Women.
Partido Manggagawa. 2010. Violations of workers' rights in the Philippines in 2010, June 10. http://partidongmanggagawa2001.blogspot.com/2010/10/violations-of-workers-rights-in.html.
US Department of State. 2010. 2009 country reports on human rights practices, March 11. https://2009-2017.state.gov/j/drl/rls/hrrpt/2009/eap/136006.htm.
Velasco, B. 2021. *Decent work and freedom of association in the ecozones: A preliminary study of the Philippine situation*. Nagkaisa: Federation of Free Workers and the Danish Trade Union Development Agency.
Velasco, B. 2023. On the ILO probe into labor rights in the Philippines. *Rappler*, February 7. https://www.rappler.com/voices/thought-leaders/opinion-international-labour-organization-probe-rights-philippines/.
Worker-Driven Social Responsibility Network. 2017. WSR success story: Accord on fire and building safety in Bangladesh, July 27. https://wsr-network.org/wp-content/uploads/2017/07/WSR-Success-Story-Bangladesh-Accord.pdf.

Worker-Driven Social Responsibility Network. 2022. Historic new accord for Pakistan's garment workers expands worker-driven social responsibility model, December 15. https://wsr-network.org/resource/historic-new-accord-for-pakistans-garment-workers-expands-worker-driven-social-responsibility-model/.

Worker-Driven Social Responsibility Network. 2023. Worker-driven social responsibility fact sheet, July 13. https://wsr-network.org/resource/worker-driven-social-responsibility-fact-sheet/.

Benjamin B. Velasco is an assistant professor at the School of Labor and Industrial Relations at the University of the Philippines (UP) Diliman. He is currently completing his doctoral dissertation at the College of Social Work and Community Development of UP Diliman. The dissertation examines the enablers and barriers to worker-driven social responsibility based on the experience of export zone workers in the Philippines. He was formerly co-convenor of the Alternative Development Program of the UP Center for Integrative and Development Studies.

Part VII
Conclusion

Chapter 22
Prospects of Alternative Regionalism in a Globalizing Southeast Asia

Ananeza P. Aban and Jeroen Adam

Abstract The different cases discussed in this volume serve as examples of development alternatives that question and go beyond the homogenizing claims of modernist approaches to development. Rather than starting from universalist, topdown concepts, these initiatives tend to be rooted in local cultural values and norms, subaltern identities, and everyday practices. From these cases, this chapter interrogates the possibility of a regionalism based on 'real regions' that become spaces of emancipation, self-realization, and empowerment.

Keywords Globalization · Southeast Asia · ASEAN · Nation-states · Alternative regionalism

22.1 The Supremacy of Global Capital Over Southeast Asia

The alternative practices of Southeast Asian communities extensively discussed in this volume are valuable evidence that inform the importance of critically rooting regional challenges and the need to organize a solidarity movement grounded in peoples' diverse but transformative alternatives. The different cases discussed in this volume serve as examples of development alternatives that question and go beyond the homogenizing claims of modernist approaches to development. Rather than starting from universalist, top-down concepts, these initiatives tend to be rooted in local cultural values and norms, subaltern identities, and everyday practices. One example that has been discussed in this book is the concept of *Tara Bandu* in Timor Leste, functioning as a context-specific customary law putting common access rights to natural resources above individual private ownership. A similar example in the

A. P. Aban · J. Adam (✉)
Ghent University, Ghent, Belgium
e-mail: jeroen.adam@ugent.be

A. P. Aban
e-mail: ananeza.aban@ugent.be

J. M. C. Sy and A. P. Aban (eds.), *Reimagining Development in Southeast Asia*, Asia in Transition 21, https://doi.org/10.1007/978-981-97-9140-8_22

field of natural resource governance is the Southern Peasants' Federation of Thailand that actively supports community rights to land and explicitly positions itself as an alternative to the hegemonic development discourse of commodification and the universalist blueprint of individual land titles. These, and other cases, also need to be understood as forms of resistance, or more profoundly, a retort to the compelling regional geopolitical situation under the supremacy of global capital. In Loktak Lake in Northeast India, the indigenous Meitei community resists the threats to their livelihoods posed by large infrastructural projects. The Payoga-Kapatagan Multipurpose Cooperative in the Philippines starts from a set of community values that oppose the hegemonic concept of individualized wage labor. Rosalinda Pineda Ofreneo and Josephine Parilla discuss extensively the concept of a solidarity economy that is centered around life values and democratic inclusivity instead of profit values and economic growth. These, and other cases discussed throughout the book, are disclosures of how another life beyond the forces of capital can truly be possible, or in fact, is already in the making. These alternatives offer proposals for better governance and more democratic societies.

Prior to exploring the prospects of regionalizing, what these ground-based experiences in Southeast Asia imply is the overriding imperative to re-examine the undercurrents of 21st-Century capitalism or neoliberalism. Robinson (2019) suggested that a clear and sharp analysis of global capitalism and its crisis must foreground the making of a peoples' movement. Following Robinson's prescription, a constant mindfulness of the power relations within this neo-liberal matrix and the impact to peoples in the Global South must be the dialectical frame. This line of thinking can be a foundational standpoint toward a deeper, solidarity-based regionalism project that addresses structural transformations in the long run.

Capitalism, the dominant economic model characterized by the relentless drive for capital accumulation, has dynamically ventured into great heights by crossing transnational borders (Robinson 2017, 2019; Sklair 1997, 2002; Al-Fadhat 2020). Global capital mobility has effectively re-organized production worldwide to ensure profit-making opportunities. The agents of global capitalism engineered this substantial restructuring through the consolidation of nation-states and national market integration into a single world market (Robinson 2017; Sklair 1997, 2002).

This economic transformation has facilitated the rise of the transnational capitalist class (TCC) (Robinson 2017, 2019; Sklair 1997, 2002), with its enabling arm, the transnational state (Robinson 2017, 2019). This is expressed by liberalizing the movement of capital, free trade agreements (unregulated markets), and solidifying the needed integration processes and infrastructure (Robinson 2017; Keet 2007; Sklair 1997, 2002).

Meanwhile, Southeast Asia's deeper global economic integration "is manifested through burgeoning cross-border company mergers, acquisitions and joint ventures" (Al-Fadhat 2020, 178). This notion of a singular market-based economy has been repeatedly stated in ASEAN documents as a trajectory of its regional integration process (ASEAN Secretariat 2015, 2019).[1] ASEAN's capitulation to global capital

[1] The documents are: ASEAN 2025: Forging Ahead Together and ASEAN Integration Report 2019.

is well captured in its ASEAN Economic Community (AEC) narratives. Such goal has been set up when ASEAN created the ASEAN Free Trade Area in 1992, as a strategy to attract foreign direct investments to a single, geographical economic space that is Southeast Asia (Nesadurai 2008, 2012). Later and as reflected in its Vision 2025 that imagines a "global ASEAN," this regional body formed its AEC that aspires to create a singular market and production base integrated within the global economy (ASEAN Secretariat 2015, 2019; Nesadurai 2008, 2012). In other words, neoliberalism is the cornerstone of ASEAN.

The AEC is the realization of the region's end goal of economic integration. It envisions ASEAN as a single market and product base, a highly competitive region, with equitable economic development, and fully integrated into the global economy.[2]

It is crucial to know further the nature of the TCC as it impacts lives and the future of Southeast Asia. For Sklair (1997, 521; 2002, 160), the TCC has four major components: (1) executives of transnational corporations; (2) globalizing bureaucrats (the states that fulfill a governance function for the global capitalist system); (3) globalizing professionals (the technical people who perform the services for the TCC); and (4) consumerist elites (the merchants and media as key players that strengthen the consumerist culture-ideology and hence help maintain the global capitalist hegemony). As a class, the players within these components have interlocking directorates or cross-membership or may take the lead on important tasks in more than one of the functions.

The TCC, with its resources, apparatuses, and enhanced structural power, eased the axis-formation of states and their allied transnational institutions to operate as a collective authority to advance the global capitalist agenda (Robinson 2017; Sklair 2002).

In particular, the states as enablers of global capital have been transformed to bolster capital growth outside territorial frontiers. Their role through national institutions is to manage economic policies and maintain the conditions supporting capitalist accumulation (Panitch and Gindin 2004, 17 cited in Al-Fadhat 2020). Moreover, they needed "to facilitate the increasing flows of investment, production and trade" (Al-Fadhat 2020, 183). In this manner, the TCC has undermined the ability of states to redistribute surpluses for the survival of the majority or has reconfigured public finance for government debt servicing and austerity measures (Robinson 2017, 2019).

This TCC-dominated global village is becoming increasingly polarized between the rich and the poor, and between the strong and the weak (Robinson 2017; Keet 2007). Two opposing poles aptly describe this quandary of endemic instability where at one end is wealth accumulation and on the other is misery and poverty (Robinson 2017). Right here unfolds the spiraling crisis of hegemony. Such maneuver has resulted to extreme global inequality that eroded social cohesion and fueled unrest which demonstrated the states' decreasing legitimacy before the people (Robinson 2017, 2019).

[2] See https://asean.org/our-communities/economic-community/#:~:text=The%20ASEAN%20Economic%20Community%20envisions,as%20the%20global%20supply%20chain.

Global capitalism is in itself highly contradictory as witnessed between the interest of business and that of the marginalized. For Sklair, this system in many respects exudes vulnerability because it has to constantly "struggle to create and reproduce its hegemonic order globally" (1997, 514). This explains why choosing a sustainable life with ecological equilibrium is incongruous with capitalism's expansive and destructive logic (Robinson 2017, 11). Keet (2007) further pointed out that the imposition of a singular set of rules and the creation of one dominant paradigm in favor of stronger economies and corporations created an imbalance instead. Keet (2007) by unmasking the myth of a strong integration among states explained that the interest of business is actually to take advantage of or exploit the disparate levels of development, capacities, and regulations between countries within regions.

22.2 State Military, Democracy, and the Global Capital

This very exploitative nature of global capitalism has generated increasing resistance from below. As a reaction, people from the margins of global capital have also "transnationalized" their struggle (Robinson 2019, 290) by organizing global peoples movements against global capital (Bond 2003 cited in Bond 2019). Throughout the book, different cases have been elaborated. A first textbook example of a highly successful transnational peoples' movement is La Via Campesina as discussed by Raphael Baladad. This organization, pushing for food sovereignty in opposition to corporate agriculture, has managed to unite different peasant movements, not only in the Southeast Asian region but even on a global scale, including Western organizations. Another example of successful alliance building that goes beyond the confines of national boundaries is the chapter by Ofreneo and Parilla that provides background to the emergence of the Asian Solidarity Economy Council which is tying together different national organizations working on solidary economy. Still, other authors delve into examples of alliance building on a national level such as the chapter by Bianca Ysabelle E. Franco that shows how different associations of women garment workers in Cambodia unite in the Workers' Information Center that is then a member of the United Sisterhood Alliance in Cambodia. And Benjamin B. Velasco's discussion of labor union organizing of export zone workers in the Philippines explains the value of mobilizing workers and alliance building as pillars of a repertoire of actions to improve working conditions and defend workers' rights. Still, it also needs to be mentioned that there is room for many of the alternative development initiatives discussed in this book to upscale their activities through building up alliances with similar organizations in other Southeast Asian countries.

In the meantime, the capitalist system refuses to concede to the transformative demands of social movements such as La Via Campesina or the Asian Solidarity Economy Council so that prevailing class power relations remain intact (Bond 2003, 8). In defense of the interest and security of transnational capital, states impose repressive means of social control and domination, as demonstrated in the use of military power to demobilize these social movements (Robinson 2019).

The issue of militarism and authoritarianism is not new in Southeast Asia. In fact, it is a prevailing phenomenon (see ACSC/APF 2017). Previous critical and scholarly assessments of ASEAN have indicated that even the conception of ASEAN in 1967 as an inter-governmental state, was already a posturing for state security as a strategy to safeguard this "Association for the elite" against the anti-capitalist movements of that era (Collins 2008; Aban and Sy 2022). Civil society groups that engage in the official ASEAN process noted in their 2019 Statement: "Power shifts and regime changes in some ASEAN countries have led to the undermining of democratic processes with the rise of authoritarian and right-wing populist leaders resulting in widespread violations of human rights" (ACSC/APF 2019).

This illustrates that the repeated plea of many social movements for the democratization of Southeast Asia requires empirical and persistent discussion and debate by integrating the element of the TCC in the regional political spectrum. One of the points that unites the different initiatives and organizations discussed throughout the book is their ongoing call for democracy, transparency, and inclusivity vis-à-vis notions of top-down governance and even outright authoritarianism. Ma. Victoria R. Raquiza and Roja Estrelita Salvador give the example of citizens' participation in budget processes in the Philippines. As they state, the inclusion of subaltern groups in these processes is not only a matter of democratic values but, in the end, also results in better governance and decision-making. Another is Honey Tabiola's chapter about the local energy planning in Vietnam that features the less explored opportunities and advantages of bottom-up energy planning to address the many challenges of the energy sector in the country, one of them is energy import dependency.

Fighting for genuine democracy, Robinson (2017) noted, is necessarily a fight against the TCC. The alliance formation against global capitalism requires the leadership of the working class and the marginalized sector. They should generate from the local to the global, geo-politically emerging from the Global South (Bond 2003), or grounded in the struggles from peoples in the peripheries of the world system (Amin 2018 cited in Bond 2019) where the impact of global capitalism is drastically hostile. This battle requires mass mobilization on a transnational scale (Robinson 2017) that is community rooted but globally oriented (Bond 2019). All these ideas, as vantage points, have shown that the process of building regional solidarity projects has to intersect the local, regional, and the global.

In retrospect, a regional solidarity, or a regionalism project within Southeast Asia, needs to look and reach beyond the narrow boundaries of nation-states, territorial demarcations, and ethnic distinctions. This movement must assert common histories, aspirations, and a vision of a region that will transcend the narrow confines of territories, nationalities, ethnic identities, and citizenships (Tadem et al. 2020, 5).

22.3 Nation-States, Peoples, and Regionalisms

As earlier explained, and as illuminated in the case studies in this volume, the fluxes within Southeast Asia's changing geopolitical landscape are deeply embroiled in this web of transnational capital. From a regional standpoint, there were in fact several expressions of trans-nationalizing the peoples' struggle against global capitalism.

This concluding chapter also seeks to clarify a distinctive regionalism that is organized by peoples in struggle and that deflects from what is normally understood as the mainstream paradigm, greatly influenced and manipulated by the global capital. This is going beyond an inter-governmental approach that prefers regional institution building as a process of interstate bargaining (Hoffman 1995 and Moravcsik 1998 cited in Beeson and Stubbs 2012).

This chapter argues that there are multitudes of regionalism across regions that are not contingent on the idea of nation-states. This does not have to imply that the concept of the state all of a sudden disappears or becomes irrelevant. Different cases in the book demonstrate how they proactively make use of democratic spaces that are being enabled by formal state legislation or transnational agreements. For instance, the chapter on the Thai civil society organizations such as the Sustainable Agriculture Foundation (SAF), the Alternative Agriculture Foundation (AAF), and BioThai is pushing for alternative systems of agriculture that recognize the importance of consciously engaging with government and state functionaries. Another example is the Peace House Shelter as a safe space for women in Vietnam that is enabled by particular national and transnational treaties on human trafficking, domestic violence, and support to victims of gender-based violence. A similar example is provided in the book showing how the Association for Development of Women and Legal Education (ADWLE) that is battling violence against women in Lao PDR, makes use of transnational treaties as a rallying point to put pressure on government officials. These examples show how organizations working from an emancipatory and people-centered perspective proactively search for democratic openings and venues, even when notions and models of authoritarianism remain dominant.

In this epoch of twenty-first-century global capitalism, Mushakoji (in Tadem et al. 2020, 4–5) contested that the 17th-Century Westphalian[3] paradigm of authority and rule that begets the notion of nation-states (also see Cutler 2001) "may no longer be that relevant where porous boundaries and labor migration patterns have created dual and sometimes multiple identities of peoples transcending ethnic and cultural lines." Mushakoji therefore posits that the concept of "multi-ethnic, multi-identity, and multi-cultural societies" has effectively obscured the concept of a distinct and

[3] Acknowledged as a principle in international law, scholars traced that the Westphalian model of sovereignty began in 1648 with the Treaty of Westphalia closing the Thirty Years' War. By reifying state sovereignty as a global governing principle, it presents that each state has exclusive sovereignty over its territory. The principle of non-interference was developed based on this framework (*See A.C. Cutler, 2001: Critical Reflections on the Westphalian Assumptions of International Law and Organization: A Crisis of Legitimacy:* https://www.jstor.org/stable/20097724 *and Wikipedia.org:* https://en.wikipedia.org/wiki/Westphalian_sovereignty#:~:text=the%20standard%20interpretation%20of%20the,argues%2C%20is%20substantially%20a%20myth).

singular national identity (ibid, 4–5). Thus, the emphasis here is a regionalism that captures people-to-people interactions and crosses the boundaries of the nation-states.

This chapter likewise supports the argument that regionalism is a political project, and that regionalism building is not confined as being a state-led affair (Bøås et al. 2003, 21 cited in Söderbaum 2012; Beeson and Stubbs 2012). There are non-state players who can potentially work within and around a regional project such as the grassroots and social movements. The aforementioned examples of La Via Campesina and the Asian Solidary Economy Council serve as quintessential examples of such social movements.

At this juncture, it is best to deconstruct first the previous and prevailing regionalism projects that can enlighten the effort to establish the parameters of a people-led, alternatives-based solidarity movement that stands as a counter-hegemonic model.

Beeson and Stubbs (2012) observed that the regionalism project in East Asia is actually a "collection of relatively weak states." With that, there is always a need to pay attention to history and placements of states and regions within the global political economy from which distinctive processes and patterns of regionalism arise (Söderbaum 2005, 223 in Beeson and Stubbs 2012).

Interestingly, in many scholarly analysis of contemporary regionalism, the versions of civil society remain significantly overlooked as a result of an overwhelming preoccupation with state-led and market-driven regionalism projects (Söderbaum 2004). By acknowledging this fact, the emerging project of Southeast Asian communities should also function to contribute to the body of knowledge about the evolving regionalism discourse.

22.4 ASEAN's Open Regionalism in the Web of Transnational Capital

In Southeast Asia, the regionalism that is commonly known is the ASEAN processes of interstate interactions, dialog, and cooperation (Nesadurai 2012, 166). One major critique of ASEAN's regionalism is the centrality of economics, trade and investment, putting primacy on competition and productivity, but without integrating the principles of human rights, social justice, and environmental sustainability (Nesadurai 2008, 2012).

The pursuit for a "global ASEAN," also coined "open regionalism" (ASEAN Secretariat 2019) faced public criticism to have brought intensified poverty, widening inequality, corporate and state-led grab of land and resources, debasement of the environment in the region, corruption due to inefficient economic governance, pressures by the increase in migrant labor, as well as inadequate social protection in terms of education, health, housing, living wages, public services, especially for the poor and marginalized population (ACSC/APF 2017, 2019, 2020). Söderbaum (2004) explained that external dependence (to global capital) becomes increasingly the cause of the complex regional problems as these external forces make or unmake a regional project.

The delusion over ASEAN's regionalism can also be discovered behind its positive messaging of the "ASEAN Way"' which essentially supports "a narrow elite-centered and sovereignty-bound framework of regionalism confined to intergovernmental contracts" (Acharya 2003, 380). The norms of sovereignty, non-interference, and consensus-based diplomacy that overshadow this ASEAN Way prevent this elite body from openly discussing "sensitive issues such as democratization, marginalization, exclusion, rights, social justice," (Nesadurai 2012, 166) which are core issues implying domestic political-economic outcomes that are results of the intrusion of global capital to public welfare. Otherwise, this is tantamount to intervention on ASEAN members' domestic affairs and politics (Collins 2008).

22.5 The Limits of Participatory Regionalism

The state-centric, top-down type of ASEAN regionalism only heightened the frustration and resistance of social movements based in the region. The urgency to call ASEAN's attention on member states accountability to the numerous accounts of the adverse impact of their "open" regionalism saw the creation of another platform that bridges ASEAN and civil society, primarily led by academics. Labeled "participatory regionalism," it tried to connect ASEAN governments and civil society (Acharya 2003). However, this initiated ASEAN engagement project received criticisms from civil society later on.

The platform was the ASEAN People's Assembly (APA), where dialogs and meetings with ASEAN officials and leaders were organized in several occasions. But it was afterward observed to have become a compromised space. This led to accusations of academics becoming bureaucratized and co-opted by the ASEAN state elite by acting as concierges who restrict civil society access rather than bridging and opening channels for dialog. Reservations of further joining APA increased when civil society organizations (CSOs) noticed the support of some APA organizers for the AEC creation which is at odds with the critical positioning among many CSOs on AEC as an indicator of capitalist globalization. (Collins 2008; Chandra 2006, 77 in Gerard 2013).

22.6 Alternative Regionalism as a Common Front

As it turned out, ASEAN's elite and open regionalism remains the dominant type in the region. The current condition conveys a resolve for an alternative model that fundamentally confronts ASEAN and the hegemony of global capital.

After APA, a civil society-led regional formation emerged that holds an annual conference somehow parallel to the official ASEAN Summits. This is the ASEAN

Civil Society Conference/ASEAN Peoples' Forum (ACSC/APF) that was conceived in 2005 and since then is acknowledged as the largest gathering of civil society, inclusive of various peoples' movements that engage ASEAN concerns. While this has been recognized as an authentic social forum that works for alternative regionalism (Nesadurai 2012), the organizers and participants across Southeast Asia saw impending challenges in maintaining this regional process with the onset of a shrinking democratic space in the region. In response to the challenges, in 2017, organizers crafted its vision to chart a more robust direction of this forum that will move beyond heavy preference for interface or engagement with ASEAN:

> This situation brings up the need for civil society and social movements to present an alternative to what ASEAN is now and will be for all time. The pressing agenda is to show what true regional integration among Southeast Asian peoples is. The task therefore is to search for an alternative model of regional integration that challenges the ASEAN paradigm—one that is based on what Southeast Asian peoples are already doing on the ground and which is guided by cooperation, solidarity, mutual benefit, the commons principle, and joint development, and which rejects cutthroat competition, the insatiable thirst for profits and narrow patriotism, ultra-nationalism, and chauvinism. (ACSC/APF 2018; Tadem et al. 2020, 13)

In both 2019 in Thailand and 2020 in Vietnam and via online video conferencing, the ACSC/APF adopted the "Resolution on Alternative Regional Integration for Southeast Asian Peoples" to undertake a process for an alternative peoples' regional integration based on the alternative practices of communities, networks, peoples organizations across the region (See Tadem et al. 2020, Appendix 48–51). In 2019, this resolution was first adopted by the thematic group called Convergence Space on Life with Dignity and annexed to the official ACSC/APF conference statement (See ACSC/APF 2019). In 2020, it was adopted by the plenary and was part of the conference statement (See ACSC/APF 2020). Consequently, the Convergence Space on Alternative Regionalism, where several workshops are held on this theme, has started in ACSC/APF 2020 and has become a significant element of the program in the proceeding annual meetings.

As a concept, alternative regionalism can be appreciated as a meeting of streams (a multi-stream instead of a mainstream to emphasize the importance of diversity of processes flowing) of communities, or a "pluriverse" (Kothari et al. 2019) of peoples engaged in varied transformative initiatives. These initiatives are reactions to the dominant, ecologically destructive and socially inequitable development model imposed by capitalists, statists, and patriarchal forces. These are likewise re-assertions of ancient ways of being that remain very relevant in the contemporary context (Kothari 2020). Examples in the book include, among others, the chapters on the revival of the lost heritage of Sorghum as a forgotten staple food in Nusa Tenggara Timur, Indonesia, the rediscovery of the Tara Bandu as a customary and communal system of resource governance across Timor Leste, and the culture-sensitive community management of the indigenous Ayta-Mag-indi in the Philippines that is rooted in their defense of their ancestral territories. Other initiatives discussed pay central attention to a reassertion of indigenous identities that have often been suppressed through coercive colonial and post-colonial state formation. Cases discussed in the book are the Lumad or indigenous schools in the Philippines or the manner in which the indigenous Meitei attempt to preserve their traditional livelihood strategies in Northeast India. These practices are resistance formations in themselves that challenge the structural foundations of injustice brought by global capitalism.

This regionalism should develop as a space, a platform, and a process where alternative practitioners converge, share their experiments and social practices, examine the sustainability of every case, and link this endeavor to the wider and bigger networks opposing global capital (Tadem et al. 2020). Similarly, Bond (2019) recommended highlighting the importance of concrete struggles that are at play across the world, and comparing notes about diverse ideas and experiences.

The backlash to the "predatory mode" of capitalist development in place in Southeast Asia that is governed with "authoritarian state behaviour" (Gerard 2013, 7) rests on the strength of communities to carve their own spaces of a radical, emancipatory, and transformative movement that supplements the current global weaving of alternatives toward a better world (Kothari 2020). Throughout the book, different examples are given of initiatives that have become spaces of emancipation, self-realization, and empowerment. Specific attention has been paid to organizations fighting gender discrimination and violence against women and the instalment of safe spaces for women. The piece by Nathaniel P. Candelaria illustrates how the organization of women cooperatives by the Pemberdayaan Perempuan Kepala Keluarga (PEKKA) in Indonesia became a means to regain confidence for women abandoned by their husbands. A similar dynamic is described by Angeli Fleur G. Nuque for the Lepo Lorun Women Weavers Cooperative in Flores, Indonesia that has become a space of empowerment for women. The same counts for subaltern indigenous communities where initiatives such as the Lumad Schools in the Philippines serve as spaces of self-realization and the gaining of a critical consciousness. Meanwhile, Andrew Aeria talked about the Rural Internship Training Programme (RITP) as a venue to respond to the overly market-oriented pedagogy of Malaysian universities by exposing social science students to rural Sarawak so they will learn about the diversity of livelihoods, cultures, and value systems there. The testimonies of Igting, collated by Karl Hapal and Regie Placido emanating from the stories of mothers, revealed how forming associations and collective leadership address exclusion and poverty toward building a stable organization with a growing social enterprise. In the process, what should emerge is a counter-hegemonic regionalism (Nesadurai 2012) that emphasizes people-to-people relations, not state-to-state interactions, and detached from corporate intervention and manipulation.

Doing all these means a paradigm shift from the colonizing and exploitative tendencies of capital and rather shaped by different, equitable, and sustainable development paradigms that will benefit all humanity and the planet (Keet 2007).

Ultimately, the success of the overall work on alternatives has to be inextricably enmeshed with the work on resistance and solidarity in order to withstand the threats imposed by the dominant system (Kothari 2020). While this volume showcased **20** cases, Southeast Asia definitely has a myriad of examples.

22.7 Decolonizing the Geographical Concept of Southeast Asia

On a related discourse, this regionalism should contribute to the demand of decolonizing the region and the concept of a region. In saying that Southeast Asia is not confined to the ten ASEAN-member states (Tadem et al. 2020), it critically questions the disputed claim of the regional geographic boundaries coming from the singular narrative of colonialism and past invasion. It confirms the knowledge that all regions are deemed to be socially constructed and therefore politically contested (Söderbaum 2012). Beeson and Stubbs (2012) explained that "Asian regionalism" is itself disputed. Its perceived borders are challenged because what is considered Asia is generally thought to have colonial origins. Instead, Keet (2007, 257) proposed "real regions" that "often include same natural resources, same ecological or topographical features, and the same peoples and communities who have been arbitrarily cut up by the boundaries drawn by colonialism." Another approach is called "bioregionalism" where people organize societies or maintain traditional landscape and seascape governance based on the understanding that geographic, climatic, hydrological, and ecological attributes of nature support life and that should be given utmost respect. In fact, there have been transboundary conservation projects, peace dialogs, and indigenous bioregional initiatives that offer optimism to this new kind of regionalism from below (Bajpai et al. 2022).

These contestations over territories also found light following the insights of medical doctor Dr. Cynthia Maung[4] who heads the Mae Tao Clinic that provides health care to stateless peoples and refugees in Mae Sot at the Thai-Burma border. She mentioned that border communities in this part of the stretch between Thailand and Burma believe that before colonial demarcations of state boundaries, there was peaceful co-existence and border sharing between the Thai people, and the Karen ethnic communities located on the other side of the border now belonging to Burma/Myanmar. This signifies that the borderland is a meeting area, a melting pot of cultures and ways of living, instead of positioning as a conflict zone. This image provides a stark contrast on how this particular border has eventually become a witness of communities fleeing to the Thai border for safety from the atrocities of the Myanmar military junta, but only to end up as stateless peoples in this part of Thailand who almost cannot avail basic services and decent employment due to their statelessness, a status of having no legal Thai or Myanmar identity.

The value of this local perception represents that peoples, not the states, were and are sharing common histories, cultural identity, and resources. However, these are under-reported stories or under-studied cases to the extent of being excluded by the dominant narrative that is still cripplingly bounded to the idea of a nation-state. This

[4] This is based on the interview conducted by UP CIDS AltDev researchers Ananeza Aban and Venarica Papa during their field research in Mao Sot, Thailand to document the alternative practices of Mae Tao Clinic, an alternative medical service provider and medical training center that primarily serves stateless peoples and refugees living along the Thai-Burma border.

fragment of history also ratifies that it is only through people-to-people solidarity where an alternative form of regional integration can emerge.

22.8 Addressing the Gaps in Building the Alternative Regionalism Movement

Weaving these peoples' stories of initiatives remains a challenge. Most of them are still largely disparate and somewhat disconnected. Many local and national groups are unaware of their similarities with their neighboring societies. What this alternative regionalism should try to mount is the connection, the bridge. Available research and documentation in partnership with progressive scholars should guide grassroots organizations and practitioners in improving and scaling up their pioneering or laudable activities (Tadem et al. 2020).

Except for a few, such as the Global Tapestry of Alternatives, Vikalp Sangam of India, Crianza Mutua of Latin America, Kothari (2020) explained that albeit there are various global peoples' movements that aim to bring peoples' power together, they did not provide a sharp focus on alternatives. The proceeding task is to popularize or inform the public about these innovative practices and eventually connect them to the global fabric to make sense of the progress of these initiatives (Tadem et al. 2020).

In 2022, social movements, grassroots federation, and scholar activists in the region conceived the Movement for Alternatives and Development in Southeast Asia (MASSA).[5] There have been hurdles, crevices, and clogs in attempting to build this regionalism project. While it engages the regional civil society process- ACSC/APF, it also nurtures a life of its own by organizing an annual conference which kicked-off in Manila, Philippines in 2018. It also conducts a people-to-people exchange, which is primarily a field-based integration of community-based alternative practitioners, that started in West Java, Indonesia in 2019. Meanwhile, the trajectory of this organization is regularly assessed in order not to lose sight on its core principles of cooperation, participation, solidarity, mutual benefit, the commons principle, joint development, and autonomy based on subsidiarity that allows local communities to manage their own lives and affairs independently. MASSA is a new force to fight against neoliberalism that is grounded on the unique and diverse alternatives of communities of practice. Additionally, it actively maintains a network with other global solidarity movements such as the 11.11.11 Flemish solidarity movement based in Belgium and the French solidarity movement CCFD-Terre Solidaire,[6] one of the longest-running non-government organizations in France. It also engages with the Global Tapestry of Alternatives (UP CIDS 2024).

---

[5] *Massa* in Bahasa and Tetun or *masa* in Filipino, means the masses or the people.

[6] CCFD-Terre Solidaire stands for Catholic Committee Against Hunger and for Development-Solidarity of the Earth.

With the birth of MASSA, Southeast Asia is indeed a fountain of alternative regionalism projects. But what is important is for a project to navigate new modes of integration outside and beyond the ASEAN and capitalist box, under the leadership of the people long been oppressed and marginalized.

References

Aban, Ananeza P., and Jose Monfred C. Sy. 2022. "Queering solidarity: Civil society at the fringes of ASEAN regionalism and alternative for the LGBTQ+." *Philippine Journal for Public Policy: Interdisciplinary Development Perspectives*, 139–170. https://doi.org/10.54096/QVGW2925.

Acharya, Amitav. 2003. "Democratisation and the prospects for participatory regionalism in Southeast Asia." *Third World Quarterly* 24 (2): 375–390. https://doi.org/10.1080/0143659032000074646.

Al-Fadhat, Faris. 2020. "The internationalisation of capital and the transformation of statehood in Southeast Asia." In *The political economy of Southeast Asia: Politics and uneven development under hyperglobalisation*, ed. Toby Carroll, Shahar Hameiri, and Lee Jones, 177–193. Switzerland: Palgrave Macmillan.

ASEAN Civil Society Conference/ASEAN Peoples' Forum (ACSC/APF). 2017. "ACSC/APF 2017 statement," November 10–13. Quezon City, Philippines.

ASEAN Civil Society Conference/ASEAN Peoples' Forum (ACSC/APF). 2018. ASEAN Civil Society Conference/ASEAN Peoples' Forum 2017 conference report, ed. Ananeza P. Aban, Raquel Castillo, Eduardo Tadem, Jelen Paclarin, Galileo Castillo, Maris dela Cruz, Jan Gabriel Castaneda, Clarisse Culla, Marc Batac, and Ivan Phell Enrile. Quezon City: Philippine National Organizing Committee, ASEAN Civil Society Conference/ASEAN Peoples' Forum 2017.

ASEAN Civil Society Conference/ASEAN Peoples' Forum (ACSC/APF). 2019. "Advancing Peoples' movements for justice, peace, equality, sustainability and democracy in Southeast Asia: ACSC/APF 2019 Statement," September 10–12, Pathum Thani, Thailand.

ASEAN Civil Society Conference/ASEAN Peoples' Forum (ACSC/APF). 2020. "Southeast Asian people's solidarity for an inclusive, cohesive and responsive community: ASEAN people's forum 2020 joint statement," November 5–7, Hanoi, Vietnam.

Association of Southeast Asian Nations (ASEAN). n.d. "Economic community." *Association of Southeast Asian Nations*. https://asean.org/our-communities/economic-community/.

Association of Southeast Asian Nations (ASEAN). 2015. *ASEAN 2025: Forging ahead together*. Jakarta: The ASEAN Secretariat.

Association of Southeast Asian Nations (ASEAN). 2019. *ASEAN Integration Report 2019*. Jakarta: The ASEAN Secretariat.

Bajpai, Shrishtee, Juan Manuel Crespo, and Ashish Kothari. 2022. "Nation-states are destroying the world. Could 'bioregions' be the answer?" Open Democracy, March 7. https://www.opendemocracy.net/en/oureconomy/nation-states-are-destroying-the-world-could-bioregions-be-the-answer/.

Beeson, Mark, and Richard Stubbs. 2012. "Introduction." In *Routledge handbook of Asian regionalism*, ed. Mark Beeson and Richard Stubbs, 1–7. London: Routledge.

Bond, Patrick. 2019. "Truncated 21st-century trajectories of progressive international solidarity." *Globalizations*, 1–10. https://doi.org/10.1080/14747731.2019.1654285.

Collins, Alan. 2008. "A people-oriented ASEAN: A door Ajar or closed for civil society organization?" *Contemporary Southeast Asia* 30 (2): 313–31. https://doi.org/10.1353/csa.0.0029.

Cutler, A. Claire. 2001. "Critical reflections on the Westphalian assumptions of international law and organization: A crisis of legitimacy." *Review of International Studies* 27 (2): 133–150. https://www.jstor.org/stable/20097724.

Gerard, Kelly. 2013. "From the ASEAN people's assembly to the ASEAN Civil society conference: The boundaries of civil society advocacy." *Contemporary Politics* 19 (4): 411–426. https://doi.org/10.1080/13569775.2013.835113.

Keet, Dot. 2007. "Alternative regionalisms: Why and how?" *Global Social Policy* 7 (3): 255-258. https://doi.org/10.1177/14680181070070030202.

Kothari, Ashish. 2020. "Earth Vikalp Sangam: Proposal for a global tapestry of alternatives." *Globalizations* 17 (2): 245–249. https://doi.org/10.1080/14747731.2019.1670955.

Nesadurai, Helen E.S. 2008. "The Association of Southeast Asian Nations (ASEAN)." *New Political Economy* 13 (2): 225–239. https://doi.org/10.1080/13563460802018588.

Nesadurai, Helen E.S. 2012. "The ASEAN People's Forum (APF) as authentic social forum: Regional civil society networking for an alternative regionalism." In *Routledge handbook of Asian regionalism*, ed. Mark Beeson and Richard Stubbs, 166–175. London/New York: Routledge.

Robinson, William I. 2017. "Global capitalism: Reflections on a brave new world." Great Transition Initiatives. http://www.greattransition.org/publication/global-capitalism.

Robinson, William I. 2019. "Capital has an Internationale and it is going fascist: Time for an international of the global popular classes." *Journal of World-Systems Research* 25 (2): 289–296. https://doi.org/10.5195/JWSR.2019.954.

Sklair, Leslie. 1997. "Social movements for global capitalism: The transnational capitalist class in action." *Review of International Political Economy* 4 (3; The Direction of Contemporary Capitalism [Autumn]): 514–538. https://doi.org/10.1080/096922997347733

Sklair, Leslie. 2002. "The transnational capitalist class and global politics: Deconstructing the corporate-state connection." *International Political Science Review* 23 (2): 159–174. https://doi.org/10.1177/0192512102023002003.

Söderbaum, Fredrik. 2004. "Civil society regionalism." In *The Political Economy of Regionalism: The Case of Southern Africa*, 115–136. New York: Palgrave Macmillan.

Söderbaum, Fredrik. 2012. "Theories of regionalism." In *Routledge handbook of Asian regionalism*, ed. Mark Beeson and Richard Stubbs, 11–21. London/New York: Routledge.

Tadem, Eduardo C., Karl Arvin F. Hapal, Venarica B. Papa, Ananeza P. Aban, Nathaniel P. Candelaria, Honey B. Tabiola, Jose Monfred C. Sy, and Angeli Fleur G. Nuque. 2020. "Deepening solidarities beyond borders among southeast Asian peoples: A vision for a peoples' alternative regional integration." In *UP CIDS Discussion Paper Series 2020–04. Quezon City: University of the Philippines Center for Integrative and Development Studies.*

University of the Philippines Center for Integrative and Development Studies. 2024. "Advancing alternative regionalism from below: People and planet first." In *UP CIDS Proceedings 2024*, October 30–31. Manila, Philippines.

Ananeza P. Aban is currently a Ph.D. researcher of the Department of Conflict and Development Studies and a recipient of the Special Research Fund doctoral scholarship of Ghent University in Belgium. She also serves as a senior research analyst of UP Center for Integrative and Development Studies Program on Alternative Development. Among her major tasks are coordinating the field research across Southeast Asia, co-editing the Alternative Practices monograph series (2020, 2023), and ensuring the international linkages of the program. She finished her Master of Community Development degree from the University of the Philippines in Diliman.

Jeroen Adam is an associate professor at the Department of Conflict and Development Studies, Ghent University. His work revolves around the politics of land access. He obtained his Ph.D. in 2009 at Ghent University based on a dissertation about the renegotiation of land access in a context of displacement in post-conflict Ambon, Indonesia.

For his post-doc, he has worked on, amongst others, the emergence of informal land markets after and land dispute settlement in the Philippines. Between 2020 and 2022, he coordinated the research component of an EU funded Horizon 2020 project on mental wellbeing among European farmers for the Flanders Research Institute for Agriculture, Fisheries and Food. Since his return to Ghent University in 2022, his work is focusing on changing land uses, access mechanisms to rural space and public lands in Flanders, Belgium. His work has been published in major academic journals such as Development and Change, Journal of Rural Studies, Asia Pacific Viewpoint and Sociologia Ruralis.

The manufacturer's authorised representative in the EU is Springer Nature Customer Service Centre GmbH, Europaplatz 3, 69115 Heidelberg, Germany. If you have any concerns regarding our products, please contact ProductSafety@springernature.com

Printed and bound by CPI Group (UK) Ltd, Croydon, CR0 4YY
01/10/2026
02225941-0011